Mobilized Marketing and the Consumer: Technological Developments and Challenges

Gonca Telli Yamamoto
Okan University, Turkey

BUSINESS SCIENCE REFERENCE
Hershey · New York

Director of Editorial Content: Kristin Klinger
Senior Managing Editor: Jamie Snavely
Assistant Managing Editor: Michael Brehm
Publishing Assistant: Sean Woznicki
Typesetter: Michael Brehm, Kurt Smith
Cover Design: Lisa Tosheff
Printed at: Yurchak Printing Inc.

Published in the United States of America by
Business Science Reference (an imprint of IGI Global)
701 E. Chocolate Avenue
Hershey PA 17033
Tel: 717-533-8845
Fax: 717-533-8661
E-mail: cust@igi-global.com
Web site: http://www.igi-global.com/reference

Library of Congress Cataloging-in-Publication Data

Telli Yamamoto, Gonca.
Mobilized marketing and the consumer : technological developments and challenges / by Gonca Telli Yamamoto.
p. cm.
Includes bibliographical references and index.
Summary: "This book explores technological developments and widespread issues concerning the explosion of mobile devices in the information age"--Provided by publisher.
ISBN 978-1-60566-916-8 (hardcover) -- ISBN 978-1-60566-917-5 (ebook) 1. Electronic commerce. 2. Cellular telephones. 3. Direct marketing. I. Title.
HF5548.32.T45 2010
658.8'72--dc22
2009034817

British Cataloguing in Publication Data
A Cataloguing in Publication record for this book is available from the British Library.

I would like to have great thanks to my family for their patience, love and support.

Thanks Kifayet (mum), Mürsel (dad), Tatsuya (husband) and Eda (daughter)

Table of Contents

Selected Readings
from the Author

Preface

The tastes, soul and perception of life of the human being are open to an infinite set of demands. The active lifestyle of the 20th century has revealed the requirement for mobile communication systems. In our era, the Internet is considered one of the most important benefits of technology and it has provided very significant means in terms of reaching individuals, corporate communication and information since the very first time that it took its part on the new global order and the internet network has expanded continuously finally reaching to a level that provides the opportunity of unlimited access to almost any part of the world. In the last twenty years, when internet and information technologies have developed, serious changes and developments have occurred with regard to the circulation of information in the world. As a result, internet created the base for information that would expose wider systems than internet itself.

Through successive developments related with information technologies; many, once upon a time, imaginary instruments and devices have now gained dominant status that affect our entire lives and time. Among them, the most important instrument portable by a person is the mobile phone. The human being began to use this mobile environment of communication at any time, location and under any conditions. This phenomenon has caused certain genuine behavioral changes on societies. So, communication networks have turned into elements that strongly influence the human life.

Two systems come forth when mobile devices, which evolved in different ways based on the technological aspect, are evaluated according to the communication networks where they are used. One of these two systems is the Internet and the other one is the mobile phone. The mobile telephone and Internet are gradually turning into developing and interlocking elements.

The use of these systems especially in the business management and marketing issues, which are open to development, not only causes the further development of the systems but also serious changes and developments in marketing and business management approaches that are fed from these systems.

Quite naturally, both positive and negative consequences have arisen in communication due to both personal and corporate use. These consequences are as important for the user as they are for the system. Societies are deeply influenced from communication values that guide the social life and seriously affect the life frame.

In our era, internet and mobile instruments are the most important components of the communication, entertainment and information systems' industries. Manual machine operated-based industries have also changed as a result of electronic computers transforming into more machinery and instruments that provide faster and better means of industry, management and marketing in a more suitable manner in response to demands. There is a correlation between the appearance of special demands and the development of machines that employ the new technology. It is possible to assign multi-functions to devices and machinery with high technology. According to Bayne (2002, p. 12), mobile instruments are the electronic devices that are owned and used by personal and professional users.

For example, the merging of the photo camera with the mobile phone is one of the most apparent characteristics of new mobile phones. In addition to other features such as sending and receiving e-mail on mobile phone and video recording, and the support of high speed data services in this respect, significant works are being carried out towards developing new areas of use and new styles of communication for mobile phones. At present, mobile phones even if they do not have e-mail sending and receiving features have been enabled to carry out such functions thanks to "push e-mail" service provided by third party service suppliers. www.emoze.com is a nice example in this regard. In one sentence, certain developments and advantages may also be created with interim solutions before dwelling on main solutions.

Mobile instruments are getting to be involved in every aspect of daily life. Pictures taken with a mobile phone at the time of a traffic accident accepted as legal evidence and the ability to make an appointment from a hospital via mobile phone and the receipt of warning messages sent to mobile phones are just a few of the high number of examples. Indeed, examples are gradually increasing and mobile phones are even involved in legal services. (Nevertheless, traffic accident rates are increasing due to calls on phone or text messaging while driving).

Also, experienced corporate changes, on the one hand, facilitate life with technology that triggers different structures and trends. On the other hand, they turn into a more complex form. The elements that influence and change our lives to such an extent should be taken into consideration from every point. In addition to the opportunities it has provided, the difficulties should also be better understood in order to make a clear evaluation of our present life.

Meanwhile, new born infants come into the world in this entire technological development. Not only generations of technology but also generations of people are also developing. For both, getting used to certain surprises in terms of the interaction and communication of the next generations will almost become an exigency. On the other hand, people have now become able to explore and evaluate the world from many different profiles. They even create many new and also virtual profiles for themselves. Paper-free document management provides up to significant cost savings (Backus, 2001). The add-ins in the legal field by virtue of e-mail, e-order, and e-state also provide time saving. Many cost-increasing processes are eliminated and savings are further increased. With these types of systems, the protection of the document against transportation, fire and similar risks is provided. E-document is developing both in the private sector and public, thus creating a brand new document supplier sector. As a bigger sized example, the European Union obtains $243 billion savings per annum with e-document (http://ec.europa.eu/information_society/eeurope/i2010/docs/studies/eei-3.2-e-invoicing_final_report.pdf retrieved 4/19/2009). The legal procedure will be completed with the acceptance of e-signature, which is to come into force in 2011.

Factors such as the increasing and aging population in the world, the gradually decreasing importance of brand and the increasing number of price-sensitive clients indicate that more different instruments will be required to reach these clients. In addition, companies now face a different environment and a stronger customer in terms of searching, selecting and bargaining. As for speed, efficiency and efficient management play critical roles, technology and information management also provide strategic superiorities. In this context, technology and information management has begun to create the infrastructure of all business practices.

According to Gilmore, the transformation of the printing press into a suitable form for circulation is defined as the most radical change that affected the status of the intellectual life in the history of Western Civilizations, and he expresses that sooner or later, its influences would be noted on every point of human activities (Postman, 1992, p. 22). A similar transformation is being experienced in our era with the mobile revolution brought by the information and communication technology.

Communication was provided via internet in recent years while the information and communication technology (ICT) was defined as the life of "network connected society". With developing new technologies, the means of communicating via mobile systems have arisen and they have become influential on changing values through becoming a determinant on the social acts.

With internet, cable broadcasts, digital broadcasts and communication channels are established among workplaces (banking services, etc.). With the expansion of tele-sales methods, marketing strategy and practices have also experienced a certain transformation. It has created an understanding of information-based contemporariness, and set foot in the fields of production, sales and services. Applications such as fiber optic and mobile phone have created the continuity of the mentioned change in personal attitudes and behaviors, and made individuals associate with every person or organization from every point.

In ordinary life, people need products in order to meet their requirements and desires. The production of any product with superior qualities is both an element that develops marketing, and it is also important for meeting the general requirements. The conversion of production into a commercial scale can be expressed as a process where marketing is also included in the production. So, various efforts are required in commercial production. Actually, this requirement was already structured by the industrial revolution. The industry has created the technology and it was eager to use the technological means; thus, creating a background and an environment to develop technology.

It is not a coincidence that the phrase "assembly line" was even contained in song lyrics once. These songs reflect the industrial revolution and its influences. Mobile communication, the new phenomenon of our life, has now shown its influence on art and culture, and it has been subject to performance arts, articles and columns and even to caricatures.

Especially the innovation invented in the production of technological products has started a new production period based on product differentiation through changing and developing the features of the product day by day. So, the idea of continuously using changing features of the product arose in contemporary marketing. The means of marketing that put forth the changes on the product currently keeps the market brisk. Meanwhile, these types of changes have created differences in the relations between the capital and consumer goods with innovations in components, materials and programs. Compared to the goods of the industrial period, various means are available at present in production, supervision and marketing. The approach towards cost has also begun to change. On the other hand, the distances between the producers and consumers have become shorter and the means of direct selling and social network have become more effective. As a result, it caused changes in the functions of the intermediaries between the seller and the buyer. It has influenced processes related to delivery and supports the idea of the customer to obtain the item and leverages purchasing trends. It also provides adjustment according to speed through organization and coordination characteristics.

Technology both provides clarity in terms of fluency and urgency while stability in personal life arises. This situation, which may reflect on the personality as a conflict, may cause ambiguities in consideration of the future. In this context, doing the correct thing will yield in the advancement of humanity with a higher standard of life.

Technology is among the main elements for a good and high standard of life. For example, obtaining information services through mobile phones all over the world may add quality to personal communication. As the demands increase, the business field dealing with technological production expands further and the operators, publishers and content providers concentrate further their searches for successful business models to support the mobile platform. The number of registered users who use 3rd generation wireless technologies in the world has exceeded 488 million people in 135 countries (as of 9/13/2007).

This new generation technology supports the communication systems with small and handy screens. The screens of the latest technological changes seriously influence the lives of people. The screens;

which were initially the screens of the cinema, are not the screens of TV, computer, gameboy, beeper, microwave oven, cardio-monitor device, alarm, mobile phone, etc. and every next day a new screen appears and gives a new direction to our lives. These screens have entered our lives following a trend from farther to nearer. Even, A. Postman suggested a new term: "screen time". This term expresses the time we spend while using computers, videos and mobile phones (Postman, A., 2005, p.ix).

It is evident that people are gradually turning towards the structure named as the "3rd Screen". TV, which is regarded as the 1st Screen, left its place to computer as the 2nd screen approximately 10 years ago. According to Postman (1994), TV is a tool that teaches consumerism to people. Likewise the way TV has changed the concepts such as "news", and "public reaction", the computer has changed the definition of "information" as the 2nd Screen (Postman, 1985). In recent years, the screens of mobile devices (3rd Screen) have appeared in front of us. Similar to the way the TV taught consumerism to people, mobile devices direct them to individualism.

This type of devices also caused further changes in the definitions of concepts such as "freedom", "reality", "information", and "memory". The small size of these screens has been preferred as they provide individuals with flexibility and comfort in terms of use. From time to time, the structures that put forth the mixture of these three screens similar to three eggs in a basket also appear and enable realizing effective promotional activities. Such tendencies are influential in integrating the progresses in technology and mobile technology with an integrative attitude, and increase the significance of the need for the effective use of promotions as an important instrument in marketing. It was in question at the beginning that operators would work with the existing TV programming system; however, mobile area-specific programs and systems have been required at present.

The spread of mobile instruments to the world was as rapid as the spread of certain viruses. The rapid adoption and internalization of these instruments by individuals at every point of the world is highly interesting. Together with this internalization, certain social behaviors have also changed. New technologies can even be more influential especially for those who desire to create a differentiation between the goods and services in this way. For example, the messages in the SMS structure influence the systems of thinking, reduce decision making times and convert them into rapid thinking and rapid acting-based processes. Those who divide their target masses into correct sections and prefer mass promotion, will also desire to be involved in a system that could also put forth individual selections. Opportunities such as mobile sales and advertising also bring along great opportunities for corporations. And mobile marketing is both an action and a lack of action between individuals and corporations. In this sense, it encompasses conflicts, which is a very important feature of the human body.

According to present technology, the "mobile human" era has begun and such systems are now being designed according to new trends. For example, the internet-based operating system Midori, which was developed after Windows, is being designed as a system that could be used on every computer connected to Internet. And high speed mobile Internet should be considered as an environment that could create many different opportunities for users. Many valuable books have been written so far with regard to mobile marketing and communication, and some of them have been directing technologies while some have dwelled on how their uses might be, and some others on the social influences.

The purpose of this book is to inspect the mobilizing marketing and customer-oriented changes in the changing world, and to discuss how we are affected by the mobile technology developments as consumers and the present situation of companies and customers, again under the light of technological developments. Because, 'mobile human' indicates a more different approach compared to normal human and new generations are more easily included in this group. 2000s may be expressed as the years when mobile instruments have expanded to the world. This book covers the generic concepts of mobile marketing to people who give importance to mobile marketing, to the present or future mobile marketers or to the people who are concerned with mobile marketing to learn some of the details such as students of information systems of business schools or IT managers intended to be the target audience.

Although the years between 2008 and 2010 will not probably be exactly the years for mobile marketing, but for the development of the iPhone and similar developments, this indicates that these will be the years for major experiments in order to put on line mobile marketers, and it is apparent that a book written in this period will guide for a sound basis making more effective work a reality, and enlighten the future.

Therefore, in the first section, technological developments in the world and general issues in the information age are being taken into account and some of the technical details are tried to be given. Mobile communication technologies' tremendously wide scope are examined and mobile services, installation systems, and the devices used on such a system's evolution and technological change is given in the second chapter. In chapter three, mobile marketing and the impacts of technology on marketing were investigated and the current situation is revised.

In chapter four, the most employed mobile marketing practices of today are clarified and practices such as mobile wap, mobile services (data, business, etc.), mobile messaging, mobile portal, research, broadcast, content, handsets and terminals are also elaborated. In chapter five, the changing customer structure and the transformation of individuals into mobile customers, individualization and the culture are detailed, generations and preferences are emphasized. Chapter six contains mobile promotions. Mobile advertising and the methods of consumer reach involving SMS and MMS are discussed in detail. The types, user profiles and some current activities are stated. In chapter seven, mobile distribution and sales are asserted in general. LBS applications and Bluetooth marketing are also taken into account in regards to their content delivery, convenience, and disparity. The present conditions of major important topics and some new trends in the world of mobilization are scrutinized in chapter eight such as the mobile family, mobile entertainment, mobile health, and mobile payment. Chapter nine is about socialization and change after the intrusion of the mobile phone and computer into our lives such as social networking, mobile dating concepts and the concept of Buzz Marketing. The book ends with chapter ten with the problems that have emerged after mobilization such as privacy concerns, terror, etc.

It is a fact that we are running towards the days that even our existence will not have any meaning if we do not have mobile phones. In a very short time, we are talking, exchanging texts, taking photographs, watching TV, listening to music, measuring our blood pressure, learning about the news, organizing our accounts, paying our bills, spending money and doing many other things with this device which we have accepted as if it was another limb. This book has been considered as a necessary source related with mobile marketing and the customer. Through the merging of many separate concepts and elements, this book somehow aims to fill in the gap in the literature in this regard.

Gonca Telli Yamamoto
Okan University
Istanbul, Turkey 2009

REFERENCES

Backus, M. (2001). *E-governance in Developing Countries*. Retrieved 4/19/2009 from http://www.ftpiicd.org/files/research/briefs/brief1.pdf

Bayne, K. M. (2002). *Marketing Without Wires*. New York: John Wiley & Sons.

http://ec.europa.eu/information_society/eeurope/i2010/docs/studies/eei-3.2-e-invoicing_final_report.pdf (n.d.).

www.emoze.com (n.d.).

Postman, A. (2005). Introduction to the Twentieth Anniversary Edition. In N. Postman, *Amusing Ourselves to Death.* New York: Penguin Books.

Postman, N. (1992). *Technopoly*. New York: Vintage Books.

Postman, N. (1994). *Disappearance of Childhood*. New York: Vintage Books.

APPENDIX: LIST OF ABBREVIATIONS

0G: Zero Generations

1G: 1. Generation Technologies

2G: 2. Generation Technologies

3G: 3. Generation Technologies

4G: 4. Generation Technologies

ALI: Automatic location information

AMPS: Advanced Mobile Phone System

APSC: Alabama Public Service Commission

ASP: Application Service Provider

B2B: Business to Business

B2C: Business to Customer

BREW: Binary Runtime Environment for Wireless

C2B: Customer to Business

C2C: Customer to Customer

CDMA: Code Division Multiple Access

cdmaOne: Wireless system based on the TIA/EIA IS-95 CDMA standard, including IS-95A and IS-95B revisions

CRM: Customer Relationship Management

CSF: Critical Success Factors

DAB-IP: Digital Audio Broadcasting-IP

D-AMPS: Digital AMPS

DMA: Direct Marketing Association

DMB: Digital Multimedia Broadcasting

DRM: Digital Rights Management

E-commerce: Electronic commerce

EMS: Enhanced Messaging Service

ETSI: European Telecommunications Standards Institute

EU: European Union

FDMA: Frequency Division Multiple Access

FOMA: Freedom of Mobile Multimedia Access

GNSS: Global Navigation Satellite System

GSM: Global System for Mobile Communications

GPS: Global Positioning System

HSDPA: High-speed downlink packet access

IM: Instant Messaging

IP: Internet Protocol

IR: Infrared

IS-95: a standard for (CDMA)

ISPD-T: Integrated services terrestrial

J2ME: Java 2 Micro Edition

Kbps: Kilobits (or thousands of bits) per second

KHz: Kilohertz

LAN: Local Area Network

LBS: Location Based Systems

M-B2B: Mobile B2B

M-commerce: Mobile Commerce

MediaFLO: Media Forward Link Only

MEF: Mobile Entertainment Forum

M-marketing: Mobile Marketing

MHz: Megahertz

MIT: Mobile Internet Technology

MMA: Mobile Marketing Association

MMS: Multimedia Message Service

MP: Mobile Payment

MP3: MPEG-1 Audio Layer-3.

MVNO: Mobile virtual network operator

NFC: Near Field Communication

NMT: Nordic Mobile Telephone

NPI: Non-public personal information

P2P: Peer to Peer

PDA: Personal Digital Assistant

PDC: Personal Digital Cellular

PC: Personal Computer

PPV: Payper review

PKI: Public key infrastructure

RF: Radio Frequency

RFID: Radio Frequency Identification

RFP: Request for Proposal

ROI: Return on investment

SAR: Special Absorption Ratio

SCM: Supply Chain Management

SMIL: Synchronization Multimedia Integration Language

SMS: Short Message Service

SPIT: Spam over Internet Telephony

TACS: Total Access Communications System

TCAR: Three carriers

TDMA: Time division multiple access

TRA: Theory of Reasoned Action

UGC: User Generated Content

URL: Uniform Resource Locator

UWB: Ultra Wideband

WAP: Wireless Application Protocol

WAN: Wide Area Network

W-CDMA: Wideband Code Division Multiple Access

Wi-Fi: Wireless Fidelity

Chapter 1
Information Age

In this section, the processes leading to the information age and mobile technology are examined. The affects of certain inventions are discussed through having a look at the period which began when human kind gained the ability to direct energy with the invention of fire, and continued with the construction of simple machinery, and then reaching the Industrial Revolution with the influence of the modern scientific approach. Here, the appearance of technological development and technique in order to provide ease for people are being inspected. In the development and advancement of the human being, it is suggested that the work that initially required raw power, was transferred to simple machinery, and then to normal machinery. Many technological innovations with many contributions to living are taken into account from fairy tales to films made through the use of virtual means as the products of the creative mind related with the passion of invention. Passage to the Industrial Revolution has been triggered with the appearance of the required ideas for the invention of the work performer and caller forces or intelligent machines.

With the invention and offering for the use of people the elements such as radio, telephone, TV, computer, and satellite during and after the industrial revolution; the beginning of the steps to a certain age have also begun. In this development, values that would be influential in the struggle of human kind are evaluated on three fundamental points; human, information and technology. In this context, the importance of mobile tools and instruments as the latest examples of the systems nested in the people is evaluated.

The information age and the current density of information production and consumption are indicators that the industrial age has ended as the herald of the entrance to the "third wave" period. This is a wave to come with the transformation of technology through evolution to nano technologies, space technologies, and advanced biotechnologies; and this wave brings in the mobile human, who is intertwined with mobile technology in the technological sense.

DOI: 10.4018/978-1-60566-916-8.ch001

The demands that caused the companies and organizations to develop technical instruments during the industrial period were stemming from the war powers and they were organizational-sourced while the demands in the development of mobile technologies are considered as personal-sourced, and the experienced change is also examined. The change of values takes into account the mobile human and the effects of mobility on human life, and is oriented towards developing the styles that push forward individual priorities in the sectors. In the societies which direct to good life philosophies, it enables individuals to form groups and create societies based on the characteristics. Even virtual lives that are based on the virtual values have become valid in our lifetime and after the prediction is that it will reach further stages. Finally, the mobile service system in the information age is explained in general.

UNTIL THE INDUSTRIAL REVOLUTION

When the human being lit fire thousands of years ago, that human being also learned to control a certain type of energy. The human being concentrated on the fire, which was explored through coincidental methods according to a rational backwards perspective. In time, the human being glorified the fire and equalized it with deistic powers. Indeed, fire is a very significant element in mythology written through rumors passed through generations. Zeus hid the fire in order to punish the mortals and their protector Prometheus. However, the wise Prometheus fools Zeus. He climbs up to Olympus; there he steals a flame from the blazing wheel of sun and secretly takes it to people. According to Sophocles, Prometheus is known as the Titan god that "carries fire" in the Ancient Greek belief. The fire created very significant means after it entered under the service of human kind. It was used in defense, protection, heating, nourishment and in almost every aspect of life, and became a very important part of life.

The wheel, which was found centuries later than the discovery of fire, was the first important tool in transportation and travel, and is one of the most important instruments of communication. According to the old tablets, when the wheel was invented by the Mesopotamians in 3500 BC, the extent of its contribution to human kind was probably not imagined clearly at that time. (http://www.ideafinder.com/history/inventions/wheel.htm 4/19/2009)

In time, with its contribution to the construction of machinery, the wheel was used in production and transportation, thus enabling the achievement of major developments throughout civilization. Besides the ability to travel and carry freight and goods over long distances, the new vehicles of communication also enabled trade to obtain more comprehensive dimensions.

Evolutionary changes were experienced in a span of thousands of years long affected communication and marketing at the time of constructing machinery that would enable transformation from simple machinery to the machinery of the industrial revolution.

It was noted that with the scientific advancement in the 18th century and in the early 19th century that industrial developments were connected with each other. Among classical economists, Smith (1776) mentioned especially this feature (Freeman & Soete, 2003, p. 229). It is known that technology interacts with activities in practice, research and development in industries associated with modern science. Technological developments, which had a very slow development tempo until the industrial revolution, managed to register the previous development of thousand years with a single leap only in the last twenty years with the invention and personalization of computers. In our age, which has integrated with circumstances that appeared during this leap, new developments are still continuing to be experienced. The complexity of technical information and data processes affects technological development and changes, and speeds up the route of development. Computers that put into service experienced warehousing, utilization and retrieval systems necessitated making changes in marketing, sales, management and training methods for industrial goods which appeared with the 18-19th century industrial revolution which could be obtained with fewer efforts.

Manufacturers and companies initially used classical marketing methods and gradually oriented towards the use of technological equipments in marketing in a more effective manner. Through the use of the technological devices, new types of competition have arisen in the market. Actually, the invention of the print press, electricity and vapor machines in time respectively have caused the industrial revolution and the societies directed towards industrialization to experience changes accordingly.

Technology has appeared in order to facilitate things for people and a transformation from the initially raw force based works to simple machinery and then to complex machinery has been experienced. As technical devices developed, the issue of doing work in an easier way was intrigued. The idea of using tools and instruments was common in that period. The issue of providing more force either while doing or getting the work done was of interest. The powers of the machine used in technology were evaluated through comparison with the power people obtained from animals. The measurement of the engine power with horse power is its most important proof. Motorized machinery had a technical infrastructure that supports the industrial revolution. The industrial period had shown integrity with its instruments, machinery, types of energy used, styles of thinking and standard of life. The technical background obtained in the industrial period changed direction in time and only the performance of work was not sufficient. Necessary ideas for new forces which would order and have the work done namely intelligent devices appeared.

The engagement of elements such as radio, telephone, computer, and satellite which were invented during the industrial revolution and offered for use of people became the precursor of the new age. With the invention of the computer, similar to many other fields, marketing also began to benefit from new technologies. The telephone provided opportunity of voice communication with people from faraway places and became one of the essential communication instruments of marketing. Now that the satellite systems have also started to be used in sound and image transmissions, the invention of the mobile phone that uses these systems brought a new dimension to communication. This invention is considered as the invention of the new millennium. Star Trek (www.startrek.com retrieved 4/19/2009), approximately 40 years ago, was suggesting that many different and interesting instruments could be used according to the star calendar. The travelling of the characters in that TV series with mobile devices in order to communicate with another world from very long distances and even from their space shuttle was an interesting approach. Just 40 years after such movies (Jensen, 2007), the audience began to use the similar devices themselves. Yes, extraterrestrial communications are only in the service of the science for the moment, but serious changes have occurred in our lives due to technology.

The most important aim of the secret services of societies that adopted the understanding of development through strategies based on the superiority of information and that gained a superior position in this way during the Cold War era was to obtain rare information and documents in the world and to protect the people with featured information, etc. Besides, the delivery of such information from one place to another had been subject to many books and movies.

The most important contribution of movies that were shot with virtual devices as based on the stories produced with creative thinking were related with the passion for discovery heralding that many technological innovations will rapidly appear. The thing that made the legendary movies of the fifties on the British secret service interesting was the supplement of tools and instruments that were created with imagination. For some movies, there is a correlation between the successes of the directors and what they do in the virtual environment with the images pushing the limits of imagination. The construction of tools that have many additional functions in addition to their actual functions has become to be perceived as quite natural.

Mobile phone is the leading one of such multi-function devices. These devices, which are tried to be produced in such a way to either illuminate even very dark environments from time to time, or according to the unlimited desires of the imagination, are also equipment as required to carry out mobile operations. Mobile phones are continuously developing. In some countries and regions, they may be used for further functions than only telephoning; for example, for mobile banking transactions. In addition to audio-visual communication, they have become so complicated that they can shoot pictures, use radio waves, and perform functions such as a clock, calendar and chronometer, etc.

Communication through Internet has exceeded borders and brought a new dimension to global relations and made great contributions to business life. Similar to other instruments in this system, developments with regard to mobile phones continuously reflect on life with positive and negative aspects. The mobile phone and Internet have become the stars of communication tools.

For example, the 2002 year's movie "The Minority Report", which deals with technological development, clearly expressed and deeply questioned the issue in which mobile instruments could be used in the future, and how the wrong approaches in this regard would affect people. So, these types of movies leading someone to self-questioning have taken the place of movies which foresee what the interactions in the digital world would be. In the following movies, the effects of technological devices of accessing to personal information were questioned. The use of eye, hand, fingerprint, etc. personal properties in obtaining personal identification data has now begun to be perceived as natural in life. These data are sufficient enough to provide the personalization of all promotion and advertising related with marketing. A new world and a brand new type of generation is presented with digital technology, which facilitates the supply and evaluation of much information that help to describe the target mass in marketing. They can be considered as the preliminary draft work in realizing more frenzy projects of the future that speak, find out the direction, communicate and pave the way.

Mobile phones, which were brought on the agenda with their imaginary uses thirty years ago, are currently being used by 80% of the population over 5 years old in the US (Mc Guire, 2007, p.xv). There is a similar trend to that in the US in all developed and developing countries, where there is a rapid increase.

The new advancements with regard to technology lead to the abandonment of previous technology and advancement so as to create different generations. The human being, affected by technology, continuously desires newer generation instruments at a much higher speed than technology itself and thus supports the continuity of innovation. In recent years, the changes between technological generations show their influences in terms of marketing in a very short time as two to three years. It requires researchers, producers and marketers to get prepared according to each generation, to develop special and different attitudes, to act by taking into account new styles and to use obtained new skills.

The indications given by technology in becoming faster, more effective and more efficient in marketing are at least as much important as the means they suggest.

Communication is involved in every emotion that appears during a time shared by a minimum of two people in the world. Communication actually provides existence in a determined relationship, and therefore it requires a shared environment. Technological instruments have virtualized these environments. The purpose the relationship serves for should be clearly identified in the relationship which occurs through the face to face meeting of existing people. In marketing, the communication network established between the customer and the company is very wide. The most important focal points on this network are the customer, manager, colleague, boss, employee and the people who affect the customer.

In order to understand if relationships are conducted properly and whether relationships are continuously expanded or not should be sought. If relationships grow, then communication also grows. There is a virtual environment in the relationship established through the use of technological instruments. Therefore, decisions made regarding the expansion of the relationship or every created relationship may not always come right. In this respect, mobile communication revealed a communication style which was not known before the twentieth century.

The destruction of the routinely defined paradigms in the technological environment (virtual environment) may create positions that lead to problems or intensify existing problems between theory and practice during the presentation of creative suggestions. There are certain difficulties in terms of managing people in this environment, which is suitable in terms of giving an opportunity for the customer and team work.

The use of devices to reduce lack of time, work stress and load in an abrupt manner may cause a busier and more stressed environment. In e-communication, the future should not remain limited with an analysis of the immediate time. The analyses of the future time should also be made and the continuity of the relationship should be maintained.

Corporate strategies, regulations, standards, accreditation and certification, measurement, evaluation, psychological and sociological aspects should be in a configuration so as to enable the correct use of the virtual environment.

The human pleasure, spirit, and understanding of life are open to infinite desires. In our era, the concept of communication has created unlimited facilities through continuously working on developed technologies. Mobility is a social revolution that designs living individualism beyond collectivism in the human life, and this revolution deeply influences marketing. It requires the necessity of different organizations with the conditions that cause the changing of traditional marketing methods which enable reaching the entire information on communication channels. Actually, mobility has always been in question since the beginning of intercultural trade. But that mobility can be described as one that emphasizes long time and long distance. The instruments used for communication are only the assistants to such acts.

Each invention includes references to the future and covers the indications of future inventions to follow it. The transition from abacus to the calculator in 1671 with Gottfried Wilhelm Leibniz, and the patenting of the pocket watch by Christian Huygens in 1675 created a difference in terms of the use of technology but these developments can be expressed as interim developments (http://inventors.about.com/library/inventors/bl1600s.htm retrieved 3/30/2008). Each of the innovations that appeared with the industrial revolution is included among the infrastructural elements of present-time technology.

The invention of the printing press, electricity and vapor machine have caused the industrial revolution, and in turn, innovations. For example, one of the most important technological developments that affected communication is the invention of telegram in 1844. Telegram has a specific importance as it is the first communication tool that uses electrical current (http://www.soylenasil.com/elektronik/telgraf2.htm retrieved 6/20/2008). By laying telegram wire over a distance, Samuel F. B. Morse enabled the transfer into a new communication period from a period when only messengers and letters were used for communication in personal production and sales processes, and he had been the first person to undersign the communication age. With a message to Baltimore, one of the Supreme Court chambers in Washington DC, Samuel F.B. Morse did not only introduce his telegram device, but also initiated a new type of communication with the message "What hath God wrought?" (http://www.pbs.org/wgbh/amex/telephone/timeline/timeline_text.html retrieved 3/30/2008.).

Innovations in communication tools do not require a total elimination of the previous tool; for example, the telegram still maintains its use today when required, but communication via telegram is gradually becoming nostalgia.

This method of communication transported messages to carry out distant communication in every field until the invention and widespread use of the telephone. Marketing is also one of the important fields where the telegram is used. It is the communication tool of the period when slow but a comprehensive communication is carried out via the machine. Therefore, it has characteristics which became identical with its time. Many business areas related with the telegram appeared; the system rapidly expanded in terms of sending and transmission, and it was evaluated as an important element of marketing and sales in parallel to the mentioned growth and expansion of the coverage area.

Alexander Graham Bell (3 March 1847, Edinburgh – 2 August 1922, Baddeck), is a Canadian scientist of Scottish origin, inventor and industrialist. He made the patent of the fixed telephone on March 7, 1876. He founded Bell Telephone Company, the first telephone company of the world, in 1877. Bell Telephone Company is currently one of the biggest companies in the US (http://performans8.blogcu.com/2082215/, 2008). Telephone companies, which wired the entire world by a communication network in 120 years, made a business out of establishing a bilateral voice communication.

The passage from telegram to telephone took 32 years. Compared to the lifetime of the human being in that period, this is equal to the average lifetime of people (http://en.wikipedia.org/wiki/Life_expectancy) and it indicates that one generation changed during the transition from telegram to telephone.

After the telegram, almost one hundred years passed and AT&T began to provide mobile phone services in 1946. This service was initially provided in a region with a single antenna. In the metropolitan region where the service was provided, it was not possible to make more than 12-20 uninterrupted calls. In 1947, AT&T introduced the cellular telephone concept; however, the technology suitable for the use of the mentioned concept was not developed then (http://www.corp.att.com/history/milestones.html retrieved 4/13/2008).

The transformation of analogue-digital telephones so as to emit images became possible through the changes at the switchboard systems of voice communication. Cable TV channels and Internet connections are also operated via these systems. The quality of image and sound has been increased with the relations established between cable systems and satellite systems. The market of the tools and instruments that employ this type of systems has grown and created a special marketing field now.

Electronic communication and calculation systems began to develop in the 1950-60s (Freeman & Soete, 2003, p. 187). The electronic computer, electronic detection and identification instruments can be counted as a post-war change, and depended on process instrumentation (radar, gun control systems, missile guidance, and numerically controlled work benches) that appeared as a result of a revolutionary approach. Even, robots were employed afterwards.

Later, in the 1970s, the production of optical fiber assisted the clearer transfer of image and sound through light pulses. Fiber optical cables began to be used in the communication industry. The company titled Corning Glass announced in 1970 that they created the fiber glass structure which could transmit light signals. GTE (General Telephone and Electronics Corporation) and AT&T (American Telephone and Telegraph Corporation) also initiated the experiments of sound and image transfer that employ fiber optic cables. These developments triggered ideas that would change the communication industry. (http://www.pbs.org/wgbh/amex/telephone/timeline/timeline_text.html retrieved 3/30/2008)

The use, installation and transfer of the system were burdensome for computers developed in the 1950s as they were cumbersome. With revolutions and improvements in the 1970s-1980s, the sizes of computers were reduced down while their functions increased.

The importance of information systems has gradually increased in the current world with a fast race against the time. The necessity of becoming more active in reaching the source of information and using that information increases day by day. Besides, as computers have become more dominant in production systems at the core of the production of traditional capital goods, mobile technology will also gradually become more dominant on sales and marketing systems.

POST INDUSTRIAL REVOLUTION

The industrial revolution began to be shaped with the use of the vapor power; it continuously developed and created a situation that multi-laterally influences societies. The information and techniques obtained during the industrial revolution also prepared the fundamental thought of the technology revolution. However, with the coming forth of communication technologies, information and information processes together with the produced machinery and devices reached above all other techniques. According to Yücel, as the use of the information increases, the production structure also changes. Information is included in the production as a more important factor than the labor and capital. The societies that practice such a production style can be defined as information societies. In the information society, the most important element is information itself. Information is obtained with labor, financial resources and time allocated on research & development activities (http://ekutup.dpt.gov.tr/bilim/yucelih/biltek03.pdf). This also originated new ideas and improvements for collecting and using information to have better conditions for life.

Marketing was also affected as much as the technology of the transition to the information age through the comprehension of the importance of accumulation of knowledge and the value of information which gained a new value in the industrial society. In the supply of goods, the idea of customer satisfaction is held prominent, and customer satisfaction is aimed from the production to the delivery to the end user. In order to obtain customer satisfaction, information with regards to what, how, where, when and why the customer wants it should be known beforehand. Information technologies are constantly used during the procurement, storage, interpretation and use of the mentioned information. The use of the telephone and mobile phone during the communication with the customer is only possible with having customer information. In the information age, general trends in industry were shaped as stated here below: (http://www.bilgicagi.com/index.php?option=com_content&task=view&id=284&Itemid=137 retrieved 4/17/2008):

- The most important capital in production and marketing is information.
- Creativity is the essence, and what is new today can be commercialized tomorrow.
- Markets are continuously globalizing further and competition is increasing.
- Global size technology cooperation and access have become ordinary.
- Consumer demands are continuously increasing.
- Products are getting more complex and production is becoming based on higher performance and system.

Table 1. Free War Forms (Source: Toffler Associates, Dia Workforce of Future: Creating the Future of the Defense Intelligence Agency White Paper, May 15, 2003, p. 20. http://www.dia.mil/thisisdia/DIA_Workforce_of_the_Future.pdf retrieved April 17, 2007)

Militaristic War	Cold War	Non-Militaristic War
Nuclear	Diplomatic	Financial
Conventional	Internet	Commercial
Bio-chemical	Information processes	Natural Resources
Ecology	Psychological	Economic Assistance
Space	High Technology	Laws and regulations
Electronic	Smuggling	Preventive measures
Guerilla	Drugs	Media
Terrorism	Dissuasiveness	Ideological

- Laws are getting stricter and new laws and regulations are required in the new systems under new conditions.
- A change of paradigm means changes from mass production to mass customization and the final target of which is "single order production" is being experienced.

The information age, in a sense, is a concept used in defining the period when information has become the prominent factor in creating fortune. So, material capital is replaced with intellectual capital. The structure of intelligence capital, which is not limited in a certain place, has changed all management and social relations that also creates new models for marketing such as social media marketing.

From the point of macro marketing, a struggle continues among those in need, the producer and the interest groups. These struggles first began with display of power on the war fields and then moved towards different directions. According to the Toffler Associates' 2003 Dia Workforce Report white paper, the struggle has been directed towards information-intense fundamentals, media and ideological issues in economic, financial, and commercial areas (See Table 1). This tendency has exposed societies to the approaches of struggle other than gun fight and took place in the essence of social upheavals. There is a demand for technology in any type of struggle and therefore technology has been supported continuously.

At present, values that could be influential in any type of struggle are evident: human, information and technology. In this context, the importance of mobile tools and instruments as one of the latest examples of the systems intertwined with people has appeared once more.

A NEW WAVE (AFTER THE 3RD WAVE)

Alvin Toffler dwelled on the former opinion styles, old formula, and doctrines in his book titled 3rd Wave, and expressed that ideologies despite their high benefits in the past, are not suitable for current time's reality. He emphasized that with the impact of new values and technologies, new geopolitical relations, new lifestyles and communication methods; new ideas, new eulogies, new classifications and concepts should be developed.

According to Toffler, the period from the beginning of humanity to the industrial age, was the "agricultural" age or "the first wave of change". The "second wave" rose in the 18th century which is the beginning period of advancements in the industry. Our present time with hectic information production and consumption is evaluated as the "third wave" when the industrial age has ended. However, this wave is also the indication of a new, rapidly forming wave. And it is a wave to come with "the evolution of technology to nano technologies and advanced biotechnologies". The period of creating anything desired is through the use of nano, space and advanced biotechnologies which will begin. In this wave, mobile technology will remain as an interim technology but yet, it will maintain its value as one of the most evident and functional instruments that can be used in a personal environment. When evaluated in terms of place, the first wave is a flow towards "village", while the second wave is a flow towards the "city", and the information age that emphasizes individualism is evaluated as a "global" flow. Individual and societies' variations also influenced the marketing idea and the conditions of the markets.

INFORMATION AGE

Information age or Digital Age has taken its place in the literature as one of the terms that express our present life. These terms imply the concentration of the global economy from physical product production to the management of information (http://en.wikipedia.org/wiki/Information_Age retrieved 6/20/2008).

The information society tries to create an environment where value given by people to information multiplies, where changes caused by the technology and innovations are more easily accepted and where people evaluate them as a social stage. Some of the expectations here are standard of living, changing of minds through support with information and the formation of new life styles in accordance with this change. These explicate a new form of customer.

According to Gouldner (1979) and Stoiner (1983), these societies are those which appeared as a result of the changes that occurred with the new class and economic conception that occurred after industrialization (Dura & Atik, 2002: p. 39-40). It is observed that a highly serious stage has been reached in terms of obtaining information, producing and sharing in modern societies together with the developments through the targeted direction. Besides, modernization and the adoption of differences in a basket of cultural change can be correlated. Certain cracks may occur on social values as a result of the society's raising information oriented values due to technology and advancements. The social, cultural, technological and structural changes in the post-industry period are interrelated.

The meaning of information society is related with the possession of a complex communication network that could develop and change information rapidly (oregonstate.edu/instruct/anth370/gloss.html retrieved 6/20/2008).

Accessing information has always been regarded as highly valuable throughout the development of humanity. The industrial revolution occurred with approaches based on innovative thoughts that were explored with sharp changes, and they created their own age. The third wave initiated a communication age that targets the individual itself. In this period, the divergence of civilization from massiveness has caused a great increase in the amount of information exchange among people.

With the increase of the means that pave the way for the information age, everybody understood that access to information would bring along superiority. Information existed in certain societies much enough to be the leaders, and in some others, as only enough to make them followers. Some societies did not have the information at all. Besides, those who have information aimed to keep their information disguised for a long time in order to obtain a value. Even, superiority was realized through such disguises. When the step was taken into the information stage, the slogan of 'unlimited access to information anytime, anywhere' arose, and this slogan began to guide the way of thinking, deciding & behaving. Modern societal transformations are also alined by Beniger (1986) & Godin (2008).

This became to turn into a reality with technological instruments that support thinking. The information disseminated via Internet has been broadcasted without any limits. In our era, even the idea that having the required and non-required information included in this system is nothing else than getting mixed up has blossomed. For its arrangement, the concept of "Information Economy" was suggested (Godin, 2008). By dint of that, the ideas of processing the information and adding it other information; thus, facilitating the procurement of new information and the creation of new values are becoming widespread.

Internet is a joint universal network, built for the interaction of computers with each other. On the Internet, communication is built via computers and computer networks in a manner which encompasses the entire of social, psychological and cultural interactions between people and the societies. Internet, which expands the sharing area of information and accelerates communication through new communication technologies, can also be described as the golden key of the information age. It is a system that relates people with each other via communication in general, and facilities the exchange of information. Internet communication is a communication model that especially creates advantages in terms of speed, time and economy in fulfilling the information requirements of people, and supports it with the use of computer and similar technological devices. This communication style shows a continuous development through benefitting from the most important technological developments of the new millennium. Communication via Internet has many positive and negative effects. Especially, Internet creates a new field of negotiation in the world and suggests a common language and communication culture. Some new words and concepts appeared after a short time because of the entrance of Internet to our lives: e-mail, site, surf, anti-virus, Trojan, spam, chat, e-book, hacker, upload-download, virtual library, forum, blog etc.

Internet became widespread as a relative system; whereas, the computer became mobilized, personal and portable. The mobile instruments which used a different system in the communication process and the limits of which have continuously expanded, showed a development parallel to that of computers. This could be appraised as the platinum key of the information age.

The machines, which were produced as mobile phones, entered in the market rapidly, and found a wide area of use. Besides, mobile phones were supported with different accessories and fitted into the mission of entertaining instruments, and even passed beyond them and brought a brand new perspective to the entertainment concept. The transformation of PDA, mobile computer and telephone into a single device is also among the features that could increase the options in terms of market. Their functions include internet message reading, calendar, address book, to do list, memo or notebook. They are at a size as small as that could be to fit in a pocket. However, mobile systems will inevitably be further equipped with new instruments that could meet most of the demands based on imagination. Most of them will be changes that aim to facilitate living.

In the Information Age, an extreme specialization and detailing of subjects within themselves is possible in the sense of gaining information and using it. For example, with the detailing of information in medicine, fewer than two main concepts as internal diseases and external diseases led to the formation of specializations that base on a single limb now. It is understood that in time, specializations will develop on the basis of the pieces that make up the limbs: such as specialization on the finger, cell, etc. Similarly it is natural for marketing to evolve towards the invention of specialization and to become more influential with such specializations as screen decoration, copy writing, planning, choreography, modeling, mobile content creation, etc. In public relations, specialization on subjects such as bulletin release is seen and the perspective narrows down as information increases. Thanks to new technologies, narrow communication networks and structure are gradually expanding. The appearance of new specialization areas can also be expected in the expansion of this structure.

Demands, firms and organizations that developed the technical tools in the industrial period were stemming from war forces and they were organizational-sourced, while the demands that develop mobile technologies are mostly personal-sourced.

CHANGES OF THE EXISTING VALUES

With the transition to the information age and the transformation mentioned in our century are not only valid for mobile technology. Always, latter technologies destroy the values of former technologies with a slightly flying away impact. The mentioned change is a differentiation in economic, social and cultural values.

As mobile systems on economic terms influence business and marketing fields and all fields that are related with the use and change of capital, they also influence industries such as those concerning books, films and music that enable the production and distribution of information. They especially narrow down the profit margins in the media. The supervision of personal rights in the publications becomes more difficult as the sharing and copying of files on the system is so easy. However, the issue of what kind of a value could be put instead of the decreasing value could not be identified and clearly resolved yet. To do that, a system is required to enable end users to make payment during the time of copying on the uses related with intellectual and artistic works on mobile systems. If it could be done, then mobile systems will both effectuate values based on human efforts and prepare a technological environment that produces value.

On the Internet and mobile systems, the main income of the content provided is in general based on the marketing income. To obtain a positive value on the system, a higher volume of commerce is required on websites (Yamamoto & Karaman, 2005). In the earlier developmental years of mobile communication technologies, the limits on Internet technologies also caused the borders to be limited and the scope to be tremendously low (see 1G). On that basis, the environment has not been enriched and it is gradually turning into a more important instrument. Even customer types have begun to be formed now. A labor-respecting customer portfolio is being created and individuals are becoming ready to develop ideas in accordance and to also pay money for the content. They even support it for the creation of a safer and better communication environment. Therefore, new values and value measures have been developed and started to be used. The payment for a jingle music as +SMS or use and obtain, etc. means of payment have now reached to such an extent as to create an important material source. In this context, content protection technologies are essentially required to be re-evaluated with the consideration of legal bases that could be influential within the issue of mobile systems.

From another point of view, mobilization welcomes new communication environments and affects the formation of new values in the communication environment. One of the newly formed values is work at home (home offices), the complete creation of remotely administered business places and a flexible work environment. Before this period, the Mobile Age, people for whom taking part in production was very difficult has become able to produce values. This also reduces the requirement for transportation and in a period when fossil fuels such as oil have begun to deplete and become more expensive, it may add value to the environment as a positive development with regard to humanity. It was stated in the previous comparison with human lifespan that previous technological innovations were equal to the average lifespan of a human. However, present technologies step into the human lifespan with almost annual and even semi-annual periods. In short, all innovations in the transformation of values can now be rapidly accepted, used and turned into a functional state.

The preferences and priorities change due to the usage of mobile systems, and the opportunities brought by the Internet indicate the danger of alienation from socialization. With mobile technology, the formation of new social gatherings is noted instead of the actual societies, and these societies are passing through a period of structuring all advantages and disadvantages of virtual association.

The reactions of people about adapting to developments and changes in the world differ greatly. It is quite natural (due to age). Old people either reject using technological instruments or try to obtain the skills to use such means very simply. The present generations are born into a life where devices such as computers and mobile phones were already included. Therefore, they are eager to use technological means on every issue. For example, in most of the developing countries, while the people in the 30-40 age group and higher prefer to use different means of communication, they also have more different understandings of entertainment, etc. As the computer and mobile phone entered their lives in a later period, the change is most severely felt by the middle-age group. Some of the people in this age group even have their mobile phones adjusted by their children and they can benefit only from certain functions of these devices. Then, the existence of certain behavioral differences and differences of habits between those who are born in the environment of computers and mobile phones and those in the previous period becomes inevitable. In 2005, the number of GSM uses exceeded 100 million subscribers. In some countries most of the users are young because of the census distribution. For example, 45% of the telephone subscribers in Turkey are young people. According to Wireless intelligence regional users of GSM between 2002-2005 increase from 2.958 million to 6.115,2 million. This means GSM market has growth two fold approximately in three years.

Also according to the same report, the number of GSM technology users in the Asia Pacific region increased by 39% in the first quarter of 2007. It is followed by Western Europe by 19% and Eastern Europe.

With these data taken into account for a developing country like Turkey, the mobile phone subscriber density is estimated to reach 96.3% in 2008 and the density of wideband subscribers to 10% (SPO Turkey 2008 program pdf).

According to the analysis on the wireless industry, service developers, content providers and gatherers will obtain continuous income increase in the next five years. This will be enabled through the global mobile data increase. The market share is expected to increase continuously and increase from 81 billion dollars to 150 billion dollars between 2005 and 2009 (uiOne™, 2007). Even the increase of personalization in terms of number of subscribers of their own services supports such an expectation. As a result of the strategic analysis at the end of January 2006, global income to be obtained from total content personalization is $12,365 million. It is expected to increase by 14% to reach $18,363 million. As of 2006, 420 million mobile phone users have subscribed to such personalization content and this figure is expected to reach 661 million users in 2010. The distribution and sales of mobile phone have gone a very long way so far since its first invention, and have become one of the most striking and most important products in the market.

Figure 1. Mobile service system in the Information Age

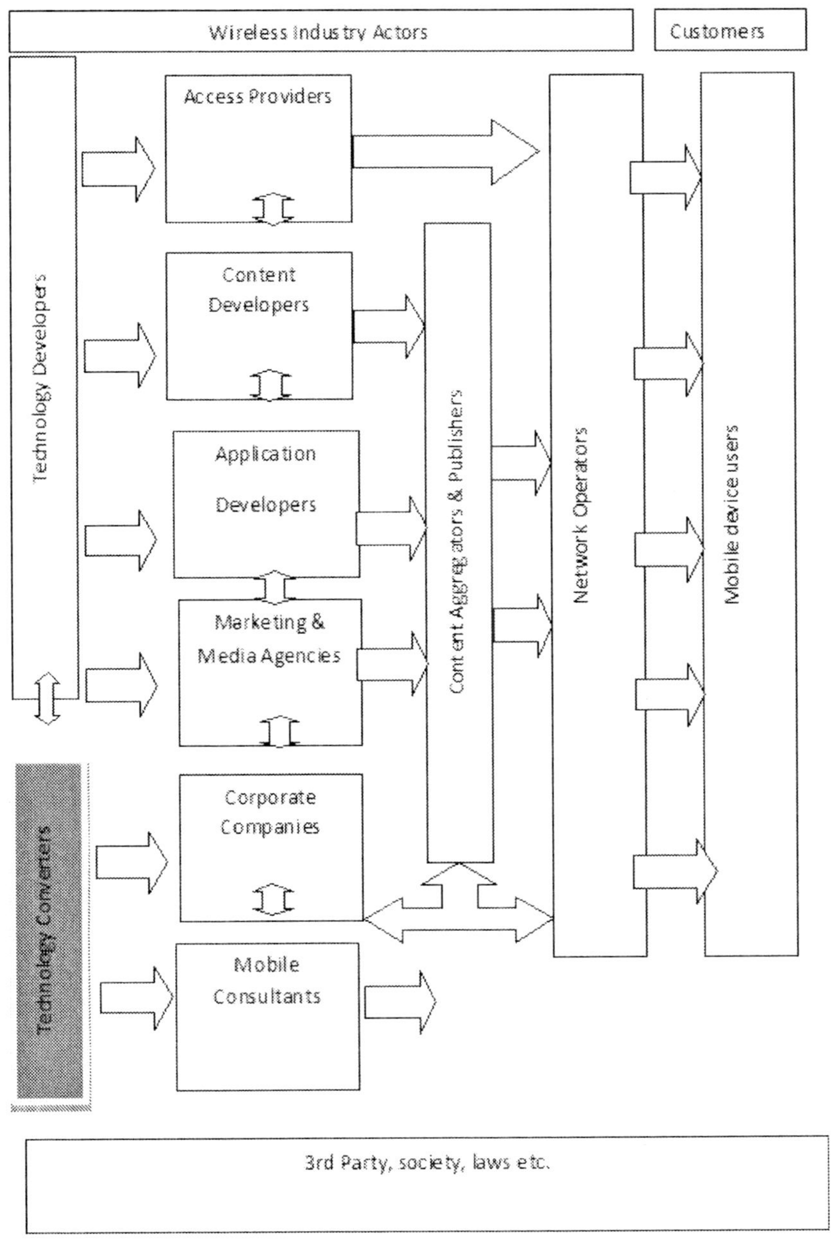

CHANGE OF VALUES IN MARKETING

The understanding of market concerns globalizing and the companies and organizations are benefiting from that while the subscription and expectation of users are proceeding. Individuals are informed about global developments as soon as possible, and they act under the influence of new technologies and new communication environment. The distance of contact between the producer and the consumer is shortened and the long distance has become very quickly surmountable in this way. The changes in logistics implementations have enabled marketing to become more effective.

The awareness of reaching information is ascending. Here, the mentioned information is conscious, correct and useful information. Brand new occupations and software such as content aggregator from content waste have arisen and they create new work environments (Yamamoto, 2004 & Yamamoto & Özbek, 2008).

The values aiming towards increasing individual priorities in the sectors will increase. The societies directed towards good life philosophies also enable the grouping of individuals and the formation of societies according to the values based on characteristics. It is already valid at present with virtual life based on virtual values, and it will reach even further stages.

In the future, such approaches may be influential on numbing and blinding the society that could expand the negative freedom area.

MOBILE SERVICE SYSTEM IN THE INFORMATION AGE

Although service providers and customers are face to face, there is also a very crowded group of players in this system. Among them, the developers and modifiers of technology create a group. On the other hand, agents such as the access providers, content providers, application developers, marketing and media agencies, mobile consulting services providers and network operators, and customs clearance have formed the actors of the mobile industry. Apart from them, third parties and the system of social and legal rules influence the system (see Figure 1).

Mobile commerce is assumed as a subset of electronic commerce where transactions are made via mobile communication network of monetary value gained through the mobile. Here, not only the performance of the transactions, but the entire market that covers electronic commerce is evaluated (Kuo & Yu, 2006). Actually, mobile commerce creates a brand new commercial environment where intense transactions are carried out in near fields.

REFERENCES

Belek, I. (1999). Post-kapitalist Paradigmalar (Post-capitalist Paradigms). Istanbul: Sorun Yayınları.

Beniger, J. R. (1986). The Control Revolution: Technological and Economic Origins of the Information Society. Cambridge, MA: Harvard University Press.

Besson, J. L. (2005). Keşifler ve Icatlar (The Book of Discoveries and Inventions), (5th Ed, Trans. N. Kula). Ankara, Turkey: Tubitak Pubs.

Dura, C. & Atik, H. (2002). Bilgi Toplumu, Bilgi Ekonomisi ve Türkiye (Information Society, Information Economy and Turkey). Istanbul: Literatür Yay.

eprints.ru.ac.za/12/03/glossary.html (n.d.). Retrieved June 23, 2008.

Freeman, C., & Soete, L. (2003). Yenilik İktisadı (The Economics of Industrial Innovation) (3rd Ed.) (Trans. E. Türkcan). Ankara, Turkey: Tubitak Pubs.

Godin, B. (2008). The Information Economy: The History of a concept through is measurement 1949-2005. Working Paper. Retrieved 4/19/2009 from http://www.csiic.ca/PDF/Godin_38.pdf

http://en.wikipedia.org/wiki/Information_Age (n.d.). Retrieved 6/20/2008

http://en.wikipedia.org/wiki/James_Bond#Films (n.d.). Retrieved 4/1/2008.

http://inventors.about.com/library/inventors/bl1600s.htm (n.d.). Retrieved 3/30/2008.

http://kelebek.hurriyet.com.tr/yazarlar/4468659.asp?yazarid=48&gid=61 (n.d.).

http://mural.uv.es/madelos/prometheus.html (n.d.). Retrieved 4/13/2008.

http://oregonstate.edu/instruct/anth370/gloss.html (n.d.). Retrieved 6/20/2008.

http://performans8.blogcu.com/2082215/ (n.d.). Retrieved 4/13/2008.

http://wiki.howardforums.com/index.php/Cingular_Acronyms_and_Definitions (n.d.). 6/23/2008.

http://www.corp.att.com/history/milestones.html (n.d.). Retrieved 4/13/2008.

http://www.davechaffey.com/Internet-Marketing/C3-Macro-environment/Statistics-on-adoption-mobile-services/image/image_view_fullscreen (n.d.). Retrieved 4/13/2008.

http://www.dia.mil/thisisdia/DIA_Workforce_of_the_Future.pdf (n.d.). Retrieved April 17, 2007.

http://www.ideafinder.com/history/inventions/wheel.htm (n.d.). Retrieved 3/30/2008.

http://www.pbs.org/wgbh/amex/telephone/timeline/timeline_text.html (n.d.). Retrieved 3/30/2008.

http://www.visualtron.com/gsm_topic01.htm (n.d.). Retrieved 4/13/2008.

http://www.wirelessintelligence.com (n.d.). Retrieved 4/13/2008.

http://www.yrp.mot.co.jp/works_history_eng.html (n.d.). Retrieved 3/13/2008.

Jensen, T. (2007). Retrieved March 30, 2008 from http://trekmovie.com/2007/11/07/star-trek-begins-shooting-today/

Kuo, Y.-F., & Yu, C.-W. (2006). 3G telecommunication operators' challenges and roles: A perspective of mobile commerce value chain. Technovation, 26, 1347–1356. doi:10.1016/j.technovation.2005.08.004doi:10.1016/j.technovation.2005.08.004

Mc Guire, R. (2007). The Power of Mobility. Hoboken, NJ: John Wiley & Sons.

Müller-Veerse, F. (1999). In Kuo & You. Mobile Commerce Report. London: Durlacher Research Ltd.

Postman, A. (2005). Introduction to the Twentieth Anniversary Edition. In N. Postman, Amusing Ourselves to Death, (20th Anniversary Edition). New York: Penguin Books.

Postman, N. (1992). Technopoly. New York: Vintage Books.

Quo vadis mekatronik? (n.d.). Retrieved April 17, 2008 from http://www.bilgicagi.com/index.php?option=com_content&task=view&id=284&Itemid=137

Ritzer, G. (2007). The Coming of Post-Industrial Society, (2nd Ed). New York: McGraw-Hill.

Tekin, M ve Çiçek, E. (2006). Bilgi Çağında Bilgi Toplumu ve Bilgi Ekonomisi (Information Society and Information Economy in the Information Age). Retrieved April 17, 2008 from http://bilgitoplumu.blogspot.com/2006/10/bilgi-anda-bilgi-toplumu-ve-bilgi.html#_ftn3

Toffler, A. (1981) Üçüncü Dalga (Third Wave). Istanbul: Altın Kitaplar.

Toffler Associates. (2003, May 15). Dia Workforce of Future: Creating the Future of the Defense Intelligence Agency. White Paper, (p. 20).

www.buzzaboutwireless.com/cms/component/option,com_rd_glossary (n.d.). Retrieved 6/23/2008.

www.chrsolutions.com/TelecomGlossary.html (n.d.). Retrieved 6/23/2008.

www.portal.state.pa.us/portal/server.pt (n.d.). Retrieved 6/23/2008.

Yamamoto, G.T. (2004). Your New Career: E-Learning Process Manager. T+D (Training & Development), 58(4), 66-68.

Yamamoto, G. T., & Karaman, F. (2005). A Road-Map for the Development of the Content Protecting Technologies (CPT) for the Content Based E-Business Models. European Business Review, 5, 226–232.

Yamamoto, G. T., & Özbek, A. (2008). A Reality in Virtual World: E-HRM. In T. Torres-Coronas & M. Arias-Oliva (Eds.), Encyclopedia of HRIS: Challenges in E-HRM. Hershey, PA: Information Science Reference.

Yücel. 21. (n.d.). Yüzyılda Bilgi Toplumu (An Information Society in 21th Century). Retrieved from http://ekutup.dpt.gov.tr/bilim/yucelih/biltek03.pdf

uiOne™ (2007). Using Device Personalization to Promote Mobile Data Services. Retrieved from http://www.easemobile.com/uione_personalization_mobiledata.pdf

KEY TERMS AND DEFINITIONS

CDMA (Code Division Multiple Access): A spread spectrum air interface technology used in digital cellular, personal communications services and other wireless networks.(http://www.braddye.com/glossary.html) Also, a Global Mobile Wireless technology underlies on a method of sharing frequency among many users by encrypting each user's signal using a different code. A method of digital wireless transmission enables a large amount of users to share access to a single radio channel by assigning unique code sequences to each user

Content Aggregator: An organization that combines information such as news, sports scores, weather forecasts and reference materials from various sources and makes it available to its customers.

Content Provider: An organization or individual that creates information, educational or entertainment content for the Internet, CD-ROMs or other software-based products. A content provider may or may not provide the software used to access the material

First Wave: The society after agrarian revolution and replaced the first hunter-gatherer cultures (http://en.wikipedia.org/wiki/Alvin_Toffler).

Information Age: Digital Age, Wireless Age.

Information Economy: An economy with an increased emphasis on informational activities and information industry.

Information Society: A society integrated by complex communication networks that rapidly develop and exchange information. oregonstate.edu/instruct/anth370/gloss.html

Mobilizing Wave: A new wave which occurred with the development of mobile technologies after the information age. This wave also supported human mobility both in the real world and the virtual world. The mobility of information and the environments also became integral parts of life.

Post-Industrial Society: Society in which an economic transition has occurred from a manufacturing-based economy to a service-based economy, a diffusion of national and global capital, and mass privatization. The prerequisites to this economic shift are the processes of industrialization and liberalization. This economic transition spurs a restructuring in society as a whole (Ritzer, 2007).

Second Wave: The society during the Industrial Revolution (ca. late 1600s through the mid-1900s). The main components of the Second Wave society are nuclear family, factory-type education system and the corporation. Toffler writes: "The Second Wave Society is industrial and is based on masses such as mass production, mass distribution, mass consumption, mass education, mass media, mass recreation, mass entertainment, and weapons of mass destruction. These could combine with standardization, centralization, concentration, and synchronization, and wind up with a style of organization." (http://en.wikipedia.org/wiki/Alvin_Toffler)

Third Wave: The post-industrial society. Toffler would also add that since the late 1950s most countries are moving away from a Second Wave Society into what he would call a Third Wave Society. He coined lots of words to describe it and mentions names invented by him (super-industrial society) and other people (like the Information Age, Space Age, Electronic Era, Global Village, technetronic age, scientific-technological revolution), which to various degrees predicted demassification, diversity, knowledge-based production, and the acceleration of change (http://en.wikipedia.org/wiki/Alvin_Toffler)

Chapter 2
Mobile Communication

The dictionary definition of mobile is capable of moving or being moved, movable (http://www.merriam-webster.com/dictionary/mobile). Mobile technology considers the portable, movable and usable movement devices and the systems that operate such devices. Therefore, mobile technology has a tremendously wide scope with the services it presents, the installation systems and the devices used on such systems. The developments in this technology come out in dizzying speed. The realized innovations offer people the means of life that gradually mobilize and merge into very different options. In this section, first the definition of mobile communication is examined in detail and the issues and characteristics regarding how it became involved in life and popularized are discussed.

Special care is shown for systems such as the wireless internet and GPS-Galileo which support mobile communication. Their features and the means provided by these systems are explained. The technological developments in mobile communication began from 0G with the establishment of communication satellites, mobile phones and radio connections, and with the formation of global telephone networks and wideband areas, and they are now detailed up to 4G, which is to take place in our lives in the future. At this stage, especially the 2.5G system used in our present time technology and also Bluetooth, ultra wide band, ad hoc wireless systems that have been developed in transition to 3G are also interrelated. Besides, the mobile phone is the most commonly used device of mobile communication and its features are also discussed.

Finally, the development of mobile communication with user dimensions is evaluated and its various functions added compared to other means of communication are discussed.

DOI: 10.4018/978-1-60566-916-8.ch002

DEFINITION OF MOBILE COMMUNICATION

Mobile communication, most simply, is the type of communication provided via mobile devices. As a means, which cannot be limited to individual movements, activity in mobile communication is related with the changing places of the individual in the process of life. The continuity of the mobile communication arises from the continuous togetherness of the communication tool with the individual who always carries the device everywhere.

The fastest developments in terms of mobile life are related with the field of communication. Changes have arisen in parallel to the previously developed technologies in mobile communication. The obtained developments in this regard have rapidly been included in life in the last decade and a brand new communication style and technologies which support this style have arisen.

Mobilization previously showed itself with wireless telephones as the devices used by certain sections. After the use of mobile computers expanded, the opportunity of reaching information from anywhere instead of reaching it in its place has been provided. As the functions of the mobile phone interlaced with those of computers, the opportunity for active individuals to communicate regardless of place, time and location also arose. These opportunities provided the person to possess a memory where he or she can communicate and keep records related with business, and made the functions of the already available devices as usable as possible everywhere. Mobile communication is communicating via tools and instruments that are produced with mobile technologies (http://www.allwords.com/word-mobility.html retrieved 4/23/2009).

According to its general definition, communication is the exchange of ideas and information mutually between two people, two groups or within groups through talk, certain signals, writing or behaviors. The transfer of data from one system to another is also expressed as communication (books.nap.edu/html/up_to_speed/appD.html 6/23/2008). According to Fabun (1968) communication is the transfer of meaning.

By looking at technological changes and developments at present, it is noted that a transition has been experienced in the use of mobile devices mainly communication via Internet, and mobile phone technologies have gradually gained weight in communication. Indeed, communication can now be easily established and managed via palmtop computers as well as mobile phones.

The mobile paradigm is gradually developing. On implementations made with mobile instruments, very big changes can be observed in very frequent intervals. Mobile devices are rapidly transforming into multi-function platforms which can transfer sound, data or multimedia applications.

For the time being, mobile communication can be evaluated as the final point of existing technologies on communication. Mobile communication is the transfer of the communication, information, message, thoughts or meanings via mobile devices at both synchronous and asynchronous times.

Mobile communication does not only consist of speaking on mobile phone. Similar to all other technologies, the formation and development or use of self-information, produces and maintains the spread of relevant information. Mobile devices are used in countless areas (science, studies, politics, tourism, marketing, management etc). It is noted from the micro perspective that they provide active use in the fields oriented towards management and realization of corporate policy and the operation of the control mechanisms in the work environment. Therefore, it densely takes its place as a success providing the communication environment in business management and marketing. Although palmtop computers, iphones and Blackberry's have been developed and recently added into the field of communication in a different style, it is seen that the most commonly used device for this type of communication is the mobile phone. Mobile can be considered everywhere, not because it covers everything but because it comes from everywhere.

In the entire world, the expansion of mobile phones and service providers as well as the gradual decrease of prices has caused telephones to expand only at the base, and the depth of the market with regard to mobile phones has gradually increased. As of 2008, in addition to many brands sold fewer than 50 US dollars, the launching of a mobile phone below 10 US dollars wholesale price is also expected to be launched in 2009. However, as the mobile phone ownership ratio reached 50% globally, the growth of the market slowed down despite all the mentioned developments. Therefore, the concentration of mobile technologies on new areas of use and the development of income depends on the innovations to be realized with regard to the sector. For this reason, the idea that product development is the best strategy is dominant in mobile phone marketing. Because the sector will not lose interest and remain open so long as it supports development and growth with innovations.

According to the predictions, the telecom sector will continue its rapid development also in 2009 with highly impressive figures. It is estimated that GSM, which is to celebrate its 21st year on September 7, 2008, will be used by over 700 operators in more than 200 countries (http://www.ntvmsnbc.com/news/435436.asp). With approximately 546 million mobile internet users globally, IDC predicts that the number of mobile devices accessing the internet will surpass the number of online PCs by 2012 (http://www.techshowwire.com/2009/02/17/gsma-mwc-2009-eservglobal-demonstrates-pre-paid-data-roaming-at-mwc/ 4/23/2009). The mobile subscriber base is expected to cross 412.12 million by the end of 2009 with the overall mobile density reaching 38% (http://www.sitm.ac.in/files/Prevision_09.pdf 4/23/2009).

The major telecommunication forecasts before 2009 were as follows (http://www.ntvmsnbc.com/news/435436.asp);

- Following the subprime mortgage crisis, the increasing possibility of crisis in certain fields of communication is not considered as a danger but rather as certain opportunities for business executives,
- Highly affordable mobile phones will be launched,
- Mergers and acquisitions will also continue in 2008, similar to those in 2007,
- Different strategic plans shall be prepared on how to act in the market in order to become stronger than the mentioned ongoing mergers,
- The prices of GPS will decrease,
- In the telecom sector, investments targeting youth will continue,
- As the GSM sector will celebrate its 21st year in 2008, R&D activities will continue with regard to what could be done towards backward and forward,
- New technologies and new business models will continue to enter the market rapidly, and precautions will be taken for tradition technologies to sustain their existences,
- 70% of the mobile communication and mobile data transfer will be realized from inside of the buildings; therefore, the service sector will be strengthened,
- The increases of speed, which are experienced in communication and which are severely demanded, will be passed through serious questioning and investment plans for speed increases that are not found healthy will be delayed,
- The volume of digital communication will increase, diversify, refresh and become more important

These forecasts express that the plans related with mobile communication should be made for effectiveness and their importance has increased; besides, new technologies are supported and the cost of such type of technologies has decreased. The expansion of the mobile market will provide a great contribution to the service sector and the development of this sector will be owed to the young people.

In terms of social structure, mobile communication can be evaluated according to two basic distinctions. One of them is personal mobile communication and the other one is the mass mobile communication. In personal mobile communication, individuals and individual subjects come forth. The device, which provides it in the best way, is the mobile phone. At present, mobile phones are the devices with which businesses are managed owing to various properties such as use on the desired moment and at the desired place in transferring daily plans and with ease of finding places of meeting. Besides, thanks to the camera on phone, it is possible to take images and send them, and which has various other functions such as Mp3 player, and navigation. The structures such as CRM and ERP are observed from the perspective of companies to the field of communication. Based on these structures, the inclusion of traditional content management platforms is also in question. Mass mobile communication is related with communication oriented towards masses through this way. Mass mobile communication can be evaluated as the systems used by organizations for propaganda, promotion and marketing. Here, the provided content is mostly based on public or corporate means. Especially in media relations such as TV, radio and newspaper, it is clear that in addition to mass communication in PR companies, financial organizations and the general blog, mobile mass communication will also be an influential field.

Mobile communication can be realized via all wireless instruments, Internet and GPS-Galileo in terms of presentation.

WIRELESS

According to Goldsmith (2005), wireless communication is the fastest growing segment in the communication industry from every aspect. Here, the potential created by the exchange of high speed and high quality information between portable devices in any part of the world is emphasized. Cellular systems developed exponentially and the number of uses reached approximately 2 billion users by 2005 all over the world (http://searchengineland.com/070919-075231.php 4/23/2009).

Besides the potential applications, which can be created with this technology, mobile phones that can use multimedia Internet, intelligent houses and devices, automatic road systems, video teleconference and remote education and self-functioning sensor networks are the samples. Highly evident technical works are realized on wireless techniques that support such applications (Goldsmith, 2008).

The devices where wireless communication is used are mobile phones, multimedia devices and palmtop computers. Besides, recording and transferring protocols and messaging technically, individually and socially covering issues such as communication and its calculations, the controls of networks direct and monitor mobile uses location-free mobile personal services, personal effects, personally created data bases, social factors, the access of channels with each other, traffic management, satellite systems and global personal communication, micro cell based personal communication systems, cable free communication industry and technological developments and applications. In this work, mostly wireless communication technologies and the social change and development it brought are taken into consideration. This will be accelerated from the development of networks.

The most significant change in the overall wireless network in the next few years will be the merger of Voice-centric cellular, with circuit-switched delivery in real time and data-centric WLAN, and with IP-based, best-effort packet-switched delivery networks into a single data network (Gatherer & Kumar, 2008).

INTERNET

In our era, as technological skills increase and devices develop, the number of those who have access to the Internet also increases in greater numbers and Internet is becoming nested into life itself. Yet, Internet has interlaced with mobile and the entry of the mobile Internet concept into life has become possible. According to Saarikoski (2006), in the first scientific article, Minges (2005) deals with this issue with the question, "Is Internet mobile?" Apparently, this question has begun to be asked further. Actually, in terms of the ability to reach everywhere, Internet is mobile but in terms of carrying everywhere. It does not quite fit into the meaning of mobility yet.

As the capacities and network speeds of devices increase, a large content begins to be supplied via mobile Internet. Even, according to the marketers, mobile Internet can turn into a platform through which the target masses can be accessed. Thanks to this platform, corporations are now able to provide any messages they desire to their clients. Applications related with mobile Internet are especially concentrated on mobile advertising, and mobile Internet users express that they are more open to mobile advertisements compared to other mobile contents (http://www.nielsenmobile.com/documents/CriticalMass.pdf retrieved 4/23/2009). For example, as of the year end of 2007, 29% of the mobile data users in the US expressed that they could remember some advertising forms on their telephones. These forms are other forms of mobile video advertising, mobile game advertising and mobile marketing. A 26% ratio among mobile Internet users remember that they see certain advertising forms during use. It is stated that this figure is 28% in the UK and 34% in Spain, which is coherent. (http://www.nielsenmobile.com/documents/CriticalMass.pdf)

According to Bayne (2002, p.12), the main difference between the Internet and wireless are the devices in use. According to researches, the number of mobile phone users which is 2 billion in 2005 will increase to approximately 3.3 billion in 2010. Just in the USA, there were 233 million registered users in the end of 2006 corresponding to 76% of the country's total population (http://searchengineland.com/070919-075231.php).

Mobile internet users regard mobile advertising as 60% more acceptable compared to an average data user. Likewise, mobile video users also have a similar tendency with regard to the reception of advertising messages in the immobile world via various media types. (http://www.nielsenmobile.com/documents/CriticalMass.pdf).

These first applications of the usage of mobile Internet are supposed to be the conventional marketing applications for the companies that want to reach the large masses as stated before. But there will be several improvements and changes with the applications and the users' preferences.

GPS-GALILEO

Named for the Italian astronomer Galileo Galileo, the positioning system is officially referred to as just "Galileo". It is also sometimes colloquially described as the "Galileo Positioning System"; however, since this abbreviates to GPS, the shorter name is preferred to avoid confusion with the U.S. GPS. (http://en.wikipedia.org/wiki/Galileo_positioning_system retrieved 4/23/2009).

GALILEO is the global navigation infrastructure under civil control. It is a system established with the cooperation of the European Union (EU) and European Space Agency (ESA). This system consists of 30 satellites, the associated ground infrastructure and regional/local augmentations (http://ec.europa.eu/dgs/energy_transport/galileo/faq/index_en.htm retrieved 4/23/2009).

It provides many facilities to GPS users all over the world; for example, in aviation, road, marine and rail transportation, telecommunication, emergency situations, exploration of resources, mining and construction, financial transactions, etc. All users and governments have contributions in the future of GPS. In the USA, the works on the functioning of civil global navigation satellite systems and international cooperation, as well as the prevention of the deliberate distortion or breakage of signal systems is supported. (http://useu.usmission.gov/Dossiers/Galileo_GPS/Sep1807_GPS_Degradation.asp).

GALILEO is also a new technologic progress, developed for the provision of new generation global services. The role played by satellite global positioning systems in our everyday lives is set to grow considerably. The real impact of satellite global positioning on society and industrial development, as is the case for all major technical innovations, will become clear only gradually even though many practical applications are already possible. While there is no question that the future of guidance systems involves satellite radio navigation, there are sectors other than the transport sector which are already dependent on this new technology even if they are not aware of the fact. This is true of the financial sector when it comes to determining the exact time of bank transactions. Some analysts regard satellite radio navigation as an invention that is as significant in its way as that of the watch. In the same way that no one nowadays can ignore the time of day, in the future no one will be able to do so without knowing their precise location (http://ec.europa.eu/dgs/energy_transport/galileo/intro/challenge_en.htm).

The €3.4 billion project is an alternative complementary to the U.S. Global Positioning System (GPS) and the Russian GLONASS. On November 30, 2007, 27 EU transportation ministers involved reached an agreement that it should be operational by 2013. (http://news.bbc.co.uk/1/hi/sci/tech/7120041.stm).

When in operation, it will have two ground operations centers, one near Munich, Germany, and the other in Fucino, 130 km east of Rome, Italy. (http://www.physorg.com/news115630526.html. in http://en.wikipedia.org/wiki/Galileo_positioning_system#cite_note-1 1/14/2009). Galileo is intended to provide more precise measurements than available through GPS or GLONASS (Galileo will be accurate down to the meter range), better positioning services at high latitudes and an independent positioning system upon which European nations can rely even in times of war or political disagreement, since Russia or the USA could disable use by others (through encryption). Like GPS, use will also be free for everyone (http://doe.sd.gov/octa/ddn4learning/statewide/gps/galileo.htm).

Development of Europe's GALILEO Global Navigation Satellite System (GNSS) began in 1999. With the system to be created, a network of 30 satellites will continuously transmit high-frequency radio signals, containing time and distance data that can be picked up by any GALILEO receiver, allowing the user to pinpoint their position anywhere on Earth. It will allow for precise positioning for air, sea and road traffic management, urban and in-door positioning as well as off-shore applications (http://ec.europa.eu/dgs/energy_transport/galileo/intro/future_en.htm).

Besides, global satellite navigation is an exciting technology and in addition to providing many industrially developed productivity and coherence, it also puts forth a new pleasure and superior level of safety for cruising, and various sports and entertainment activities.

GALILEO is designed to meet the needs of a wide range of positioning and navigation applications with unparalleled precision and a guarantee of signal continuity and reliability even in extreme latitudes. As a key component of the second phase of development of the Global Navigation Satellite System (GNSS), GALILEO is scheduled to be fully operational in 2008.

The Future of GNSS: Galileo + GPS

According to the data of the European Union, the vast majority of satellite navigation applications are currently based on GPS performances, and great technological effort is spent to integrate satellite-derived information with a number of other techniques in order to reach a better positioning precision with improved reliability. (http://ec.europa.eu/dgs/energy_transport/galileo/intro/future_en.htm)

According to the scenarios, EGNOS, the European regional augmentation of GPS, will start to provide its own services and then, with the advent of GALILEO, the service and quality of global satellite navigation system (GNSS) will be doubled. The availability of two or more constellations, more than doubling the total number of available satellites in the sky, will enhance the quality of the services, increasing the number of potential users and applications.

Also according to EU data, significant enhancements are expected to be realized with GALILEO's specific characteristics. Firstly, for urban areas or indoor applications, the design of GALILEO signals will improve the availability of service (broadcast of data-less ranging channels, in addition to the classical pseudo random ranging codes). Secondly, the high-end professional market will also benefit from the GALILEO signal characteristics. Three carrier phase measurements will be essential for the development of specific "TCAR" algorithms, leading to centimeter accuracy over large regions. These kinds of improvements are creating worldwide opportunities especially foremost the communication and transportation sectors and then spread to all other sectors.

WORLDWIDE OPPORTUNITIES

GALILEO services will be a primary component in future high-tech industries, comprising exciting new markets and opportunities. With countries including China, Canada, the Ukraine and India already expressing a concrete interest in GALILEO's space-based navigation technology products and services, the time is now to invest in what will undoubtedly become a massive worldwide market.

The range of potential applications with GALILEO is extremely wide. Looking beyond the transport sector, where it will enhance safety, efficiency and comfort, GALILEO's advanced technological features and its commercially oriented services will make it a valuable tool for nearly all economic sectors. Integration with other technologies such as mobile communication or traditional navigation aids will further increase its potential.

Useful applications will benefit both industrialized countries and the developing world. These will include infrastructure and public works management, agricultural and livestock management and tracking, coordination of external staff and even e-banking and e-commerce authentication. But the value of GALILEO is not limited to the economy and companies. GALILEO will also be a key asset for the provision of public services. In addition to the use of the system for features such as rescue operations and border control, specific user groups will benefit greatly. For instance, it can guide the blind and people suffering from reduced mobility, monitor Alzheimer's sufferers with memory loss, help protect the environment and guide explorers, hikers or sailing enthusiasts.

DEVELOPMENTS OF MOBILE COMMUNICATION

Mobile communication was initially born with technologies aiming to make defense industry more effective, and was developed to enhance intelligence, and gradually, it expanded over the society and came into our lives. Handheld radios can be regarded as its first and simple form.

When mobile phones went into an uptrend in 1996, they were regarded as a fashion accessory for everyone and its beyond was not thought (Michael & Salter, 2006, p. 2). Because, similar to every new fashion element, it came in as an element that facilities everyday life with certain changes in life, human and comfort. However, it developed much further when values obtained with its use were realized. While mobile devices started to be trendy, the networks were advancing.

Global telephone networks took a different shape with the use of communication satellites, mobile phones and radio connections, and a wide means of services appeared with the creation of the wide band area.

At present, mobile phones have expanded almost from the deepest rural areas of Africa to the regions on the North Pole, anywhere people live. According to Saadhna Panday from the South Africa Human Sciences Research Council, "More people have access to a cell phone than a land line" (http://www.csmonitor.com/2008/0115/p20s01-woaf.html). Indeed, the opportunity to access mobile devices without requiring a cable line has arisen at any point of the world. And that presents brand new and wide opportunities in terms of marketing.

The telephone industry is among the branches that are mostly influenced by technological developments and new changes. According to Borstorff, Featherstone & Haymon (2006), the telephone industry was relatively developing in the 1950s, and it was very simple. Bell system had the highest market share in the US. In addition to that, there were also small companies such as Farmers in the same market. Only simple services were provided with regard to the telephone and prices did not change until 1982. The profitability of small companies was also affected from the inflation and the increase in operating costs in 1980. The same was true also for farmers. Although The Alabama Public Service Commission (APSC) made some attempts to increase prices, it could not be very effective. With such an approach, the government failed to respond to the problems of the telephone companies on time. Therefore, small companies were urged to make a decision to reduce costs and dismiss staff. Finally, prices increased in 1982 and remained stable until 1995. The Bell system dissolved in 1983. With the dissolution of the actor with the biggest slice of the market, the telecommunication world went into a big change and deregulation occurred in the sector. In 1995, APSC ratified a law that requires independent telephone companies to charge prices according to the conditions of each individual state taking into account the inflation rates. The law, validated in 1996, opened the telecommunication industry to competition at every level.

Because of the open price system, the systems employed under telephone gained a competition-based understanding. The companies and organizations, which tried to make the market more effective, were not satisfied with the means in their hands, and they introduced new marketing instruments and methods such as telemarketing and telesales. Individuals conceived the rational of shopping and paying

Table 2. Most commonly used mobile functions at present

1. Call	2. Mobile game
3. SMS	4. Mobile search
5. Subscriptions	6. Use of operator's login
7. Calculator	8. Surf on WAP sites
9. Camera	10. Ticket
11. Light	12. Coupon

via telephone and lines. This rational turned into a habit in a very short time and influenced the social attitude towards receiving the service at its door. These kinds of efforts also set up conditions for today's marketing environment and customer habitude.

Technological life criteria should be formed in the societies before the transition into mobile life. These criteria mainly include the increase of technological tendencies (technology creation and utilization tendencies), the increase of the number of mobile instruments and users, and the turning of mobile instruments into a necessity. As technological progresses in the world are realized stage by stage, mobilization period also appeared gradually. First, analog systems were put into operation; these systems were vastly used especially in the Cold War period for intelligence and preparedness for war, etc., and portable systems began to be preferred. And finally, technical details that brought out the transformation into the present time when mobile systems were dwelled upon. The following is Table 2 that contains general information regarding the systems that found a wide area of use in the world. These systems are classified from 0G to the presently evolving 4G (see Table 1).

THROUGH MOBILE COMMUNICATION SYSTEMS

These systems should be collocated with their formation dates and their features. Therefore, we classified the systems from analog to current most advanced 4G.

First Generation (Analog Mobile Phone systems)

As also mentioned earlier, mobile phones as one of the most important products of our time's communication technologies were not easily reached. The developments followed each other stage by stage and the products with common properties began to increase on period by period basis with obtained facilities. Following the facilities of the period of rotary printed media, the radio was invented in 1920s and so, a new media instrument was brought on the agenda. After the press, radio was the first media instrument addressing to wide masses, for the first time delivering the voice. The period of radio and the period defined as 0G in telephone developed in parallel to each other. Zero generation (0G) can be defined as a telephone which is not cellular which has few users and which can work as an analog system.

1G - First Generation mobile phone networks were the earliest cellular systems to develop, and they relied on a network of distributed transceivers to communicate with mobile phones. First Generation phones were also analogue, used for voice calls only, and their signals were transmitted by the method of frequency modulation. These systems typically allocated one 25 MHz frequency band for the signals

Table 1. Major mobile communication systems with year of first widespread use (Source: M. Steer, 2007. "Beyond 3G" IEEE Microwave Magazine, p.77)

Generation / System	Year	Description
0G		**Broadcast, no cells, few users, analog modulation**
MTS	1946	Mobile Telephone Service, half duplex, operator assist to establish call, push-to-talk
AMTS	1965	Advanced Mobile Telephone System, Japan, full-duplex, 900 MHz
IMTS	1969	Improved Mobile Telephone Service, full duplex, up to 13 channels, 60–100 km (40--60 mile) radius, direct dial using DTMF (dual tone multi-frequency) keypad.
0.5G		**FDMA, analog modulation (also Autotel)**
PALM	1971	Public Automated Land Mobile radiotelephone service used digital signaling for supervisory messages, technology link between IMTS and AMPS.
ARP	1971	AutoRadioPuhelin (Car Radio Phone), obsoleted in 2000, used cells (30 km radius) but not hand-off. 80 channels at 150 MHz, half-duplex and latter full duplex.
1G		**Analog modulation, FSK for signaling, cellular, FDMA**
NMT	1981	Nordic Mobile Telephone, 12.5 kHz channel, 450 MHz, 900 MHz
AMPS	1983	Advanced Mobile Phone System, 30 kHz channel
TACS	1985	Total Access Communication Systems, 25 kHz channel, widely used up to 1990s, similar to AMPS
Hicap	1988	NTT's mobile radiotelephone service in Japan
Mobitex	1990	National public access wireless data network, first public access wireless data communication services including two-way paging network services, 12.5 kHz channel, GMSK.
DataTac	1990	Point-to-point wireless data communications standard (like Mobitex), wireless wide area network, 25 kHz channels, max bandwidth 19.2 kb/s (used by the original BlackBerry device)
2G		**Digital modulation**
PHS	1990	Personal Handyphone System, originally a cordless phone, now functions as both a cordless phone and as a mobile phone
GSM	1991	Global System for Mobile Communications (formerly Groupe Spécial Mobile), TDMA, GMSK, constant envelope, 200 kHz channel, max. 13.4 kb per timeslot (at 1900 MHz), 2 billion customers in 210 countries
DAMPS	1991	Digital AMPS, narrowband, (formerly NADC for North American Digital Cellular and prior to that as USDC for U.S. Digital Cellular), pi/4DQPSK, 30 kHz channel
PDC	1992	Personal Digital Cellular, Japan, 25 kHz channel
cdmaOne	1995	Brand name of first CDMA system known as IS-95. spread spectrum, CDMA, 1.25 MHz channel, QPSK
CSD	1997	Circuit Switched Data, original data transmission format developed for GSM, max. bandwidth 9.6 kb/s, uses a single timeslot
2.5G		**Higher data rates**
WiDEN	1996	Wideband Integrated Dispatch Enhanced Network, combines four 25 kHz channels, max. bandwidth = 100 kb/s
GPRS	2000	General Packet Radio System, compatible with GSM network, used GSM time slot and higher-order modulation to send 60 kb per time slot, 200 kHz channel, max. bandwidth = 171.2 kb/s
HSCSD	2000	High-Speed Circuit-Switched Data, compatible with GSM network, max. bandwidth = 57.6 kb/s, based on CSD, higher quality of service than GPRS
2.75G		**Medium bandwidth data—1 Mb/s**
CDMA2000	2000	CDMA, upgraded cdmaOne, double data rate, 1.25 MHz channel
EDGE	2003	Enhanced Data rate for GSM Evolution, compatible with GSM network, 8PSK, TDMA, max. bandwidth = 384 kb/s, 200 kHz channel

continued on following page

Table 1. continued

Generation		
System	Year	Description
3G		**Spread spectrum**
FOMA	2001	Freedom of Mobile Multimedia Access, first 3G service, NTT's implementation of WCDMA
UMTS		
WCDMA	*2004*	*Main 3G outside China*
OFDMA	*2007*	*Evolution to 4G (downlink high bandwidth data)*
1xEV-DO	2006	(IS-856) Evolution of CDMA2000, max. downlink bandwidth 307 kb/s, max. uplink bandwidth 153 kb/s.
TD-SCDMA	2006	Time division synchronous CDMA, China. Uses the same band for transmit and receive, base stations and mobiles use different time slots to communicate, 1.6 MHz channel
GAN/UMA	2006	Generic Access Network, formerly known as unlicensed mobile access, provides GSM and GPRS mobile services over unlicensed spectrum technologies (e.g., Bluetooth and WiFi.)
3.5G		
HSDPA	2006	High-speed downlink packet access, high download speeds up to 14.4 Mb/s, incorporated in UMTS
3.75G		
HSUPA	2007	High-speed uplink packet access, high upload speeds up to 5.76 Mb/s, incorporated in UMTS
4.0G		Low latency (e.g., for VoIP) + MIMO + OFDM + wireless broadband (WBB, > 100 Mb/s) + software defined Radio

to be sent from the cell base station to the handset, and a second different 25 MHz band for signals being returned from the handset to the base station. These bands were then split into a number of communications channels, each of which would be used by a particular caller. (http://www.mobile-phone-directory.org/Technology/1G_-_First_Generation/ retrieved 4/13/2008).

Following the radio, developments which were also noted in the field of cinema and voice could be added to silent motion pictures. In the mid 1950s, television was invented as a mass media instrument and the opportunity of watching on screen broadcasts without having to go anywhere came into the lives of people (Sharma, Herzog & Melfi, 2008, p.32). Instruments such as radio and television took their places in mass communication as the instruments that affect society on social terms as the media, and influenced the world.

Analog telephone systems were initially developed in the US in the 1970s. The first mobile phone service began to be given via an analog system in Chicago in 1983. In Japan, this system was commissioned by Japan Nippon Telegraph and Telephone Public Corporation in 1979 (http://www.yrp.mot.co.jp/works_history_eng.html retrieved 3/13/2008).

In case of AMPS, the first 1G system in the US began to be used in July 1978. There, each channel is distributed through separation of adjacent channels, each of which covers a 30 kHz area. As this system restricts the number of calls in a time period, it was not regarded as sufficient and it was not used efficiently especially in terms of suitable radio waves. Nevertheless, the system had enabled multi-users on the same channel to use the channel simultaneously. This system was defined as "frequency division multiple access" (FDMA).

As the power outputs of cellular transmitters were limited and they were designed for a certain coverage area, it is not possible for the frequencies on other cells at a further distance to block each other. So, the system is considered as one, where the frequencies are reused and the capacity of network increased. The cellular structure of the network can also be used for other properties on cellular phone conversations. Mobile telephone provided a kind of change in passage from one cellular region to another. And this has enabled the utilization of a certain group of frequencies by telephone during the time of change.

The systems that were being newly developed in that period showed a peculiar development as there was not a technology-based certain accumulation of knowledge, and the product's developmental stages were realized with the effects of the changes that arose in short terms. Such developments may sometimes be in the form of shock waves or make the previous systems unusable.

NMT 450, the Nordic Mobile Telephone System using the 450 MHz band, was the first cell phone network to start operating in Europe in 1981. Later, in 1985, the United Kingdom began operations with its TACS (Total Access Communications System). With the introduction of 2G networks, the 1G phones were destined to become obsolete, as they were not adaptable to the new 2G standards. (http://www.mobile-phone-directory.org/Technology/1G_-_First_Generation/). The new 2G standards created a more secure conversation environment and therefore, it turned into a standard in demand by mobile phone users.

These systems initially created a serious benefit in terms of corporate communication, and brought a brand new approach in terms of sales. Salesmen began to visit their clients through taking appointments, and even to take many orders on phone and gain a great deal of time. The savings were not only limited by time, but also corporate costs were significantly reduced. Besides, customers were remembered and called, and the replacement of certain face to face sales efforts with those on phone has reduced the costs of companies.

After mobile systems were included in marketing, a sales style other than catalogue sales arose and it started to be used effectively in new marketing. Although new type of sales organizations that are defined with the names such as telesales and telemarketing had arisen with the development of analog telephone systems, these methods became more usable.

Meanwhile, as mobile phones enabled a highly efficient use with regard to sales or marketing (especially for researches) and ensured benefits from the means of technology, it has turned into a permanent attitude. The relations established on these systems were generally in the form of inbound/outbound calls. Meanwhile, it is also known that during access to the customer through these systems, the more aggressive calls of the manufacturers or the intermediary companies will create a reaction and an irritating situation for the customer.

Second Generation Digital Portable Phone Systems

The developments in digital technology realized a transformation with a dizzying speed in the last twenty years. In the early 1990s, with the configuration of PDC digital cellular systems and their application to the telephone systems, the ground for a different development has been prepared. As of May 1990, there were 42 million PDC subscribers in Japan. This technology was followed by the IS-95 digital cellular telephone system developed in the USA in 1998. In front of the IS-95 and IS-136 digital cellular system formats in the US, the GSM formats are found and started to be used in Europe and Asia.

It will be rational to consider 2G mobile phone networks as the first mobile system that used purely digital technology. The demands placed on the networks were identified on the networks and particularly in the densely populated areas within cities, they handled a large number of calls, reducing the risks of interference and dropped calls. Although many of the principles involved in a 1G system also apply to

2G, 2G networks also entered the market and received a demand. There are differences in the way that the signals are handled on these two systems. For example, the 1G network are not capable of providing the more advanced features of the 2G systems, such as caller identity and text messaging. Equipped with better features, 2G telephones triggered and expanded the mobile phone market.

One of the most successful 2G digital systems in terms attaining a standard is GSM. This standard was developed during the 1980s, and is currently under the control of the ETSI. With the benefit of a well-specified system, commercial GSM services were able to start operating in Europe in mid-1991. The GSM system is able to utilize any of the three frequency bands at 900, 1800 and 1900 MHz. So, many GSM phones could operate as dual-band or tri-band phones, whereby they are adaptable to the local frequency system in the region the user travels through.

In GSM 900, in addition to various band widths, for example two frequency band widths of 25 MHz can be used for uplink and downlink-uplink connections, as well as the band 890-915 MHz for the uplink communications from the mobile station to the base station, and the band 935-960 MHz for the downlink communications from the base station to the mobile station. Each band is divided into 124 carrier frequencies, spaced 200 kHz apart, in a similar fashion to the FDMA method used in 1G systems. Then, each carrier frequency is further divided using TDMA into eight 577 long "time slots", every one of which represents one communication channel. The total number of possible channels available is therefore 124 x 8, producing a theoretical maximum of 992 simultaneous conversations. In the USA, a different form of TDMA is used in the system known as IS-136 D-AMPS, and there is another US system called IS-95 (CDMA one), which is a spread spectrum code division multiple access (CDMA) system. The CDMA technique is also used in 3G systems (http://www.mobile-phone-directory.org/Technology/2G_-_Second_Generation/).

According to Steer (2007), 'Many services in 2G were conducted under the hegemony of GSM. In this system, with the development of the code division multiple access system (CDMA), the use of mobile phone expanded unlimitedly'. CDMA enabled the conceptual development of 2G systems and as it symbolized GSM, being mostly classified as 2.5G.

On the basic 2G networks, as only 9.6 kbps is used similar to that in GSM, it was insufficient for any complex digital services.

Although the basic approach on mobile phones aims to invent complementary or more functional systems, sometimes fundamental changes are also made as the changing of the basic approaches would make the communication stronger. These changes both target the quality of speaking and communication and they also enable the use of fax, etc. instruments and increase the efficiency of the business world with the inclusion of supporting systems.

Naturally, mobile systems can be considered as an interim product in the passage to the information age at the point from where the industrial revolution is reached. Thanks to being able to pass through certain stages, in addition to the currently developing digitalization, mobilization has also begun to be used effectively and expanded. Now, the identity of the caller can be seen on phone, individuals can communicate with the person or organization they desire, and messaging through texting on manual phones can easily and directly be sent with the handheld device.

This is both supportive and preventive for marketers. But, because it prevents the calls that could cause unnecessary frictions, it can also be considered as a material and moral support. Mobile telephones provided a very rapid distribution of SMS messages. It is possible to turn the instruments into a form that could yield in faster and more economic commercial benefits through sending mass messages with devices that enable both reaching individuals and masses and the sales area in a very short time. Mobile phones, as one of the most important mobile devices, are also more practical for the user. The models are gradually becoming lighter, portability is increased or they are turned into more ergonomic and healthy devices with the addition of various pieces (bluetooth etc.) in order to expand its use.

3rd Generation Mobile Phone Systems

According to Michael & Salter (2006, p.11), 3G mobile phones are simple but functional devices that receive travel or direction warnings, and they can pay bills and parking tickets.

3rd Generation (3G) mobile phone systems, provides its users with an access through data on an expanded bandwidth, that switches on an optimum system (Steer, 2007).

According to Qualcomm's report, 3G technologies provide over 488 million subscribers in the world with enhanced data and sound services over wide band. According to the same report, on 4 September 2007, there were over 400 3G operators in 135 countries all over the world.

3G - Third Generation mobile phone networks are defined as the latest stage in the development of wireless communications technology. The main features of 3G systems are that they support much higher data transmission rates and offer increased capacity, which makes them suitable for high-speed data applications as well as for traditional voice calls. In fact, 3G systems are designed to process data, and since voice signals are converted to digital data, these results in speech are being dealt with in much the same way as any other form of data. Third Generation systems use packet-switching technology, which is more efficient and faster than the traditional circuit-switched systems, but they do require a somewhat different infrastructure to the 2G systems. (http://www.mobile-phone-directory.org/Technology/3G_-_Third_Generation/).

Besides, according to the mentioned source here above, high data processing speed may provide a wider scale data service for the subscribers of 3G mobile phones. Mobile internet access and multimedia applications can be shown as its best example. Compared to previous mobile phones, new features presented by 3G telephones will expand their limits with certain new services to be offered. They may be listed as watching TV, multimedia, videoconferencing, Web browsing, e-mail, paging, fax, and navigational maps.

In this open-ended development process, the first country to adopt 3G systems was Japan. The reason was the sharp impact on the Japanese PDC Networks on the severe competition in the big digital mobile phone market in Japan. Unlike GSM systems, the 2.5 G stage in Japan was slurred over with some sort of enhanced services and the gap between the 2G and 3G was filled in with such services. Passage to the new standard is considered as a starting point towards solvings capacity issues.

Recently, an explosion in mobile phone sales is observed in Japan. One third of Japan's population uses mobile phones. According to the estimates, the registered number of subscribers was 60 million in the beginning of the new millennium. IMT-2000 based commercial services were started in Japan in 2001. These services provide some more functions to the devices and some more features to the users.

Third generation mobile phones have also become suitable for global roaming especially in Europe, North America and Japan. However, due to the problems arising from different standards, there is still a need for wide band and multi-modal telephones. There are problems in terms of providing 3G services to the regions that are obliged to use 2G or 2.5G networks.

In mobile, operating frequencies of the 3G system use 2GHz (the IMT-2000 core band) radio spectrum and by the agency of that, they are separated from crowded frequency bands used by 2G and 2.5 G networks. UMTS systems may also enable different data ratios depending on the situation of the user. For example, a maximum of 144 kbps capacity is used for moving vehicles, while the capacity for pedestrians is up to 384 kbps and that for domestic or stationary users is 2 Mbps (http://www.mobile-phone-directory.org/Technology/3G_-_Third_Generation/).

As the 3G systems encompass more different features, the data exchange can be provided easily; therefore, it can be considered as a big step with respect to the intertwining of the mobile phone and internet. Due to that, mobile systems can serve more different aspects such as entertainment, watching, and relaxing compared to the telephone, and the demands for entertainment and videos may change the tendencies and attitudes in the marketing area.

Bluetooth Technology

Bluetooth is a specification for the use of low-power radio communications to wirelessly link phones, computers and other network devices over short distances. In other words, it is the name of the low-power radio signals that enable wireless connection over short distances to telephones, computers and other communication devices. Bluetooth technology was initially designed to support mobile phones, PDAs, wireless earphones, simple wireless networks and peripheral units. Wireless signals of Bluetooth cover short distances, such as 10 meters (30 feet). Bluetooth devices generally communicate at less than 1 Mbps. (http://compnetworking.about.com/cs/bluetooth/g/bldef_bluetooth.htm)

Bluetooth is one of the technologies that have been developed for wireless communication since the 1990s. With this technology, a signal can simply be sent, for example from a computer to a printer as wireless (Haig, 2002). However, in such an environment with such highly developed wireless communication, Bluetooth created a tremendously impressive and new opportunity for marketers in both sales and marketing environments.

Bluetooth is a technology that provides totally cable free and synchronous transfer of data from cable networks to the other side, other networks and the Internet, and which provides a transparent access (Michael & Salter, 2006, p.65). This brings a lot of opportunities for marketers and people who have Bluetooth mechanisms in their mobile devices.

Bluetooth is an important technology that makes many things more comfortable and easier. The main aim of Bluetooth is data synchronization (Scholz, 2007). For example, people who have a Bluetooth featured communication device can use this system to synchronize their devices or to use a wireless earphone. However, due to security concerns, the users have to shut down this feature for most of the time. On the other hand, companies have begun Bluecasting, and people do not want to experience another raid of spam mails on these tools.

According to DMA (2005), Bluetooth mobile technology provides marketers with important opportunities in terms of a rich and totally free content at certain places and locations. For mobile devices that can recognize any Bluetooth device, transactions such as coupons, music download clips and access to WAP micro sites can easily be provided.

Bluetooth Devices

Bluetooth is currently included as standard technology on many products such as telephones, printers, modems and earphones. This technology provides data transfer between two or more approximate devices of short bandwidth. It is mostly used with a Bluetooth earphone, connected with mobile phone or palmtop computers.

Bluetooth facilities the installation and retrieval among devices. Bluetooth devices may sometimes advertise all services created, and enable the recognition and use of such services on many networks.

Ultrawideband

This technology is similar to that of Bluetooth technology and expresses that data can be transmitted with minimum power on an ultra wide band up to approximately 70 m (230 feet) despite many barriers of distance. This technology has certain superiorities. These are low power consumption, further travels of radio waves and the ability to surpass many barriers (Gratton & Gratton, 2004). This system is actually based on military radar studies (Engler, 1995).

UWB is a very important feature in the wireless field. It is able to provide high data transfer in 10 m areas with low power consumption (Kolic, 2004). Thanks to that, it is possible to obtain high speed communication in short-distance environments with a low force between various digital instruments. Especially the services given in multimedia instruments have reached a competitive price level against the services given through alternative wired communication systems. Owing to this, we could get rid of the highly disturbing cables at our homes and offices. These systems have also started to be considered as a technological supporting home-office formation.

The ability to transfer any record from a video recording device to a PC is an example. Another example is being able to view the photographs in a digital camera on a larger screen. Meanwhile, mobile phone also grew in complexity and began to possess various other features such as video, photo, etc. Portable MP3 players, DVD players, and digital cameras are the instruments that created a trend in the UWB market. Their addition to the functions of the mobile phone is a significant development in terms of marketing and sales, such as new type of presentations or enlightenment of the usage or the products etc.

Ad Hoc Wireless Networks

The present wireless devices are gradually becoming smaller, lighter, cheaper and faster. According to Readhead & Trill (2004), Ad Hoc networks (symbiotic networks) are complex network forms consisting of a symbiotic connection of two or more devices from one network. If such devices would be equipped with useful functions, people will also want to use these functions at their homes or even in motion. These types of networks are developed for these types of devices. Due to them, mobilization is also realized without extra cost.

These applications were initially seen in the military field like many of the other communication devices. These types of devices, distributed to different vehicles or combat vehicles (tank, truck etc.) and soldiers, continuously circulate in the war field. A high depreciation rate can be identified for a continuously circulating network. Mobile devices are also regarded as vitally critical in terms of security (Zhou & Haas, 1999).

According to Gedge (2004), Ad hoc Networks are attracting attention with their capacities to bypass the new generation technologies where personal or social networks are located, and they have highly destructive effects on manual network operators.

In Figure 1, on account of home usage, cabled communication uses computer and internet installation compositions. The connection structures can easily be seen here with consoles, mobile phones, headsets etc.

Figure 1. Wireless Home Network Diagram Featuring Ad Hoc Connection

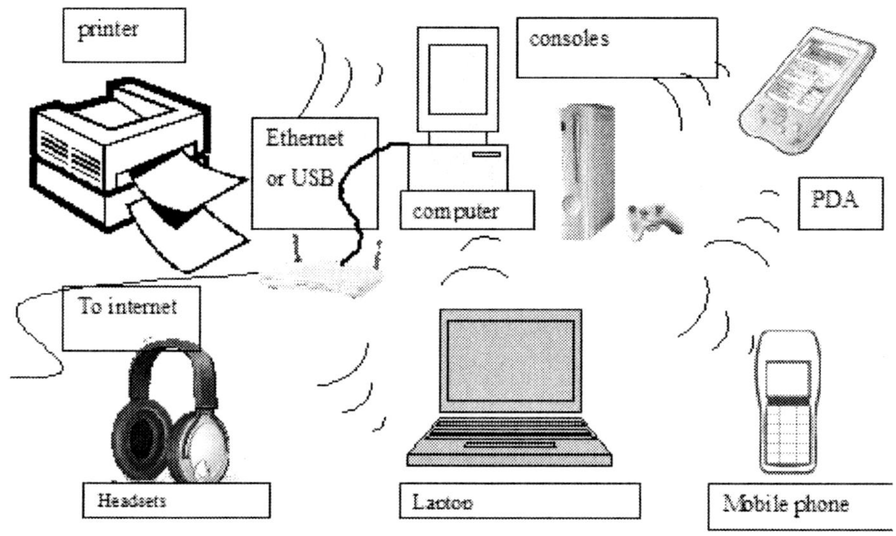

WiMAX

The WiMAX uptake has a unique set of potential, opportunities and challenges in the emerging markets. WiMAX is used as an alternate way to serve areas that cannot use traditional wire line networks. And the target market for WiMAX is lucrative enough to have boosted expectations for technology. However WiMAX deployments have been slower than what the hype suggests owing to factors like regulatory issues and the low buying power of consumers that stand to benefit the most from technology. There are some examples of commercial uses of this technology although some of them are only as access technology and not as a mobile platform (http://www.sitm.ac.in/files/Prevision_09.pdf 4/23/2009).

4th Generation Mobile Phone Systems

The fourth generation mobile systems are becoming much lower in cost and much more effective in terms of capacity. It brings together the idea of using new technological advantages in fields that may come into mind easily. It is important to recognize the more effective use of the scopes of existing 3G systems. Regarding 4th Generation telephones, the creation of an entire mobile work environment in terms of the customer is focused with all details as security, etc. All productions towards this aim can fully benefit from multi-media and they can be realized at very high speed. Communication can be realized with an extremely low delay. These systems can also use semantics, robotics and new forms of mobile technology.

WiMAX is the 4G wireless technology designed to enable pervasive, high-speed mobile internet access to the widest array of devices including notebook PCs, handsets, smartphones, and consumer electronics such as gaming devices, cameras, camcorders, music players, and more. WiMAX also delivers low-cost, open networks and is the first all IP mobile internet solution enabling efficient and scalable networks for data, video, and voice (http://www.intel.com/technology/wimax/ retrieved 4/26/2009).

Figure 2. Four generations of mobile. (Source E. Bohlin. (2006). "Future Mobile Networks: Business Models" Presentation held at 1st Caspian and Black Sea Regulatory Conference, Istanbul 25 - 27 May 2006)

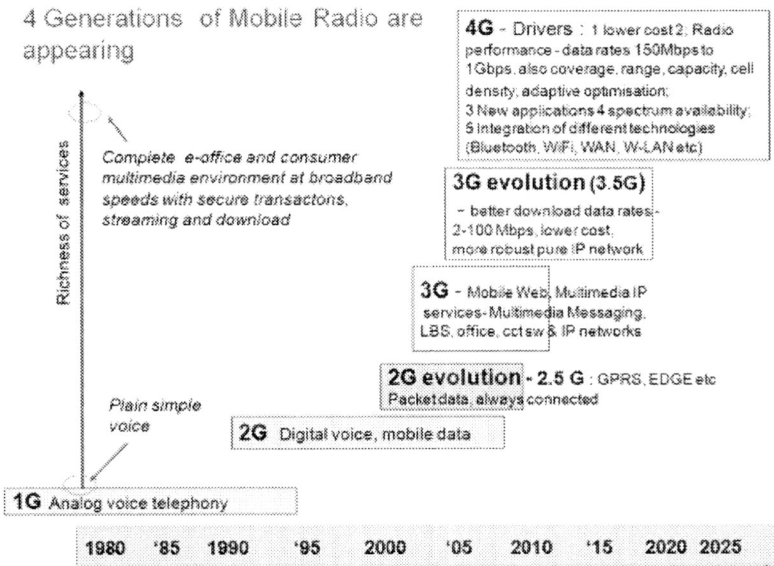

Source: E. Bohlin. (2006). "Future Mobile Networks: Business Models" Presentation held at 1st Caspian and Black Sea Regulatory Conference, Istanbul 25 - 27 May 2006.

With the 4G networks setup, customers would receive improved coverage for voice calls across countries and a network for mobile internet with speeds that would be 10-15 times faster than today. The launch of high-speed mobile broadband services based on LTE-technology is expected at the end of 2010 (Orr, 2009).

MOBILE PHONE AND ITS PROPERTIES

Mobile telephone is a personal communication device (http://www.infonu.com/downloads/HowToGo-Mobile.pdf, p.1) that has several abilities for its users. It is also known with other names such as the cellular phone or wireless telephone. It is individual and not shared with others like land phones. If there had already been such sharing, mobile phone would not be much different from land phones. Briefly, mobile phone is one of the personal instruments of people like their underclothes.

According to Chokvasin (2007), a real mobile phone has features that enable it to recognize its owner; for example, that person's phone number, the ring melody, the image on the telephone, the telephone book and notes, etc. of that person. Mobile telephone is a factor of everyday life, which sticks on the person and cannot be rejected or neglected.

Mobile telephone is a media with quite high interaction which enables the receiver to immediately reply after receiving the message (Bauer et al. 2005, p. 182). The ability to establish uninterrupted bi-lateral communication is the main feature which differentiates mobile channel from others.

Mobile phone brands try to remain in the market with emphasis that they are the single channel that provide 7 days and 24 hours access to the actual target mass. Mobile communication turns became an important part of life with its own means.

In a very short time after its entry to the market, mobile phones nested into the lives of people and their users became almost not able to live without them. For most people, the mobile phone started to be considered as a new limb added to the human body with its confidence increasing and life simplifying features. So, mobile phone began to be used in every moment of life and mobile communication in every point of life.

As an interesting product in terms of marketing, although it is included among luxury products by virtue of its features, it is also included among the easily sold items in the market. Price, availability, and the distribution at the infrastructural organizations make this possible. The presented benefits are continuously developing. As benefiting elements continuously develop, its continuous consumption and depreciation of models are in question. Therefore, they are produced with very fast consumption-based thoughts.

The basic features of mobile phones are:

- Synchronous voice, image, text exchange and saving
- Speed
- Economy
- Bilateral Feature (Interactive)
- Ease of Usage

These features increase the importance attached to mobile communication. Because of mobile phones, even the definition of proximity has changed. It enables reaching the desired point anytime, anywhere and by anyone. These features have great importance in terms of mobile communication.

Mobile messaging in mobile communication can be made audio, visual and written while messages in the past were sent as letters, telegrams and news. Mobile messaging is an important feature, which also approximates telephone to computer. Mobile messaging includes exchanging and saving audio, visual and written messages. Therefore, the concepts related with messaging in mobile communication are mostly used interchangeably. However, mobile communication has enabled passage to messaging perception with a wider scope.

Mobile messaging can be realized with different methods. The methods used in a mobile area should be considered as pieces of mobile data services. Mobile data services include varieties such as mobile TV, mobile music, mobile games, themes, video, location-based services (LBS), business related services, mobile news, mobile advertising, "push to talk", and even "look to talk" which are continuously developed in terms of content. In the personalization of mobile devices, demands should be focused on, and related services should be provided.

For a mobile device user, phones are prepared as it is selected for different customer segments and along with the received melody on the interface, the melody that rings in redial, the wall paper, graphics and many other applications and services are presented as interesting elements, and so the telephone is completely specialized as based on personal selections. The most commonly used mobile functions are calling, playing games, SMS messaging and mobile search (see Table 2). It could also used as a light when needed.

Mobile phones are not only used for communication purposes, but also for operational purposes. As operations provide the use of the phone instead of a simple calculator, clock and time-measuring alarm, chronometer, calendar, calorimeter, and step meter; they also provide many functions such as listening to music, watching films, using it as a dictionary, determining the location, messaging on the Internet, monitoring the place the person is located and finding direction. Therefore, mobile phones are enriched devices. As the non-sound services increase in mobile communication, the use of such services expands and mobile marketing becomes more influential in our lives.

In fact, the mobilization trend in the world advances from two main points in addition to the phone calls. The first one of them is registering progress in sales and distribution, payment and delivery through the use of mobile communication and the second one is the penetration of mobile advertisement and promotions into the lives of people. This causes new structuring in the entertainment sector and creates an environment oriented towards individuality in the concept of entertainment. Because of the new features of mobile phones, it is easy to reach the customer and to send the messages (mostly advertising messages as the previous custom).

According to E-marketer, mobile advertising was 1.5 billion in 2006 and it is expected to reach 14 billion dollars in 2011. According to this forecast, the sector will grow by approximately 10 times in the next three years. Because of this system, approximately 5 billion dollars is expected from the US side. According to forecasts, the share of global mobile advertising will increase from 5.4% in 2006 to 21% in 2011.

Thanks to the mobile operators who provide services to mobile phones. In addition to the communication function, it is now possible to download songs, video clips, logos, melodies, and games to the telephone, to check e-mails, to learn scores of games, to read newspaper, to reach the news and weather forecast, to learn the conditions of traffic, to obtain the definition of a word and to surf on WAP and Web sites, and consequently to explore the Internet. In order to open and view web pages on the mobile phone, the telephone is required to have a special tariff or browser. As some web sites have mobile phone recognition systems, they can automatically view private pages (WAP pages). The mobile phone is designed not only as a device to call people, but also as a camera, mp3 player or a device for other functions that make life colorful. Many service providers serve their users to help them live by using all these features of their mobile phones, and to make them live the most pleasant experiences via the mobile Internet world (http://www.vodafone.com.tr/VodafoneLive/vodafonelive.nedir.php).

THE USAGE EVOLUTION OF MOBILE COMMUNICATION

By 1977, AT&T Bell Labs constructed and operated a prototype cellular phone system. A year later, public trials of the new cellular phone system were started in Chicago, IL with over 2000 trial cellular phone customers. Then, the sector finalized its decision towards opening to the public in general, and with the outcome of scientific tests, the mobile phone was also added among the subjects of marketing. In 1979, the first commercial cellular phone system began operation in Tokyo. In 1981, Motorola and American Radio phone started a second U.S. cellular radio-phone system test in the Washington/Baltimore area. In 1983, the first American commercial analog cellular phone service or AMPS (Advanced Mobile Phone Service) was offered in Chicago, IL by Ameritech, which expanded the mobile market. (http://www.affordablephones.net/HistoryCellular.htm)

Analog AMPS system is the first generation telephone system and it has pushed the limits with requests and demands that exceed its capacity after a while. These systems were first used on corporate communication and they created serious benefits. A brand new approach arose especially in terms of sales and marketing. In the new communication environment, people were able to call and answer calls at any time.

Most sellers began to call their clients via mobile phone for appointments, and they even began to receive many orders via mobile phone. It created a serious time saving for the staff in the work environment. Marketers, who work on behalf of the company, enabled the company to acquire very important customer information: the portfolio of mobile phone numbers. The telephone numbers of clients or callers could be entered on the system. These records could be obtained via manual telephones for a certain time period, and they could be registered on mobile books as written documents. With mobile phones, the marketing staff became able to reach their clients from their cars. Reaching the staff with mobilized telephones could be possible from the analog system at worst, when they were in their cars or at the point where they were located.

However, as the production oriented infrastructure was weak on this system, the ergonomic structure of the devices fell short and certain difficulties of use arose, which triggered the idea of searching a better version for each model produced. Insufficient technological devices caused results for the customer in certain cases, and it became a problem for the manufacturer in many aspects. Difficulties arose especially in terms of access, and technological constraints were shown in the passages which include different service providers. Despite everything, the width of the market and its readiness to grow enabled the entry of competitor forms to the market and accelerated the marketing traffic. However, mobile phones created wide coverage areas, high capacities and load distribution instruments, and gained a very significant superiority compared to manual telephones in the market. The battery life of earlier mobile phones was short and as its speed could not meet the expected standards either, the necessity of various technologies in this regard arose and highlighted the importance of R&D studies. Besides, service suppliers noticed the potential in this aspect and provided support for the advancement of mobile phones. Technological developments provided passage to digitalization, and caused a fundamental change in initial systems.

In the 1990s, mobile systems completed the establishment stage, and therefore, they began to see significant decreases in their costs. The growth trend began rapidly. The mentioned growth was mainly due to the second generation mobile phones, which enhanced digital systems. The products took their places in the market with brand new technologies such as GSM, D-AMPS, and CDMA.

With the increasing number of mobile phone functions added to the initial voice call, individuals were provided with the opportunity of texting via phone in addition to the voice call. Fast and easy communication was provided with short message systems. In addition to the immediate delivery of messages, the services to deliver the messages later to unreached numbers were developed. Messaging sales promotions were used as a means of sending advertisement messages to clients and new marketing devices were put into operation. With the engagement of additional functions such as SMS and Bluetooth, messages began to be used effectively on commercial terms. A situation of aggressive calls or messaging of companies arose instead of customer acquisition and sustainment. In the latter applications on the system, the clients who owned one of the mobile communication devices for recent applications in the system were interested in receiving personalized messages and could easily adapt into messaging. As messaging is cheaper, the young population with a restricted budget and concern on cost began to use messaging more effectively. Young people sometimes tried to reduce their mobile phone costs to a minimum with the provided promotions, and together with the promotions, messaging almost turned into a method of providing financial support for this type of devices as a mutual fulfillment of interests.

With the innovations brought by technology, the creation of a new social environment around the mobile phone has become possible. However, as the message load began to increase and successive messages began to arise from many different organizations, they also started to create a pressure on the customer and different, offensive reactions of customers appeared.

The enhanced systems with new technological means caused both speech and communication quality to increase, and enabled the use of the fax, and other devices when support systems are required. When such support systems are put into operation, written messages can be transferred to paper and turn into documents with legal validity. This which came together with the industrial revolution both enables obtaining assistance from the system during the documentation of the contracts to be made in the business world, and increasing professional efficiency and changing of conditions related with contracts. Naturally, mobile systems should be considered as the final stage of the industrial revolution and an interim point towards the information age. According to computer supported businesses, due to the stages on mobile devices, which have more functional qualities, only digitalization was not sufficient and a second step in digital utilization area towards the creation of systems for a more effective use of mobilization was put into operation.

Mobile devices with a brand new communication environment which allows to see the caller's identity on the screen and through which individuals can both orally communicate and send text messages, both provide time saving and a positive psychological impact as they prevent undesired effects.

On the other hand, they caused messages such as SMS to expand very rapidly and enabled such instruments to create value on commercial terms. Included in the social environment as the devices that can be used commonly and comfortably, mobile phones have also begun to be more practical, lighter and cheaper compared to the devices in the past.

Besides, a new symbol on present time business cards is the inclusion of mobile phone numbers in addition to e-mail. With mobile devices, it is also possible to use audio, visual and written elements for communication. The transmission of visual elements helps use the system more effectively and creates both audio and visual effects. Here, a user-dominant situation appears in general. Whether to accept or reject the received visual or written elements depends on the user.

The producers, who perceived the requirement of companies for high speed access and multi- function operations with faster data processing speeds as a result of the demands in Internet use and the increase in the number of Internet users, have focused on the development of third generation systems. 3G systems are to be enhanced with new features on mobile phones. This technology which indicates the intertwining of mobile phone and Internet as it provides a wider data exchange easily can be considered as a great leap in communication.

Owing to 3G communication devices, the understanding of using other systems for entertainment, watching and listening have evolved towards these devices and caused them to change. As a result of gradually increasing service quality, obtaining value added voice quality, messaging, high speed Internet communication, video call and enhanced security methods, will provide new opportunities for mobile trade. The compatibility of new system mobile devices with previous networks will provide a less economic load on service providers. The continuous progress of communication is not only important for production but also for a further level of system compatibility and it may provide significant contributions in terms of developing commercial applications.

Mobile phones are enhanced with new functions every day. These new functions develop new areas of use. The software and functions of use are upgraded. Although the new 3G system had just been launched not long ago, many companies had already started researching and developing a fourth generation (4G) system. As the change is based on very short periods, researchers aim to design 4G as a system that can reach a much faster speed, have a connection speed of up to 100Mb per second during connection, possess tighter network security and also bring up the quality during communication whether it is voice or video calls. The 4G system is expected to be launched in 2010 (http://wiki.media-culture.org.au/index.php/Mobile_Phone_Technology_-_Future_Development). 4G telephone devices provide connection with a tremendously low delay. So, highly efficient systems and a perfect connection will be established with a communication speed as fast as light.

Mobile phone is a multi-functioned device in essence, but this functionality is being increased continuously with the idea of making a demand for more. This is one of the determinants in the mobile phone market. Smartphone and PDA phones are already launched in the market. These devices are incorporated with computing functions which will be able to replace other devices such as laptop, PDA and even entertainment devices, and have the functions of creating music, drawing pictures, and shooting movies. (http://wiki.media-culture.org.au/index.php/Mobile_Phone_Technology_-_Future_Development) PDAs are handheld devices which can carry computer applications such as telephone/fax, Internet and network operations to the system manually. It is possible to benefit from PDAs as mobile phones, fax exchange, web scanner and personal organizer. Data can be entered in these mobile machines without requiring a keyboard like other portable computers, only with thin point pen handwriting on the screen. These devices have the function of recognizing hand-writing. Even certain PDAs use the technologies of voice recognition and communicate the oral data input as a document to the other side.

Although the keyboard is used on some PDAs, some others allow data entry with a thin point pen. Apple Computer was among the first companies to launch these devices in 1993. In a short time, many different manufacturer companies launched similar products. One of the most popular of these devices at present is the Palm Pilots series produced by Palm Inc. PDAs are currently named as palmtops, handheld computers, or pocket computers (www.mycapel.com.au/Business/SmallBusinessIT/Glossary). PDA telephones are the devices that employ both PDA and cellular phone technologies and which generally have a touch screen and a rarely used mini keyboard (http://www.webopedia.com/TERM/P/PDA.html). Apart from them, there are other models with smart phone, PDA, mobile phone and numerical keys on the mini keyboard. In this context, PDAs are also sometimes named as "Smartphones".

Actually, many mobile phones are gradually turning into PDA devices. At present, the most developed mobile phones can be synchronized with PCs that work with short range wireless technologies such as Bluetooth. Mobile telephones are gradually shrinking, while their powers of operation are increasing. Many sector analysts express that smartphone, PDA and other similar portable handheld devices are becoming more competitive with each other. With 3G networks, mobile phones can now be used via wideband instead of computers in cases that require high speed and uninterrupted connection in video conferencing, watching TV and accessing the Internet. (http://wiki.mediaculture.org.au/index.php/Mobile_Phone_Technology_-_Third_Generation_Phone).

In 2001, Japanese NTT DoCoMo launched the world's first commercial W-CDMA standard mobile phone system, FOMA. Later in 2003, Hutchison Whampoa launched a 3G mobile phone service. As stated earlier, providing high speed access to the Internet is the most important aim according to the 3G communication rule. This communication rule functions on telephones as similar to IP (Internet Protocol) and uses different communication than GPRS. GPRS provides only a limited data flow whereas

3G provides a higher data flow. So, video calls are supported. 3G's Internet access speed is up to 2MB (megabytes) per second, while that of GPRS is only 144B (bites) per second. So, the 3G system is able to provide WAN (Wide Area Network) and LAN (Local Area Network). Here, the key concept is to turn the local-lined terminal to wireless. Another service is the provision of M-Commerce (Mobile Commerce) and global positioning services with 3G. M-Commerce is associated with the idea of electronics, and thus the transaction is realized (http://wiki.media-culture.org.au/index.php/Mobile_Phones_-_Technology_-_Third_Generation_Phone).

Through mobile phones, each link on the chain of marketing began to communicate faster and in a more reliable manner with each other. PDA, Bluetooth, 3G and 4G networks and due to the ease of use of mobile phones, they managed to become permanent in the market with their characteristics based on change.

According to Bohlin (2006), the number of worldwide mobile phone users was 1.5 billion in 2004, but it will increase to 3 billion in 2010 and this figure will be saturated with 2.3 billion people in 2030. Ling (2004) for example, expresses the development in Bangladesh as with 1 land phone per 100 people,

Figure 3. Global mobile user population growth (Source, Bohlin, E. (2006). "Future Mobile Networks: Business Models" Presentation held at 1st Caspian and Black Sea Regulatory Conference, Istanbul 25 - 27 May 2006)

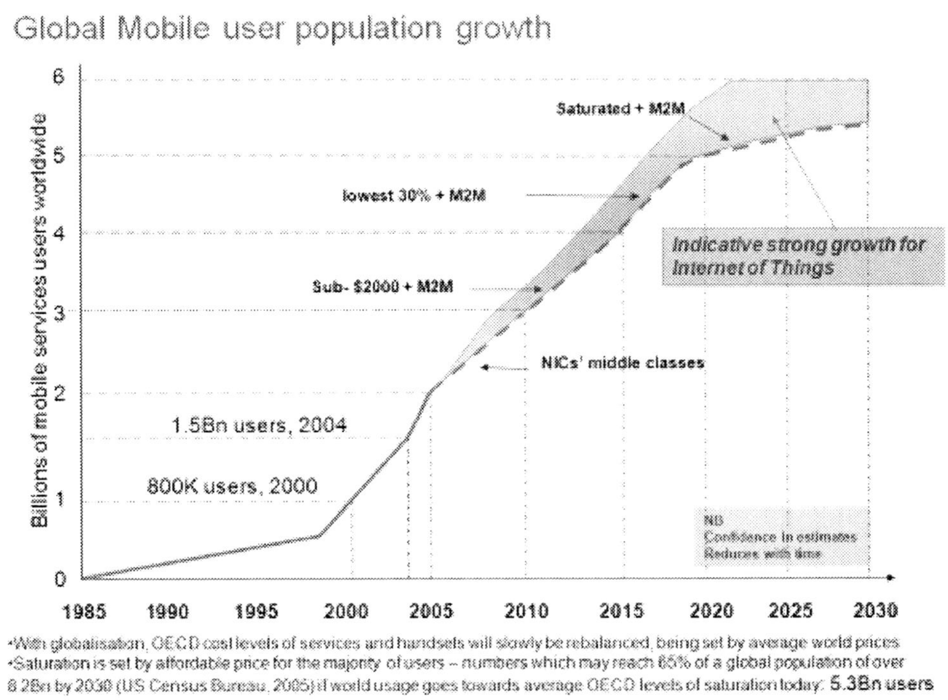

Table 4. Broadcast differences comparison from 1966 to 2007 (Source: Based on Sharma, Herzog &Melfi, p. 57. based on Piper Jaffray (2007) The User Revolution: The New Advertiser Ecosystem. (February 2007),16.)

					2007
				1	Mobile voice call
				2	Mobile message
				3	Mobile telephone ringtone
				4	Mobile music
				5	Mobile societies
				6	Mobile games
				7	Mobile TV
				8	Mobile media
				9	Mobile internet
				10	Newspapers
				11	Magazines
				12	e-mail
				13	TV broadcast
				14	Radio
				15	CD Player
				16	Cable TV
				17	Digital TV
				18	Laptop computers
				19	Satellite TV
				20	Internet call
				21	Internet
				22	DVD player
				23	Satellite radio
			1986	24	MP3 player
		1	Newspapers	25	Tivo/DVR
		2	Yellow pages	26	Slingbox
		3	Magazines	27	ipod/itunes
		4	TV broadcast	28	Blogs
		5	Radio	29	Online video
		6	Cassette tapes	30	Console video games
	1966	7	Walkman	31	PC video games
1	Newspapers	8	VCR	32	Massively Multiplayer On-line Role Playing (MMORP) Games
2	TV Broadcast	9	Cable TV	33	Film download
3	Magazines	10	Personal Computers	34	Podcasts
4	Radio broadcasts	11	Console video games	35	Instant Messaging (IM)
5	Albums and record devices & track	12	PC video games	36 37	Social networks Videocasts

while local farmers use their solar-energized mobile phones to withdraw micro loans and to learn the prices of agricultural products, and so, they increase their powers to bargain during negotiations with merchants.

In spite of positive factors such as electric power which provides the functioning of mobile phones included in the clean energy category and accumulate and use energy, the rapidly changing models of telephones have created certain effects in terms of polluting nature. Here, the producers undertake an important responsibility in terms of producing recyclable models because the scope of disposal for anything with a number of 2-3 billion units in the world would be serious.

REFERENCES

Bauer, H. H., Reichardt, T., Barnes, S. J., & Neumann, M. M. (2005). Driving Consumer Acceptance of Mobile Marketing: A Theoretical Framework and Empirical Study. *Journal of Electronic Commerce Research*, *6*(3), 181–192.

Bayne, K. M. (2002). Marketing Without Wires. New York: John Wiley & Sons.

Bohlin, E. (2006). Future Mobile Networks: Business Models. Presented at 1st Caspian and Black Sea Regulatory Conference, Istanbul 25 - 27 May 2006.

Borstorff, P., Featherstone, M., & Haymon, J. (2006). Changing of the Guard: Farmers Telephone Co-operative. Retrieved 5/25/2008 www.abe.sju.edu/proc2006/borstorff.pdf

Chokvasin, T. (2007). Mobile Phone and Autonomy. In S. Hongladarom & C. Ess, (Eds.), Information Technology Ethics. Hershey: Idea Group Reference.

DMA (2005, August). Mobile Marketing Best Practice Guidelines. London: Direct Marketing Association, UK Ltd.

Engler, H. F., Jr. (1995). Technical Issues in Ultra-wideband Radar Systems. In J.D. Taylor, (Ed), Introduction to Ultra-wideband Radar Systems. Boca Raton, FL: CRC Press.

Fabun, D. (1968). Communications, The Transfer of Meaning. Retrieved 4/23/2009 from www.eric.ed.gov/ERICWebPortal/recordDetail?accno=ED020179

Gatherer, A., & Kumar, S. (2008). The Wireless World in 2012: A Day at the London Olympics 2012. Retrieved 4/23/2009 from http://focus.ti.com.cn/cn/lit/wp/spry101a/spry101a.pdf

Gedge, G. (2004). Ad Hoc Wireless Networks. In Mobile and Wireless Communications: Key Technologies and Future Applications (pp. 137-149). London: BT Communications Technology Series, The Institution of Electrical Engineers.

Goldsmith, A. (2008). Wireless Communications. Retrieved from http://www.stanford.edu/class/ee359

Golsmith, A. (2005). Wireless Communications. New York: Cambridge University Press.

Gratton, S., & Gratton, D. A. (Eds.). (2004). Marketing Wireless Products. Oxford: Elsevier Ltd.

Haig, M. (2002). Mobile Marketing: The Message Revolution. London: Kogan Page http://compnetworking.about.com/cs/wirelessfaqs/f/adhocwireless.htm (n.d.). Retrieved 7/1/2008.

http://compnetworking.about.com/cs/bluetooth/g/bldef_bluetooth.htm (n.d.).

http://doe.sd.gov/octa/ddn4learning/statewide/gps/galileo.htm (n.d.).

http://ec.europa.eu/dgs/energy_transport/galileo/faq/index_en.htm (n.d.). Retrieved 4/23/2009.

http://ec.europa.eu/dgs/energy_transport/galileo/intro/challenge_en.htm (n.d.). Retrieved 7/1/2008.

http://ec.europa.eu/dgs/energy_transport/galileo/intro/future_en.htm (n.d.).

http://en.wikipedia.org/wiki/Bluetooth#Bluetooth_profiles (n.d.).

http://en.wikipedia.org/wiki/Galileo_positioning_system (n.d.). Retrieved 4/23/2009.

http://en.wikipedia.org/wiki/Galileo_positioning_system#cite_note-1 (n.d.). Retireved 1/14/2009.

http://en.wikipedia.org/wiki/Mobile_TV (n.d.). Retrieved 7/1/2008.

http://en.wikipedia.org/wiki/Music-On-Demand (n.d.). Retrieved 7/1/2008.

http://en.wikipedia.org/wiki/Push_to_talk (n.d.). Retrieved 7/1/2008.

http://en.wikipedia.org/wiki/Ultrawideband (n.d.).

http://eprints.ru.ac.za/12/03/glossary.html (n.d.).

http://highered.mcgraw-hill.com/sites/0072464011/student_view0/chapter6/glossary.html (n.d.). Retrieved 7/1/2008.

http://news.bbc.co.uk/1/hi/sci/tech/7120041.stm (n.d.).

http://opencontent.wgbh.org/report/glossary.html (n.d.).

http://people.ischool.berkeley.edu/~hal/Courses/StratTech09/Lectures/Google/Articles/summary-user-revolution.pdf (n.d.). Retrieved 4/25/2009.

http://searchengineland.com/070919-075231.php (n.d.). Retrieved 4/23/2009.

http://searchnetworking.techtarget.com/sDefinition/0,sid7_gci532097,00.html (n.d.). Retrieved 7/1/2008.

http://useu.usmission.gov/Dossiers/Galileo_GPS/Sep1807_GPS_Degradation.asp (n.d.).

http://webopedia.com/TERM/P/PDA.html (n.d.). Retrieved 7/1/2008.

http://wiki.howardforums.com/index.php/Cingular_Acronyms_and_Definitions (n.d.).

http://wiki.media-culture.org.au/index.php/Mobile_Phone_Technology_-_Future_Development (n.d.).

http://wiki.mediaculture.org.au/index.php/Mobile_Phone_Technology_-_Third_Generation_Phone (n.d.).

http://wordnetweb.princeton.edu/perl/webwn?s=lan (n.d.).

http://www.abiresearch.com/products/market_research/Mobile_Business_Applications_and_

http://www.affordablephones.net/HistoryCellular.htm (n.d.).

http://www.bci1.com/cctv_terms_v.htm (n.d.). Retrieved 7/1/2008.

http://www.bluetomorrow.com/content/section/10/37/ (n.d.). Retrieved 7/1/2008.

http://www.buzzaboutwireless.com/cms/component/option,com_rd_glossary (n.d.).

http://www.chrsolutions.com/TelecomGlossary.html (n.d.).

http://www.csmonitor.com/2008/0115/p20s01-woaf.html (n.d.).

http://www.deviceforge.com/articles/AT8171287040.html (n.d.). Retrieved 7/2/2008

http://www.geo.unizh.ch/publications/cartouche/lbs_lecturenotes_steinigeretal2006.pdf(n.d.). Retrieved 7/1/2008.

http://www.gsminfo.co.uk/gsm-glossary.html (n.d.).

http://www.infonu.com/downloads/HowToGoMobile.pdf (n.d.).

http://www.informit.com/store/product.aspx?isbn=0130078174 (n.d.). Retrieved 7/1/2008.

http://www.intel.com/technology/wimax/ (n.d.). Retrieved 4/26/2009.

http://www.intel.com/technology/wimax/ (n.d.). Retrieved 4/26/2009.

http://www.micro2000.co.uk/network_glossary.htm (n.d.). Retrieved 7/1/2008.

http://www.mobile-phone-directory.org/Technology/1G_-_First_Generation/ (n.d.). Retrieved 4/13/2008.

http://www.mobile-phone-directory.org/Technology/2G_-_Second_Generation/ (n.d.). Retrieved 4/13/2008.

http://www.mobile-phone-directory.org/Technology/3G_-_Third_Generation/ (n.d.). Retrieved 4/13/2008.

http://www.mobile-phones-uk.org.uk/gprs.htm (n.d.). Retrieved 7/1/2008.

http://www.mycapel.com.au/Business/SmallBusinessIT/Glossary (n.d.).

http://www.nielsenmobile.com/documents/CriticalMass.pdf (n.d.). Retrieved 4/23/2009.

http://www.ofcom.org.uk/consult/condocs/mobile_call_termination/wmvct/annexf/ (n.d.).

http://www.polyphonic-ringtones-ring-tones.co.uk/glossary.html (n.d.).

http://www.portal.state.pa.us/portal/server.pt?open=512&objID=922&&PageID=223047&level=3&css=L3&mode=2&in_hi_userid=2&cached=true (n.d.).

http://www.qeiicc.co.uk/organising_an_event/it/glossary (n.d.). Retrieved 7/1/2008

http://www.sitm.ac.in/files/Prevision_09.pdf 4/23/2009 (n.d.).

http://www.stanlake.co.uk/recruitment-candidates/recruitment-glossary.php?char=M (n.d.). Retireved 7/1/2008.

http://www.thebesengroup.com/downloads/Mobile.Data.Service.Portfolio.pdf (n.d.).

http://www.uwb.org, (n.d.). In Mobile and Wireless Communications: Key Technologies and Future Applications, (pp. 55-56). London: BT Communications Technology Series, The Institution of Electrical Engineers.

http://www.vodafone.com/start/investor_relations/shareholder_services/faq/glossary.html (n.d.).

http://www.vodafone.com.tr/VodafoneLive/vodafonelive.nedir.php (n.d.). Retrieved 4/24/2009.

http://www.yrp.mot.co.jp/works_history_eng.html (n.d.). Retrieved 3/13/2008.

Kolic, R. (2004). Ultra Wideband -- the Next-Generation Wireless Connection.

Ling, R. (2004). The Mobile Connection. Amsterdam: Elsevier.

Michael, A., & Salter, B. (2006). Mobile Marketing: Achieving Competitive Advantage Through Wireless Technology. Oxford, UK: Butterworth Heinemann.

Miller, J. L. (2007). Verizon Picks 4G LTE Technology. Retrieved 4/26/2009 from http://www.webpronews.com/topnews/2007/11/29/verizon-picks-4g-lte-technology

Minges, M. (2005). Is the Internet Mobile? Measurements from the Asia-Pacific region. *Telecommunications Policy*, *29*, 113–125. doi:10.1016/j.telpol.2004.11.002

Mobile. (2008). In Merriam-Webster Online Dictionary. Retrieved June 27, 2008, from http://www.merriam-webster.com/dictionary/mobile

Orr, J. (2009). Telenor and Tele2 to Build Joint LTE Network in Sweden. Retrieved 4/26/2009 from http://mobilebroadbandnews.com/2009/04/20/telenor-and-tele2-to-build-joint-lte-network-in-sweden/#more-1400

Readhead, A., & Trill, S. (2004).The Role of Ad Hoc Networks in Mobility. In Mobile and Wireless Communications: Key Technologies and Future Applications (pp. 137-149). London: BT Communications Technology Series, The Institution of Electrical Engineers.

Saarikoski, V. (2006). The Odyssey of the Mobile Internet. Retrieved 7/30/2008 from http://www.tieke.fi/mp/db/file_library/x/IMG/20156/file/Saarikoskivaitoskirja.pdf

Scholz, H. (2007). Women in Mobile 18. Retrieved from http://www.m-trends.org/2007/05/women-in-mobile-18-heike-scholz.html

Services (n.d.). Retrieved 7/1/2008.

Sharma, C., Herzog, J., & Melfi, V. (2008). Mobile Advertising: Supercharge Your Brand in the Exploding Wireless Market. San Francisco: Wiley.

Steer, M. B. (2007). Beyond 3G. *IEEE Microwave Magazine*, (February): 76–82.

Toh, C. H. (2001). Ad Hoc Mobile Wireless Networks: Protocols and Systems. Upper Saddle River, NJ: Prentice Hall.

Virrantaus, K., Markkula, J., Garmash, A., & Terziyan, Y. V. (2001). Developing GIS-Supported Location-Based Services. In S. Steiniger, M. Neun, & A. Edwardes, (Eds.), Proc. of WGIS'2001 – First International Workshop on Web Geographical Information Systems, Kyoto, Japan., (pp. 423–432), Foundations of Location Based Services. Retrieved 7/1/2008 from wordnet.princeton.edu/perl/webwn

Zhou, L., & Haas, Z. J. (1999). Securing Ad Hoc Networks. *IEEE Network, 13*(6), 29–30.

KEY TERMS AND DEFINITIONS

1G: In mobile telephones, first-generation systems were analog, circuit-switched. Voice links were poor, handoff unreliable, capacity and security low (www.polyphonic-ringtones-ring-tones.co.uk/glossary.html). These were known as first generation wireless. This is the elementary generation when basic analog cellular phones were introduced. The term also named as First Generation Technology that featured in this period began in the late 1970s and lasted through the 1980s (www.gsminfo.co.uk/gsm-glossary.html).

2.5G: This is the intermediate step between 2G and 3G technological generations and describes the state of wireless technology and capability usually associated with General Packet Radio Services (GPRS) that is between second and third generations of wireless technology. Each generation provides a higher data rate and additional capabilities. (www.portal.state.pa.us/portal/server.pt)

2G: Second generation wireless technology is a term used to describe digital cellular networks such as GSM which were introduced as a replacement for analogue cellular networks(www.vodafone.com/start/investor_relations/shareholder_services/faq/glossary.html) and 2G devices spectrum within 880-915 MHz, 925-960 MHz, 1710-1785 MHz or 1805-1880 MHz bands (www.ofcom.org.uk/consult/condocs/mobile_call_termination/wmvct/annexf/). This is the generation when digital voice encoding was introduced to the market.

3G: Third generation wireless is digital and includes but is not limited to such enhanced features as high-speed transmission, global roaming and advanced multimedia access. 3G will become more and more visible within the next few years and a more robust connection for mobile phones offering the opportunity to surf the Internet and send and receive large e-mails and attachments. It will also allow users to do video calls. This standard allows for high-speed, always-on data transmission and reception. Users can handle e-mail, instant messaging and Web browsing as smoothly as current wired technologies. (www.buzzaboutwireless.com/cms/component/option,com_rd_glossary, 6/23/2008)

4G: Fourth generation wireless. This stage has increased transmission rates. It is expected to provide universal access and device portability by allowing different wireless networks to inter-operate (www.chrsolutions.com/TelecomGlossary.html) and an emerging technology that promises to integrate different modes of wireless communications -- from indoor networks such as wireless local area networks (LANs) and Bluetooth, to cellular signals, to radio and TV broadcasting, to satellite communications. Although proponents of 4G vary in their approaches, most agree it will not be commercially available until about 2010 (http://wiki.howardforums.com/index.php/Cingular_Acronyms_and_Definitions 6/23/2008).

Ad Hoc Wireless Networks: Decentralized wireless network ad-hoc network tends to feature a small group of devices all in very close proximity to each other. "Ad hoc" wireless networks eliminate the complexities of infrastructure setup and administration, enabling devices to create and join networks "on the fly"-anywhere, anytime, for virtually any application.

Bluetooth Technology: provides entirely wireless connections for all kinds of communication devices that allow connection between these devices over a limited area (10 metres). Bluetooth technology is a short-range wireless radio technology that allows electronic devices to connect to one another. Generally, Bluetooth has a range of up to 30 ft. or greater.

GPRS: General Packet Radio Service is a method of enhancing 2G phones to enable them to send and receive data more rapidly. With a GPRS connection, the phone is "always on" and can transfer data immediately, and at higher speeds: typically 32 - 48 kbps. An additional benefit is that data can be transferred at the same time as making a voice call. GPRS is now available on most new phones.

GSM: Global System for Mobile Communications is a standard for digital cellular communications (in the process of being) adopted by over 60 countries (eprints.ru.ac.za/12/03/glossary.html) and using narrowband Time Division Multiple Access (TDMA) for voice and Short Messaging Service (SMS).

LAN: It is the acronym of "local area network". It is a local computer network for communication between computers especially, a network connecting computers and word processors, also a computer network within a limited area.

Location Based Services: Location-based services exploit any of several technologies for knowing where a network user is geographically located. LBSs are information services accessible with mobile devices through the mobile network and utilize the ability to make use of the location of the mobile device. (Virrantaus et al. 2001 in Steiniger, S, Neun, M & Edwardes, 2006).

LTE-Technology: Long Term Evolution Technology, comparable to WiMax, has Qualcomm-backed Ultra Mobile Broadband, with theoretical speeds of up to 100 Mbps (Miller, 2007).

Messaging: (also called electronic messaging) is the creation, storage, exchange, and management of text, images, voice, telex, fax, e-mail, paging, and Electronic Data Interchange (EDI) over a communications network.

Mobile Business Services: Includes communications, information access, computing, integrated information access/computing and business process solutions given over a communications network via mobile device.

Mobile Communication: Wireless communication with portable devices.

Mobile Data Services: Mobile data applications that include mobile advertising, mobile commerce, mobile positioning, and mobile entertainment. Mobile data technologies include EDGE, EV-DO, HSDPA, IMS, LTE, MMS, SMS, UMB, WCDMA, and WiMAX.

Mobile TV: Television service delivered to subscribers via mobile telecommunications networks, such as mobile phone carriers.

Music on Demand: A music distribution model conceived with the growth of two-way computing, telecommunications and the Internet in the early 1990s.

PDA: Acronym of personal digital assistant. It is a handheld device that combines computing, telephone/fax, Internet and networking features. A typical PDA can function as a cellular phone, fax sender, Web browser and personal organizer. Unlike portable computers, most PDAs began as pen-based, using a stylus rather than a keyboard for input.

Push to Talk: also known as Press-to-Transmit is a method of conversing on half-duplex communication lines, including two-way radio, using a momentary button to switch from voice reception mode to transmit mode.

Smart Phone: or s-phone, any electronic handheld device that integrates the functionality of a mobile phone, personal digital assistant (PDA) or other information appliance. This is often achieved by adding telephone functions to an existing PDA (PDA Phone) or putting "smart" capabilities, such as PDA functions, into a mobile phone. The applications can be developed by the manufacturer of the handheld device, by the operator, or by any other third-party software developer.

Symbiotic: A term for parasitic network where the perceived benefits of sharing your connectivity outweigh the negative aspects associated with a parasite.

Ultrawideband: A radio technology that can be used at very low energy levels for short-range high-bandwidth communications by using a large portion of the radio spectrum.

Video Monitoring: A new trend which uses video to monitor remote sites in any part of the world. Video images are transmitted through the telephone network.

WAN: It is the acronym of "Wide Area Network", a broad connection between computers or networks located in separate cities or area codes.

Chapter 3
Mobile Marketing

This section overviews marketing and the impacts of technology on marketing, and presents the current situation where certain traditional rules of marketing are not valid anymore and certain new efforts are required. Also the mechanism of exploration is taken into consideration before going into the details of mobile marketing. Mobile marketing is discussed with its narrow and wide meanings.

A great development has been experienced through approximately the last 20 years of mobile phones, and these devices have turned into significant marketing devices. Not only mobile phones, but also computers converge with mobile, and many different devices such as mp3 players and video players are also mobilized. The convergence of computers with mobile devices makes new devices more functional. It is now possible to influence the target mass in an efficient and measurable manner by using mobile devices individually or in a classical marketing mix.

The section on the features of mobile marketing provides the details of the features that differentiate mobile marketing from conventional marketing devices. Mobile devices which are turned into marketing devices are discussed. The development of mobile marketing is evaluated with the mobile marketing concept and mobile phone device, and mobile phone line marketing concepts. After a brief definition of the concepts, the characteristics of mobile marketing are presented. Advantages such as individuality, directness and reliability of mobile marketing are considered and mobile marketing models are inspected. This chapter ends with the SWOT Analysis of mobile marketing based on literature screening.

MARKETING, TECHNOLOGY AND PRESENT SITUATION

Marketing is the total of efforts aiming to meet human needs and desires. Global marketing has always been influenced by technological developments. For example, similar to the way railways transformed

DOI: 10.4018/978-1-60566-916-8.ch003

products and services from local to national and international, and thus contributed to the development of marketing and the globalization of the marketing environment concept, the reinforced concrete technology brought great changes to urbanization and structuring and triggered both movable and immovable property markets; other technologies also create their unique impacts on the markets.

New technologies have not only changed the lives of people, but they have also significantly changed business practices (Yamamoto & Telli, 2005). Every new technology brings a new dimension to the marketing concept and its interpretations with respect to both product and service. Developments in computer technologies have directed social life towards technology unexpectedly and, mobilization right afterwards has further accelerated such changes and new interpretations. The created mobile technologies enable using the marketing elements in different ways.

It is clear that e-commerce created radical changes on marketing systems and accelerated such developments before mobile development. Sheth and Sisodia (1999) indicated a change in society as they expressed the radical change in the content of marketing together with the expansion of e-commerce. At the end of the '90s, computers developed at a faster pace than mobile phones. In the later period, however, mobile phones developed much faster than computers with regard to their features and functions.

According to Sheth and Sisodia (1999), the development of electronic commerce caused a transformation in society and various points of twist occurred in the evolution of social and commercial change as a result of the transformation. With the mentioned change, sales, advertising, distribution and the promotional styles of conventional marketing cannot even create a sufficient influence, let alone a major one.

The application of conventional marketing rules is getting more difficult as the environment is getting more complicated. International networks of commerce, created in the globalization-oriented market strategies with painstaking efforts, high costs and trained labor force, transformed into new systems through Internet and mobile devices, and provided an inexpensive, simple and fast relationship between the producer and the consumer. There are almost no limitations with regard to the number of manufacturers the consumer can refer to, and literarily, surfing is possible among the product and service options of organizations on technological networks.

Due to these extensive means, differentiation became a point that should be carefully dwelled upon at every point from the beginning of launching the product to marketing and afterwards because products and services are gradually becoming more similar. In order to deliver the differences of similar products/services to customers, differentiation on the promotional instruments should be obtained first. One of the most important promotional materials is advertising. For example, at present digital advertisements have turned window shops into live marketing and promotional instruments, and mobile phones can also be used as a channel which can be accessed at any time and individually. Digital advertisements turn color, light and sound games and effects into brighter and more appealing and more entertaining form; thus making it more effective. However, in addition to providing differentiation, this and other similar promotional oriented approaches have also become easily imitable. On the other hand, the Internet placed an important role in this differentiation in terms of communication and sales; the developed web sites and internet sales have caused new business opportunities to appear.

Brand new opportunities will also appear as a result of Internet access from mobile phones. After the exchange of corporate and personal e-mails from any point where mobile phones can be connected to the Internet became possible, it has become possible to give a more "online" direction to marketing. Digital systems have undertaken missions beyond being a device only, and gained a multi-functional status where hardware, software and services are compiled together.

Actually, new technological instruments led to a shift even beyond "online". All sectors and sectorial companies serving as a mobile content technology developer, an application service provider (ASP) or similar roles are supported using variable data with new technology and software. They also introduce new systems and new means of marketing.

Customers, the most important agents of marketing, on the one hand deliver their knowledge to the system, and on the other hand they become more conscious by the information and documents circulating in the system. Customers can now be informed about anything related with products and services very easily. Even, innovation which is not attractively presented by a manufacturer is rapidly replaced by a substitute who gains advantage in this way, and an environment in favor of the customer is developed in terms of reaching the maximum level of information. The easily convinced and fooled customer has gone and instead, very conscious customers with many alternatives have appeared. On account of both corporate technological developments and satisfaction of the customer's requirement, it has been required to evaluate the existing situation which should also include mobile environments from a marketing perspective. Therefore, we can say that when a company enters into an environment of competition, it cannot reach its actual performance without making true use of mobile devices. As a result, the issue of mobile marketing gained importance and new strategies and tactics appeared with the questioning of the basic concepts related with mobile marketing.

Developing mobile instruments have more functional structures compared to the Internet or television. Therefore, there is a situation beyond competition between these communication devices and those which can only be used at certain locations. In this sense, mobile phones and equipments support a very different marketing approach based on integrated communication in marketing. Mobile phones also have the required features to perform the functions of the Internet and television in parallel to them. With other devices and components (for example, loudspeaker or screen) attached to mobile phones, the uses of devices such as TV, music player, radio can be replaced and their functions can be provided to large masses.

In social and individual approaches, increasing demand by raising the saturation point also causes the expectations in the market to increase and this approach is influential in changing marketing structures.

THE DISCOVERY MECHANISM

People have great responsibilities in this world. Everybody may be created equal, but some go into more different searches and make various efforts to facilitate life on earth and to enable and guide to the development of the humanity. Discovery, with the most general meaning, is to find something new or to realize something new. The efforts of such people can also be defined as approaching life with a kind of discovery mechanism. In fact, the human being comes face to face with this discovery mechanism as soon as he or she is born. Some of them are things that are seen and known in this world, while some others can be considered as unknown and mysterious things. Some of the unknown and mysterious things are revealed with inventions.

In addition to research and development activities, inventor entrepreneurship activities have also become more preferable. For example, the combination of academic researches with applied researches during the industrial revolution has led to an industry and university cooperation approach. Also in the new period, formations in which cooperation could be much faster and individual participants can create different pools with open-ended points or environments of ideas will accelerate the creativity mechanism.

In the present situation, international companies are expanding their influence and adding new dimensions to their differentiational efforts through mutual and cross agreements on patents, licenses and know-how. Here, both the mechanisms of new device development and the exploration of their users' characteristics are dealt with.

Mobile communication and the innovations brought by that means in a sense is the rediscovery of the human being and carry relationships to different dimensions, and engage a different understanding of the social network. However, it is highly probable that depth of thinking may be lost and philosophical shallowness may appear in the lives which are filled with momentary information which are structured in order to meet the daily requirements only. Studies are required on how to overcome that.

THE MOBILE MARKETING DEFINITION

From the point of reaching the customer, mobile marketing is defined as marketing activities related with the corporation, product, service, idea, action and persons, and where customers or potential customers are reached in remote places by mobile phones, palmtop computers and similar devices suitable for remote and wireless communication. At present, marketing efforts with various forms of data such as SMS messages, bluetooth messages, video trailers etc. have developed. Mobile marketing is also expressed with other terms such as m-marketing, mob-marketing, and m-commerce and these terms are most of the time used as intertwined (http://goncatelli.blogspot.com).

With a narrower scope, mobile marketing can be defined as the process of realizing marketing campaigns through using the mobile environment (DMA, 2005). Mobile marketing can be realized through the integration of mobile devices into marketing strategies. Besides, this definition can be evaluated as the realization of advertising campaigns via mobile devices. Nevertheless, the addition of mobile devices to commonly used methods should only be considered as the beginning.

From the marketing communication aspect, the Mobile Marketing Association (2006, p. 22) defines mobile marketing as "marketing which uses wireless media as an instrument and which can provide integrated delivery and directly response process through a cross media marketing communication program" (http://mmaglobal.com/uploads/glossary.pdf). Of course, the scope of this definition is wider than the foregoing because it includes delivery to the customer or the use of media as an instrument. However, this view requires benefitting from other media instruments rather than just a mobile device. Probably, this may be true for the beginning stage. However, it should not be forgotten that mobile marketing turns into a total instrument per se after the individualized environment turns into a habitude.

Leppäniemi & Karjaluoto (2007) exerts the definition of Mobile Marketing Association in 2006 as being the integration of mobile media in the cross media marketing communication program and its adoption for bilateral use. The understanding of cross media publishing is the publication of the same content, which is news and photographs, in more than one channel and thus making them more profitable (Arıkan, 2006). Cross media marketing communication is considered as the integral communication expected to be realized between the media elements such as TV, radio, published and printed media, and the Internet and the mobile service providers. Nevertheless, mobile marketing is an amorphous form, which exceeds the definition and also covers and changes concepts such as entertainment, health, and social life, and presents new differences into life.

Although mobile marketing was initially considered as a supporting activity for existing marketing activities of mass communication instruments such as billboard and TV, it should be considered as a brand new system that hits the target by guiding the customer's actions, triggering the customer and directly presenting innovations as well as due to covering individualism, not mass. Mobile marketing can be considered as the realization of marketing activities that are required to be realized directly with the individual via mobile instruments, and the formation of the systems in accordance. Mobile marketing is a method of marketing that benefits from mobile technology in order to activate the target mass through personal messages.

Mobile marketing has now become remarkable with its interesting practices in sales, payment, advertising, control and marketing management. Although there is not much evidence regarding the effectiveness of mobile marketing yet, marketers have begun to spend more serious amounts of money on mobile media. As a determinant factor in this regard, it simplifies the organizations' search for better marketing investments due to the continuously changing marketing communication environment (Leppäniemi & Karjaluoto, 2007) because marketing activities already made via existing systems such as TV, radio and the Internet are mostly prevented by inhibitions developed by audiences (zapping, blocking, etc.) and even ignored.

Mobile marketing is affected by the developments of mobile technologies. An important point in the development of mobile technologies is that mobile technologies are based on customer satisfaction and customer information. This situation develops in a different way than the interrupted structure: visits or calls from time to time or from the habits of land phone use because on account of mobilization, a highly interactive environment has been created between people and mobile devices. This highly interactive environment, of course, has certain advantages and disadvantages.

According to Peni (2007) in terms of promoting mobile marketing, products and/or brands should be advertised and promoted via various instruments of mobile phone technology such as SMS, MMS or video streaming. However, this definition is almost the same as that of mobile advertising.

The Internet and mobile environment create an "alternative channel". The interest of those who are interested in these channels is rapidly increasing. Because these environments can provide a direct, face to face interaction with customers. Mobile communication hardware, which is always kept by individuals, provides very important benefits to marketers.

MOBILE DEVICES TRANSFORMING INTO MARKETING INSTRUMENTS

It is possible to say that mobile phones have developed sharply in approximately 20 years since their entry into our lives. Initially considered only as an alternative to the land phone, mobile phones received an unexpected interest as we have also said earlier, and turned into a limb of the users in time. The number of mobile phones is increasing day by day as these devices have transformed so as to provide various services such as SMS, photography, entertainment, or banking services as complementary to each other, or sometimes as totally different functions. In addition to mobile phones, computers are also approaching to the mobile state and many different devices such as the mp3 players and video players are becoming mobile. Computers also become more functional as they become more mobile.

On the other hand, elements such as PDAs and portable GPS devices have exceeded the promotional dimension of marketing and they provide outstanding benefits on planning, supply and distribution. Knowing the location of a product or a truck in the world enables an exact planning of the future movements of the product or the activities related with that product, and so, the processes of the supply chain can be followed exactly. These devices have especially been very useful for sales, marketing and logistics companies. This situation, leading to the planning and production of mass tailor made products, provides a different alternative for which marketers should prepare themselves.

With the rapidly increasing mobile phone use in the world, interest towards its employment for marketing purposes is also rising. Due to their properties of being easily used even in places without an Internet infrastructure, mobile phones create an influence on a wider area and mass than the Internet. By dint of mobility, it has become easier to follow up on the efficiency of the marketing activities of brands. Owing to the mobile "call to action" applications in classical marketing campaigns, the healthy measurement of a campaign and the impact on the selected communication channel can be made. In addition to being influential in classical marketing structures, these net measurements may increase the rates of reaching customer satisfaction targets easily through the correct use of customer orientation in marketing. It is now possible to reach the target mass in an effective and measurable way by using it either on its own or in the classical marketing mix.

In terms of use, according to Juniper Research, e-marketer, M: Metrics, CMG Logica data, currently more than 200 million mobile devices are used in the US. People replace their mobile phones and buy new ones every 14 to 16 months due to increasing capacities, as well as differentiating and developing features. Now, the number of mobile phone subscribers is higher than that of the land phone. Every four of five people below the age of 30 receive a text message every day. There are 30 million active text message users in the US and the number is gradually increasing. Besides, 94% of the sent text messages are read. 24% of the mobile marketing messages are shown or forwarded to another person. Mobile marketing has increased customer return of around 1% in conventional media to 15-20%.

Mobile marketing has gradually becames very important in the world. In addition to the development in technology, the "pleasant" marketing communication between the brand and the consumer also has a role on the basis of this development. Here, a freer, more individual preference-based and personal communication has a wider place.

After being discovered by fast moving consumer goods producers initially, mobile marketing is now also expanding into other sectors all over the world very rapidly. The reason is to find a way of reaching the consumer in the fastest way possible. This effect will reach very different points when the aspect of payment is also added.

THE DEVELOPMENT OF MOBILE MARKETING

Mobile marketing employs various technologies such as SMS, MMS, WAP and video streaming. Mobile devices have quite various functions, changing from the simplest telephone to the most enhanced mini-computer. These devices are the main instruments of mobile marketing. The first entry of mobile marketing into our lives was with the SMS advertisements created on the mobile line when mobile emerged as a new channel. They also began to be used for certain attempts of various services provided to expand their services through music download, jingle download, and screensaver download. Then, they were supported with mobile-related campaigns and accelerated mobile phone sales; on the other hand, they caused the formation of very different functions addressed to the customer because of the competitive environment. Therefore, the mobile marketing concept and mobile phone device (product) and mobile phone line marketing concepts continued supporting each other in the first stage because service providers work in cooperation with mobile phone producers. Therefore, if the question "which one" is asked, mobile marketing emerges as a wide scope approach that can expand from many different points (see Figure 1).

As also mentioned earlier, many technologies such as SMS, EMS, MMS, mobile e-mail, LBS (Location Based Services), WI-FI, video streaming and broadcasting with peculiar features, target mass and different marketing scenarios may reach consumers within mobile marketing.

Among the most interesting campaigns with positive outcomes are those with easily modifiable scenarios such as promotions, advertisements and sponsored games, texted barcode and discount coupons, products reminding the brand and brand logos, operator logo, screensaver, ringtone and free games, which provide financial support to the customer and involve the customer in the environment (http://www.kolaymanagement.com/show.asp?k=Z&h=68). Among these scenarios, the game scenario is expected to develop gradually.

In addition to the Internet, which is based on local use and yet also used in mobile marketing, mobile devices and telephones are regarded as the most convenient devices in terms of a direct, interactive and targeted communication of an organization with its customers any time and anywhere, even on the move.

As the new technologies are introduced into the economy and rapidly advance in the market, new labor force profiles arise and in this way, the employees' education, information, skills, behaviors and attitudes are required to be re-organized according to the new order. Companies evaluate the data they obtain by using information systems and feel the necessity for a more convenient structuring and entrance into new researches; they take into account the forms of behavior of their competitors and customers and rapidly revise their supply chains and identify new marketing strategies with the data they obtain, and try to integrate in the global world. The basing of business management to a process-based thinking requires the use of information systems as an indispensable preliminary condition as they provide attaining the factors of information, skills, attitudes and decision making in a short time. So, corporations begin to act more coherently with the contribution of information systems in their marketing functions, management and organizational decisions.

Under the light of them, mobile marketing develops its peculiar areas of use owing to new technologies, and it can offer a means such as communication and entertainment to its customers altogether,

Figure 1. Mobile marketing

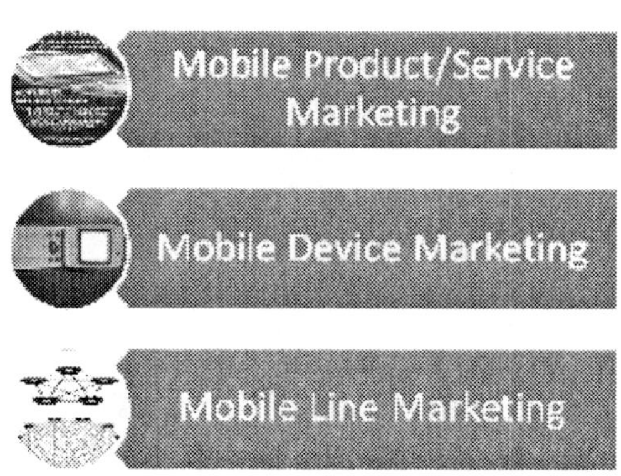

and it draws the attention with its favorable fruits of a new marketing style that can make use of the marketing information systems.

Marketing practices and its sub-functions are affected by the developments in the new generation technologies and new data networks, and thus, they face very serious changes. In the next period, the efforts to form correct mobile marketing strategies are expected to become prominent and important.

Through mobilization, the density of customer-producer interaction increases and it brings in very different applications that were not known before. Here, the clear uses of technological issues in the previous periods also steps in. For example, concepts such as data-based and permission marketing have become suitable for use in the most efficient way. Because the demands and requirements of the consumer can be much more easily monitored and identified with them. Facing a totally new consumer mass, it will be impossible for companies to ignore mobile marketing and act without it in the next 10 years.

THE SIDES OF MOBILE MARKETING

There are many interacting sides in mobile marketing (see Table 1). In mobile marketing, the main sides are the buyer and the seller. The buyer is the side that obtains a virtual product or service itself or information required for the buyer to evaluate the product through a mobile channel. The seller is the party who presents and sells the product or service based on the demands and needs of the buyer via a mobile channel. The seller organization or person may not necessarily be always presenting its own products. Here, the producers also enter into the line.

Service suppliers provide the channel that enables transferring the mobile service to the buyer. These are the organizations that can transmit the desired messages and create the interaction between the buyer and the seller. Content developers are the people or organizations who produce and evaluate the information that guides the seller and service suppliers by interpreting the information obtained from the customers in a suitable manner on a customer basis. These are the organizations that regulate the activities for virtual presentation and promotion of the product or service.

Insurance and transportation companies are the organizations that provide a safe product exchange between the buyer and the seller during mobile shopping. Information technology companies enable the

Table 1. Mobile marketing sides

Buyers	Sellers
Producers	Service suppliers
Content Managers and Developers	Advertising and Media Agencies
Insurance Companies	Transportation Companies
Information technology companies	Financial organizations
Electronic / mobile notaries	Non-Governmental Organizations
E-state	E-government
Universities and R&D institutions	Other Public Organizations
Other Educational Organizations	Approval Bodies
Supra-national organizations	Legal Bodies

advancement and effectuation of existing technologies and they create the structures of device, instrument, system, process, etc. which could meet mobile demands. Financial organizations such as banks, etc. provide the financial systems required for the safe exchange of payments, loans, monetary amounts, etc. during mobile shopping. They are also the organizations which personal payment responsibility is transferred to within the framework of law.

Electronic notaries are the organizations that receive and store the records of agreements and contracts made via mobile devices within the framework of law. Non-governmental organizations and bodies may activate social factors which influence mobile marketing. In transactions via mobile communication channels, this is the widest authorization body on legal issues that are related with the government with legal regulations binding on the parties of transactions. Other public organizations are the public organizations which influence mobile marketing in a similar fashion to non-governmental organizations. These are the bodies that can ensure social welfare and supervise the system, and events and situations in the system.

Universities are the organizations that strengthen the career with developments regarding mobile communication and marketing by producing scientific and experimental information, and that can create a qualified labor force for this market. Universities are also the guiding and directing organizations that can produce projects and supervise the implementations of projects on mobile communication and marketing. R&D organizations are the scientific organizations that can work independently and carry out transactions with regard to development of mobile devices and products, and which should receive support from the social environment and the state. Other educational organizations are those which can convey information related with the use of mobile devices to every level of society.

Approval bodies are those which hold mobile authorities in hand and permit the use of such authorities (e-signature, etc.). Legal organizations monitor legal regulations related with mobile communication and marketing, and make decisions on penal and legal measures if required (courts, experts, etc). Supra-national organizations are the organizations that undertake international legal obligations and regulations that should be formed in international mobile communication and marketing issues, and possess the power to impose required sanctions. Such an organization can be formed beyond all legal systems and with the representatives of each country.

CHARACTERISTICS OF MOBILE MARKETING

The main properties of mobile marketing include advantages brought by the devices, the created channels, and from the point of the individual or corporation, the personality, directness, immediate closeness, reliability, measurability, etc. These characteristics are mostly related with evaluations from the customer's perspective.

Openwave.com lists the characteristics of mobile marketing as follows:

- **Personal:** Mobile telephones and other mobile devices are personal elements. There are generally not lent to anyone even to family members. Elements such as logo, screensaver, ring tone, login password are chosen by the user. At present, these telephones are regarded as a kind of identity card and their numbers are becoming identical with the identity of the person. This provides marketers with unequal opportunities in terms of selection of content and address of their messages to the customer. For example, special applications can only be made to the names of mobile phone owners.

- **Direct:** The communication here is person to person or brand to person. There are not any elements such as press, distributor or retailer between the customer and the company. The user may be influenced by anything the marketer says. In mobile marketing, direct services can be provided to the customer without any intermediaries between the customer and the organization, operator or marketer.
- **Immediate or almost immediate:** The buyer receives the presented message to the target market within minutes and even seconds depending on the number of messages sent from the network and the status of the network. The message is sent and it is received by the user. It ensures the delivery of the right message to the right person at the right time. The receiver may read the message immediately or leave it to another time suitable for him or her or read it again anytime, if he or she wants to. Therefore, the acquisition of the message at the time it is sent, and the possibility of immediate and/or re-reading is higher than all other channels.
- **Reliable:** It is possible to obtain information from service providers on how and when a message is received. It can increase the comprehensive cost of the campaigns via these mobile lines because the marketer organization is also paying a cost of communication in order to learn its conclusion. In such a campaign, if measuring recognition is desired, then it is also included in the scope of the campaign. In short, the status of the marketing activity can be solidified with measurable and countable data. The undesired download of disturbing and continuous programs as spam from the mobile channel without the consent of the customer is more difficult compared to Internet and it requires permission. Therefore, it may also support and increase the reliability of the things presented to the user.
- **Bi-directional:** Wireless communication is also bi-directional similar to that in Internet. Similar to the possibility of talking to customers and listening to them, it is also possible to approach the customers with various presentations. Even, in this way, the customers may establish a relationship with the organization or the brands of the product through direct and personal interaction.
- **Measurable:** Clear and solid measurements are obtained in the campaigns via mobile devices. So, it is possible to obtain beneficial and special information through performed tracking. Measuring the rate and time of the responses to the performed campaigns are possible. Thanks to measurability, the situation of the message or marketing strategies can be evaluated and organized according to the situation.
- **Emphasizes Individuality in Mass Marketing:** Technically, mobile marketing can help organizations reach mass audiences. Such campaigns may be expensive if they are organized with unclear audiences and may fail to provide a sufficient benefit. Contacting the entire mass without determining a target or dividing the market means neglecting personal communication channels. Instead, mobile campaigns require the presentation of personally convincing messages rather than those addressing to a wide mass of audience.
- **Not Just an Advertising Instrument:** As it is well known, marketing covers a wider area than advertising. Mobile marketing brings differentiation not only because of presenting introductory messages, but also in terms of building relationships with customers, creating a dialog and providing services to its customers.
- **Not Just a Marketing Instrument:** From time to time, it may turn into a media of actual social interactions beyond marketing, and also turn social interactions into a marketing instrument.
- **Easy Integration with Other Media:** One of the most important capabilities of mobile marketing is easy integration with other media in terms of concept or implementation. Marketing efforts can

be transferred more effectively when media campaign instruments are used with integrity. Even, in order to reach the best results in mobile marketing, conventional channels may also be collocated. Besides the transfer of audio, visual and written presentations to other media with mobile instruments may be tremendously fast and easy.

- **Suitable for Complex Offers:** From the perspective of 2G and 2.5 G users, it appears that marketing with SMS is suitable for technical structure. SMS does not contain pictures and it is limited to 160 characters. This limit, also automatically determines the attention period of the audience. If the message is longer than 160 characters, then it can coherently be divided into two and maintain its meaning. Therefore, thinking short should be learned. However, these restrictions will be resolved with 3G technologies and mobile devices will turn into complete multimedia devices per se.
- **Personal Privacy:** People are sensitive and protective with regard to their personal privacies. The receipt of an undesired commercial message may prevent that brand from reaching the customer forever. It expresses the necessity of realizing mobile communication upon permission. There is always a better option for the attractive rather than the repulsive. Besides, being accused of SPIT (Spam over Internet Telephony) should be prevented. This issue will be considered in detail later.

Besides, from the point of realized marketing practices, mobile marketing benefits from virtual environments. It reaches and serves the customer in a virtual environment from the time of purchase to the delivery to the consumer, and it can also be interactively tracked down by the customer at every stage. It creates a magical environment that uses virtuality to present reality. It depends on the effective use of structures in marketing such as the supply chain. It enables the conversion of elements to be stored on the database as effective data.

Mobile marketing can be used to obtain the following (Pousttchi & Wiedemann 2006):

- To create brand awareness
- To change brand image
- To promote sales
- To bring in brand loyalty
- To create customer database
- To spread word from mouth to mouth

In addition to them, it also contributes to the creation of other elements such as social environment, entertainment, and reliability.

According to Sultan and Rohm (2005, p. 89), one of the leading points in mobile marketing is creating a value for brand-customer relationships and interacting in a manner which is certainly not coercive. Customers are willing to feel something that has informative or entertaining value for them. Information obtained by customers from mobile marketers should be workable for the customer (Peni, 2007).

The communication devices in mobile marketing reduce the decision making processes and turn them into the processes based on fast thinking and fast moving. They change the processes related with delivery and influence the customer's opinion and purchasing tendencies associated with the product. Therefore, with its new features presented during organization and coordination, it supports adjustments according to speed.

Unlike the Internet, mobile communication devices create a stronger body for the person to communicate with the organizations whose titles and identities are known, and with whom the person agrees to communicate, with other services added to Internet services.

ADVANTAGES PROVIDED BY MOBILE MARKETING

Mobile marketing provides serious advantages in terms of marketing. Considering the subject from strategic planning to policies, it is apparent that measurement, one of the most important characteristics of mobile marketing, will seriously influence strategic plans. Because the feature of measurability may enable a careful follow up of the performed mobile campaign. As it is possible to measure the response rate and time of the performed campaign or activities, it facilitates the realization of long term plans.

Even if commerce is made strategically on a global scale, information from mobile marketing activities may enable making correct decisions. With the merging of mobile data with other data in the past in corporate plans and policies, easy and healthy decisions can be taken. Because, following the requests and behaviors of the customers and acting according to their actual requirements will enable easy selection in terms of both organizations and people.

As an important part of mobile marketing, mobile communication begins from the process before production and provides the information and data entry that could be influential in the entire marketing process. Under the guidance of obtained information and data, production and supply planning may easily be realized. So, such plans may create significant cost advantages in production and supply. On the other hand, with the addition of features such as Galileo and GPS to the logistics aspect, determining the place of a product during its transportation and warehousing becomes possible. Mobile marketing with an effect that improves the process and increases the quality reaches its aim through benefitting from the features that support various them such as just in time and total quality systems in production.

Mobile marketing eliminates time limitations on shopping. It is possible for the customer to do the required thing or place the order to buy a product in 24 hours a day and seven days a week. In other words, mobile marketing provides an entire shopping flexibility.

As mobile marketing is a new marketing channel that uses mobile devices and technologies unlike classical marketing, it appears with different options for customers. It presents new lifestyles into users in terms of shopping environment and social environment with sometimes SMS and sometimes bluetooth or other features.

With regard to products and services, mobile marketing both reaches the customer at the most suitable time, and it also encompasses a kind of training function from the aspect of information the customer will regard concerning products and services for the conscious use of them. Thanks to that, the advantages of reaching customers with more transparent, effective and correct efforts can be obtained.

The organizations which understand the importance of mobile marketing take advantage of personality in mobility and initiate effective marketing. This, in turn, provides precedence to companies in competition. It is even possible to show customers some aggressive campaigns to prevent shifting of customers to competitors as beneficial for them, and to continuously attract customers to their own products and services. Briefly, owing to mobile marketing, the applications that mean hitting the bull's eye can be made.

In the present time, when the elements such as entertainment, work, information, and education have closely approached each other, the aim is to draw out activities that merge or enable their parallel functioning.

Mobile marketing personalized activities such as marketing, communication and advertising for the customer in a very short time. Besides, with the inclusion of payment systems to mobile marketing, all payment activities, which were previously realized with smart cards, can now be moved on mobile and people will be able to shop even without carrying any money on themselves.

On the other hand, the synchronous exchange of money and products also creates an almost full time change in the finance market especially where the exchange of big amounts of money is required. With such advantages, new changes may occur even in the contract structures; new legal arrangements can be made which would lead to the occurrence of e-document or m-document providing organizations in this respect or to the use of mobile devices and saving time and applications as a result.

The organizations that can act in accordance with customers may also organize their promotional activities and especially discount coupons in a manner to address to the actual requirements of the customers. The tailoring of corporate discounts and promotions to the target mass also provides another advantage with its financial benefit for the customer.

It can also help in the organization of the marketing activities that could be accommodated according to changing generations.

Corporations may also obtain certain advantages by using mobile marketing as a supporting instrument together with their former habits and styles along certain conventional marketing activities.

MOBILE MARKETING MODELS

Mobile marketing required developing different marketing models as a result of the changing environment with the influence of mobile marketing technology. Like e-marketing, mobile marketing also followed a similar development to those of the classical marketing models. However, as models exceeding the classical ones were developed in e-marketing, new models will naturally be developed also in m-marketing. Mobile marketing models begin from mobile communications and develop until mobile shopping systems and enable the creation of different systems. With regard to advertising, one of the most important functions of marketing, (models funded from advertising) and mobile purchasing models for purchasing activities are also in question. Such models are evaluated according to the structures oriented towards corporate fund or income provision.

For example, the mobile purchasing model Mobile B2B is for business to business mobile electronic shopping. Similar to all of the properties of B2B, there is also mobile support available at any point of the world. A buyer or seller, who could not be reached under normal conditions, can be reached this way. Through these transactions, buyers and sellers can easily reach each other in mobile communication environments without any physical barriers. The standard and frequently repeating commercial transactions are automated in the electronic environment and so, new values are obtained. Corporate time and labor revenue are realized and the costs are reduced with the integration of side services such as shipment, insurance and financing in this environment. Besides, the retrospective follow up and reporting of concluded transactions are also possible on time easily and free of mistakes.

In the mobile advertising model, the income is obtained from banners, sponsorships, advertisements and other marketing methods.

Models may sharply differ in terms of application, and the organizations and sectors that develop the model are based on multilateral thinking. Such models may be considered oriented towards various services such as mobile sales, mobile health, mobile finance, mobile education or mobile entertainment. These models are developed in different ways with respect to the properties and attributes of the services, and they are obliged to respond to the demand which may occur in any aspect and to the requests and requirements of the customers.

For example, in the mobile shopping model, a supply chain management close to full-controlled model is possible. A piece of the structure and application after the customer's order are listed in the following steps (See Figure 2). As can also be seen on Figure 2, the customer places the order in the first step via the mobile device. The received order is transferred to the corporate data base, and also the bill of order is transmitted to the relevant units. The data base is primarily focused on two functions. The first one of these functions is the required orders for the transformation of data to services and the second one is the analyses of the data. Data analysis is especially important for long term customer monitoring, evaluation of preferences and characteristics, as well as the segregation of the characteristics.

If the requested product is available on hand, then its compliance with the customer's order is checked and the necessary preparations for the delivery to the customer are made if it is suitable for delivery. Here, the products to be delivered are again organized according to the preferences of the customer. For example, if the customer checks the fast delivery option, then the cargo service can be arranged accordingly. Besides, the preparations also include the packaging of the product, its invoicing, and the preparation of the return document if it is rejected by the customer, which is to state the reasons, the entry and exit transactions and the warehouse order. Then, the product is shipped via postal service or cargo. The information of the cargo can be transmitted to the customer's mobile device beginning from the time the products are shipped. The final stage is the realization of delivery at the time and place indicated and requested by the customer.

If the requested product is not available, or if it does not match with the required properties with a mobile or online order message, the supplier's feedback message and information on the product and time of supply can be presented. If the customer requests fast delivery options, then information from the supplier for delivery to the customer should be obtained. The information obtained from the supplier can be transmitted to the customer. After the product is supplied, the conditions of the third step apply exactly as they are.

New marketing models will also arise according to new technological advancements to be introduced in mobile communication. For example, mobile store models and shopping systems have not been distributed on the social scale yet, and these types of models will be used more effectively with 3G and in the period afterwards. In these structures with the properties of online stores, the customer will be able to browse via mobile device and place an order. However, the native models, bluetooth marketing or LBS are used at present. At present, the existing models are the web sites that work with news, publication, information or music, etc. Purchases with the facilities provided by telephone operators and by subscription or prepaid system, the existence of these types of services depended on such revenues.

The creation of multi-channel platforms similar to those in e-commerce may be considered as another model in mobile marketing. At these platforms, all of the channels such as the Internet call center, mobile devices, kiosks etc. are used altogether and the marketing activities of these channels support each other, as also shown in the figure (See Figure 3).

Another mobile marketing model is the modular model created by using the dynamics arising from the inherent structure of the mobile marketing model itself. In this model, each activity and application

Figure 2. An order application model

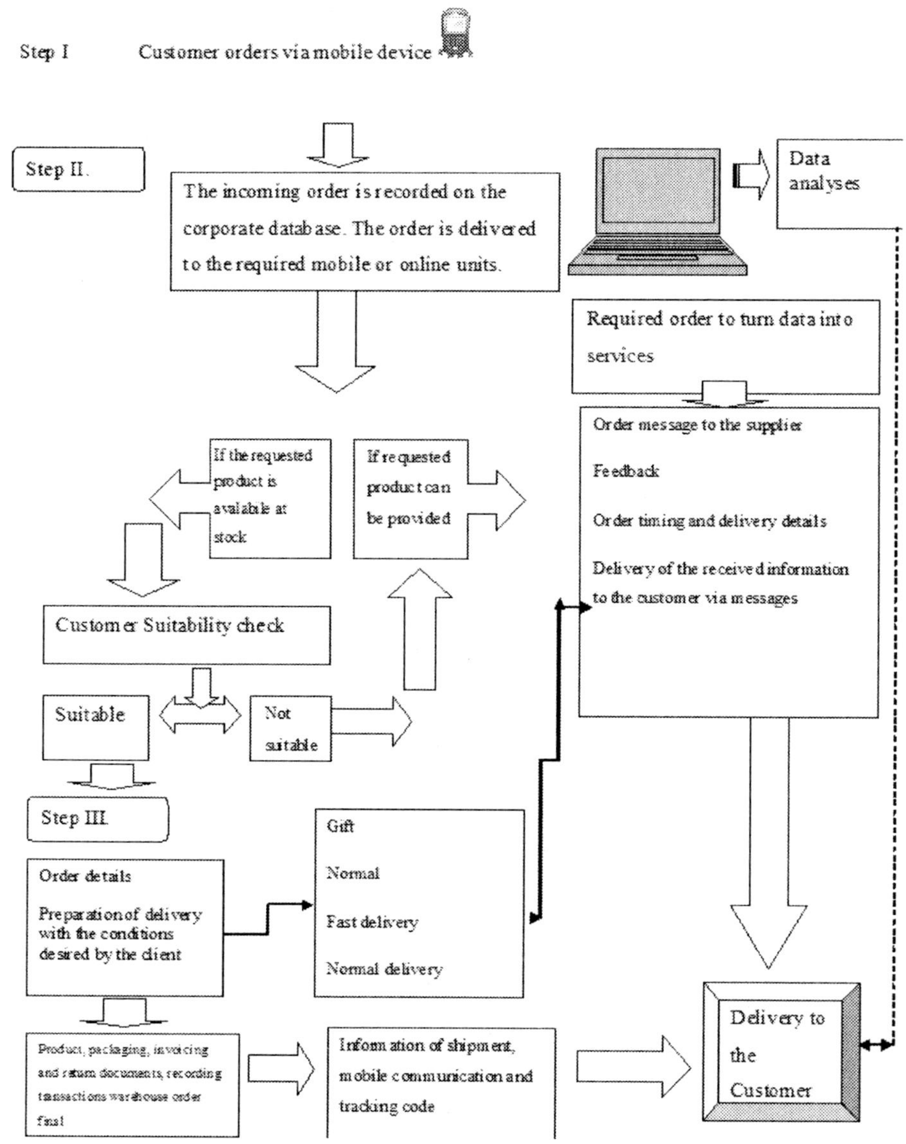

is associated and specialized between different modules and turned into the most suitable form in terms of applications. In the modular system, in addition to creating the independent functions of wireless communication agents such as service suppliers, application and content providers, the project groups which create and develop private systems for new entrepreneurs are also employed.

Mobile marketing can also be evaluated as push- and pull-based. In push-based mobile marketing, the content is sent by the advertiser or marketer to a mobile device generally once. Here, there is not a content requested with a subscription system. In push-based mobile marketing, the content is provided by voice, SMS messages, e-mail, MMS messages, TV or radio messages, picture messages, researches and similar advertisements and elements (MMA, 2006). In pull-based mobile marketing, the content

Figure 3. Multi-channel platform

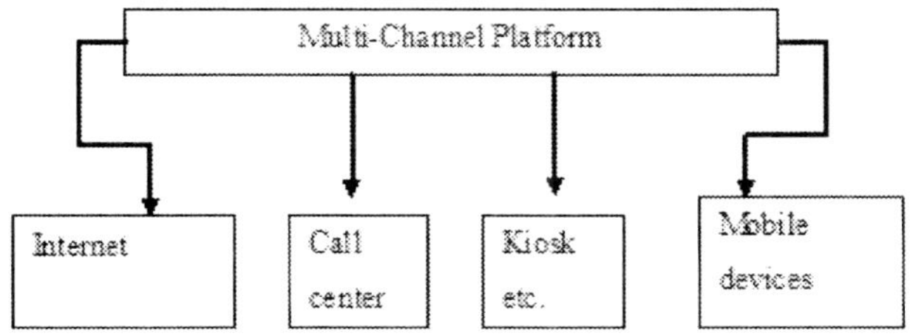

is sent to a registered subscriber upon his or her request and generally once (MMA, 2006). When the customer requests a mobile coupon or responds to marketing communication, it can be considered as an example for pull-based marketing.

A SWOT ANALYSIS ON MOBILE MARKETING

According to the SWOT Analysis based on the literature screening of Peni (2007) with regard to mobile marketing, the basic characteristics such as interaction, access to customer, and personalization are listed among the strong aspects of mobile marketing. The weak aspects are: lack of research on the issue and

Table 2. Mobile marketing SWOT based on literature screening-analysis (Source: Peni, 2007, p. 7)

Strengths Interaction Anytime, anywhere access to customer Personalization Complementary with other media or complements mobile marketing attempts of other media Can be presented to individuals, not masses One to one dialogue with the customers Person-specific addressing Means of multimedia Ability to deliver high speed messages Being fast and effective Direct accessibility	Weaknesses Few number of researches on the subject Few number of scientific cooperation and sharing of knowledge on the subject
Opportunities Consumer behavior Viral marketing Value added Value of knowledge Value of entertainment Creating loyalty Being a piece of the entire marketing strategy Prevalence of mobile devices all over the world Developed technology for devices Higher interaction with customers Customers' acquaintance to work with mobile devices A closer brand relation Sharing of cost with customers (viral marketing) Place and time in independent media Increasing use of mobile devices Increasing number of multimedia mobile phones	Threats Consumer laziness (old habits, behaviors and images) Regarding as Spam Concerns of Secrecy Technological Barriers (delivery processes, screen size etc) Difficulties in practice Resistance to change (The desire to preserve the former system by marketing and those who serve in marketing) Dependence on permission (Permission should be obtained in order to access customer and sent messages) Legal regulations Requirement of other media for its development

low amount of technological cooperation and sharing of information. The threats related with the issue are the factors from the point of both the user and the intermediary such as difficulties in practice, old habits and the lack of legal infrastructure as well as other environmental factors. Opportunities are mainly prevalence all over the world, the creation of independent media and time, increasing value of information, and changing values of entertainment causing the formation of new consumer behaviors and interaction (see Table 2).

REFERENCES

Arıkan, A. (2006, May) Bilgi Endüstrisinin Kitle İletişimine Etkisi: Değişen Medya. I. Uluslararası Bilgi Hizmetleri Sempozyumu İletişim. Retrieved from http://www.aykutarikan.info/ARIKAN-PAPER01-2006.pdf

DMA. (2005). Mobile Marketing Best Practice Guidelines. London: Direct Marketing Association (UK) Ltd.

http://mmaglobal.com/uploads/glossary.pdf (n.d.).

http://www.abiresearch.com/press/1202-2008+Global+Mobile+Device+Market+Still+on+Course+for+1.3+Billion+Units+Despite+Economic+Woes (n.d.). Retrieved 4/25/2009.

http://www.kobiegitim.com/test/A_Mobil%20Pazarlama_dosyalar/frame.htm#slide0127.htm (n.d.).

http://www.kolaymanagement.com/show.asp?k=Z&h=68 (n.d.).

Leppäniemi, M., & Karjaluoto, H. (2007). Mobile Marketing: From Marketing Strategy to Mobile Marketing Campaign Implementation. In Proceedings of the 6th Annual Global Mobility Roundtable Conference, Los Angeles, June 1-2.

Peni, H. (2007). Mobile Music Marketing: Willingness Toward Mobile Music Recommendations and Viral Marketing. Masters Thesis, Lappeenranta University of Technology, Kouvola, Finland.

Pousttchi, K., & Wiedemann, D. (2006). A Contribution to Theory Building for Mobile Marketing: Categorizing Mobile Marketing Campaigns through Case Study Research. In Proceedings of the 5th International Conference on Mobile Business (ICMB 2006), 26-27 June 2006, Copenhagen, Denmark (pp.1-8). Los Alamitos, CA.

Sheth, J., & Sisodia, R. (1999). Revisiting Marketing's Lawlike Generalizations. Journal of the Academy of Marketing Science, 27(1), 71–87. doi:10.1177/0092070399271006doi:10.1177/0092070399271006

Sultan, F. & Rohm, A. (2005). The Coming Era of "Brand in the Hand" Marketing. Sloan Management Review, 47(1), Reprint number 47115, 82-90.

Yamamoto, G. T. (n.d.). http://goncatelli.blogspot.com

Yamamoto, G. T., & Telli, A. (2005). High Technology for Marketing: New Applications & Integrated Circuits. In Science Marketing 4th International Science-to-Business Conference, October, 18-19 2005, Pretoria, South Africa.

KEY TERMS AND DEFINITIONS

Mobile Marketing: It is an emerging concept, but the speed with which it's growing its roots is remarkable with marketing on or with a mobile device such as a mobile phone or PDA.

Mobile Device Marketing: Marketing of the emerging mobile devices such as mobile phones, PDA's etc. According to ABI Research, Nokia was the world's largest manufacturer and also seller of mobile phones with a global device market share of approximately 40% in 2008. Other major mobile phone manufacturers (in order of market share) include: Samsung (15.2%), Motorola (9.3%), Sony Ericsson (8.3%) and LG (9.2%). These manufacturers account for over 80% of all mobile phones sold and produced. (http://www.abiresearch.com/press/1202-2008+Global+Mobile+Device+Market+Still+on+Course+for+1.3+Billion+Units+Despite+Economic+Woes 4/25/2009).

Mobile Line Marketing: The marketing of services that mobile operators perform to be on line competition.

Mobile Product/Service: Marketing a mobile product or service by mobile channel with the permission of the customer.

Mobile Marketing Models: Several models that try to explain mobile marketing applications.

Mobile Advertising: Advertising performances mostly served for specialized customers that could be available with/from the mobile device.

Mobile Purchasing: Realizing shopping from mobile devices.

Mobile B2B: The usage of mobile channels within business transactions.

Pull-Based Mobile Marketing: In pull-based mobile marketing, users participate upon his or her request via mobile phone. Then the content is sent to a registered subscriber upon his or her request and generally once.

Push-Based Mobile Marketing: In push-based marketing, the content is sent by the advertiser or marketer to a mobile device generally once to stimulate action to buy the product.

Chapter 4
Mobile Marketing Practices

Business types seen in e-commerce such as: e-shopping, e-store, e-tailing, e-auctioning, e-banking, e-trading, e-entertaining, e-learning, e-realtor, e-gambling, e-directory, e-news, e-zine, e-advertising, and e-logistics are now transforming into m-shopping, m-store, m-tailing, m-auctioning, m-banking, m-trading, m-entertaining, m-learning, m-realtor, m-gambling, m-directory, m-news, m-zine, m-advertising, and m-logistics.

According to Warren, Hakes and Simmons (2009), mobile marketing can be conducted in eight ways. These are:

- Mobile messages,
- Mobile portal,
- Mobile off portal,
- Mobile broadcast,
- Mobile content,
- Mobile CRM,
- Mobile local,
- Mobile handsets.

According to StrategyAnalytics White Paper (2008), more people are spending more time using mobile phones, with the phones themselves, and the way people use them are changing dramatically. Nowadays "data usage" is any non-voice application that uses the mobile handset as the delivery platform. Mostly employed mobile marketing practices are mobile wap, mobile services (data, business etc), mobile messaging, mobile portal, research, broadcast, content, handsets and terminals. These practices are shortly explained within this chapter.

DOI: 10.4018/978-1-60566-916-8.ch004

MOBILE INTERNET/WAP

As also stated earlier, mobile phones were not used for any other purposes than simple text messages and voice calls until the 1990s. However, in that decade, significant studies have appeared with regard to small screen web and similar browsing activities (Buchanan et al., 2001).

According to Saarikoski (2006), different interpretations arose in the west regarding what mobile Internet is. When Saarikoski began to his work, The Odyssey of the Mobile Internet in 2003, he expressed a definition which is quite different than what is understood in 2008. For example, according to the simplest definition, it is a direct connection to the Internet via a mobile phone or portable device. According to a highly specific definition; on the other hand, it is to receive a detailed standard document through a mobile Internet device or network. It means the use of a very clear and concrete benefit-creating ability even when the person is on the move.

Again according to Saarikoski (2006), from the perspective of the user, mobile Internet is receiving value added service through a mobile network accessed by a small and light device. The definition of mobile network is the continuity of a connection even when the user is on the move (in cases of changing place in anyway). The devices used are those with both battery life and with the feature of making voice calls all the time. In the access and use of such services, the content is required to be easy and the system should have the ability to load the mentioned content. Here, probably the user may also participate in creating the content because the content of a personalized device should also be tailored to a person's demands. These means can be interpreted as benefits to the user which can be considered as simple and easy.

According to Necat (2007), Mobile Internet Technology (MIT) is a system where the information is carried to the mobile user via various devices. Leading such devices are mobile phones, personal PDAs and laptop computers, etc. In essence, MIT provides the communication between storage or pool of information and a mobile device. Wireless networks provide communication via wireless applications that are created with various mark-up languages, communication protocols and user interfaces. This definition of MIT expresses the handling style of mobile Internet from a technological aspect. Mobile Internet application development has various difficulties (see Table 1). The leading difficulties are the device types with various capacities, multi-browser applications, mark-up languages, capacity differences, and application diverseness.

Table 1. Difficulties of developing mobile internet applications

Device types with different capacities
Multi mark-up languages
Multi browser applications
Multi applications of WAP Standard
Differences in customer processing capacities
Application of cookies
Differences in situation management
Poor development user interfaces
Security levels
Other

WAP (wireless application protocol) is briefly the infrastructure of access to the Internet via mobile phone. This system is considered as a group formed in order to access the Internet from mobile devices.

WAP systems can be considered in two categories as common as WAP systems and WAP push systems. The first one is the common system that enable accessing to websites and information from devices such as the mobile phone which were mentioned afore in general, while WAP push systems are those systems that come on stream when the customer receives a short message that contains a clickable link, not unlike a URL embedded in an email, that causes the content to be downloaded. The consumer may have opted for a service that sends such messages in areas of interest, or may have initiated the exchange in response to a print ad, or may have found something of interest while browsing the mobile web. (http://www.adobe.com/devnet/devices/articles/mobile_ecosystem.pdf).

As the mobile phone is a personal device due to its structure, it is quite natural for them to carry very large amounts of information. This information may include names, telephone numbers, recorded messages, e-mails, birthdays and dates, diary and even the number of steps the person takes in a certain time period. This can naturally be considered as logic that enables development in the social network provided by Web 2.0 applications. Web 2.0 applications may be considered as user interaction and development of the main principles on how Internet is operated regardless if it is fixed or mobile.

Internet has gone beyond being only a simple online source which is consulted, researched and acted accordingly. Internet has gradually turned into a network where information is open to common use for the social masses and databases are rapidly shared. Already the websites such as Facebook, Twister, and YouTube are among its best examples in this recent period. Therefore, Web 1.0 is only regarded as a publishing device (which distributes information and knowledge), while Web 2.0 can be considered as a content provider platform which also evaluates creative content and focuses on the subject.

In order to use the mobile Internet properties; in addition, to a mobile device with the required features, the user should also undertake the costs to be realized in this regard. These costs should not be only considered as material, but also as other social and individual costs.

MOBILE SERVICES

There is not an exact conceptual definition with regard to mobile services yet, while it expresses different meanings in telecommunication, information systems media or service markets that use or make use of the mobile. Mobile services present real time, on-demand access to content that makes it more valuable to the customers (Kleijnen, de Ruyter and Wetzels, 2007). According to Pihlström (2008), mobile service can be defined as the content services that can be provided by a mobile device. These are any kind of services that can be retrieved via mobile device and that can be delivered in interaction between an organization and customer.

Mobile Data Services

Collecting and working with data and creating and analyzing new forms of data are different kinds of services from the other. The analysis of data are in many areas such as data related new generation technologies as mobile services, business administration strategies and planning, sample case modeling for enterprises, customer segmentation and target market identification, penetration into a new market or launching a new product, developing cooperation, project management, and providing new evaluations according to customer request, and auction strategy (http://www.thebesengroup.com/downloads/Mobile.Data.Service.Portfolio.pdf), and their reduction into suitable structures and transmission to relevant organizations can be made owing to the mobile systems of our present day.

Here, on the one hand, whether certain services related with some research oriented data, and on the other hand, the transformation of obtained data is strategically in accordance with companies' requirements are in question. Mobile devices have the ability to integrate activities in every area of marketing. So, mobile devices have been involved in enterprises as an element that supports the integrated marketing approach (Yamamoto, 2003).

Operating Strategy and Planning: Acquisition of information is a service type required in creating, changing or abandoning strategic decisions. In mobile data systems, it is possible to use mobile data and reach the latest information. The up-to-datedness, coherence and scope of the data obtained for identification of operating strategies and the preparation of suitable plans is critically important. Information is an element which is always required. Literarily, the data industry is being developed on mobile systems.

The mobile data industry is changing very fast, and both the producers and customers need to obtain appropriate data in order to adapt rapidly into this change or to solve any problems which they might encounter. Companies that provide service on this issue are gradually representing a more professional approach in formulating, implementing and evaluating cross functional decisions which are in accordance with their clients' objectives. Information service providers observe domestic and foreign events (especially global tendencies) and enable being informed of the changes in the right time when the necessity arises; thus, also enabling the control of mobile instruments, services and products, production, marketing and market dynamics. The industry of mobile instrument production may also benefit from the same platform and carry out studies to meet new demands.

Business Modeling: Financial analyses, business analyses and models are produced in order to enable clients to carry out their businesses better and more successfully. In this sense, information services -in close contact with customers- aim to disclose the details requested by the customer such as financial liabilities, opportunities and the periods of work to be done. Appropriate strategies can be developed with the help of customer-tailored and flexible business models and/or instruments.

Market and Customer Segmentation: Addressing to the entire market does not seem to be a quite successful option for competing in the present mobile market. Segmentation is the process of separating potential customers into subgroups according to similar demands, requirements and characteristics (for example, similar sales behaviors). As a result, it is possible to measure and acquire the behavioral tendencies of customers. This may provide the targeting of niche markets in the mobile environment to be created, or facilitating the formation of new opportunities on the basis of customer characteristics and the physical use of the information.

Market Research and Competition Analysis: In the area where marketing elements are located in the mobile system, the evaluation of the market and the activities of competitors have the utmost importance. The purpose here is to measure market size and growth potential, to identify the trends that lead the market and to outline the competitive structure and the new competitive tendencies that arise. Creative ideas may be required as the usability of competition based on momentary changes in the market is in question. Despite fast decision making risks, it may also provide the advantages of being near the customer.

Marketing and Launching Product: The use of correct vehicles in presenting the products to the market, addressing to the correct mass and presenting the correct content and its effective and efficient use in terms of finance and marketing should be considered. Here, the key point is the inclusion of new products, services and applications in the mobile ecosystem and their presentation as prepared in accordance with all units. [Shape, form, color, ease of use, provision of access superiority, price and possession of other skills satisfactory for personal preferences, as well as the elimination of the effects of the physical environment (in water, in air, etc.)].

Cooperation and Partnership Development: In terms of maintaining the continuity of operations, finding partners that support each other for strategic development and developing cooperation and partnership is an important marketing policy. This policy provides cooperation towards growth and development by obtaining investments and contributions of partners instead of direct investment. Gaining new power based on values increase with the sharing of information held individually by companies, and creating a synergy based on it is one of the fundamental purposes of cooperation. These cooperation and partnerships could be done on a profit or solidarity basis. These kinds of efforts could uplift the industry to the stage of surveillance while creating a new mobile ecosystem.

Project Management: In addition to attributes such as information, skill, capital, and cooperation which are required in order to develop and meet the deadlines for certain projects, time and cost parameters should also be known as an important issue. Although investments oriented towards R&D are costly in terms of fast moving consumer goods, they may be influential with their contribution to the projects of the goods to be launched. In this context, the gathering, delivery, evaluation and reflection of project-management oriented data are rapidly coming into existence through benefiting from the mobile system being influential in terms of reflecting on the project in a workable way.

Some activities need to be done during the preparation of projects and conduct of marketing practices. The main one is the creation of a request for proposal or some other creative activities. The preparation of a request for proposal by most service providers in order to fulfill their customers' requests is critical in terms of clearly outlining the required attributes, responsibilities and liabilities during the sales process. Due to the request for proposal forms to be created, it will be easier to identify what the customer wants and to act fast accordingly.

3G Auction Strategy & License Application: Auctions can be held via mobile devices similar to e-commerce. Regardless of the type of auction, participation by phone in every field within the process of auction, which is among the required license applications, has become more personal. In the meantime, some of the enterprises will prefer to be auctioneers. However, the legal infrastructure should operate highly properly in this respect.

Mobile Business Administration Services

The groups which prefer to dwell on certain subjects rather than on all subjects, may desire to reach a conclusion by using mobile administration services in a similar fashion. Mobile administration services have begun to be presented densely on mobile systems that have turned into a new communication environment. They mainly refer to three basic groups as advertisers, publishers and network operators (http://mobile.yahoo.com/business). And the end group is considered to be users who provide their own customers with the services created by the above mentioned services.

- **Advertisers:** These are the groups that can reach the mobile consumer group with innovative results that are suitable for the target.

- **Publishers:** These are the groups that aim to provide their contents to the mobile consumer group and to profit from their formation and distribution.
- **Network Operators:** These are the prominent groups that put forth customer experiences and lead to the growth of mobile marketing and advertising.
- **Users:** These are the groups that receive services from the above mentioned advertiser, publisher, and network operators and present these services to their own customers in order to create an advantage.

Advertisers are required to specialize on mobile devices so as to send small scale messages with which the sender could directly influence and stimulate the receiver. Another group is the publishers. They can be considered as the individual or corporate content providers which create massive influences. In this respect, some of the classical publishers have also laid hands on mobile publishing. However, new formations such as mobile blogs can also be foreseen in publishing. Meanwhile, network operators create the group that put on stream the mentioned advertisements and publications and enable their delivery to the customers.

The organization of relational structures such as that of a sales consultant who becomes acquainted with the customer and distinguishes his or her characteristics and makes the customer feel the desire for the long term relationship through the services provided with mobile devices and operators is essential in mobile operational services.

MOBILE ADVERTISING

In big cities, people are almost constantly held to advertising raids, and every day they see 1000 advertisements on average (Postman, 1993). As this figure was related with the 1990's, this number must have increased further with the means of the present day. Despite such a massive raid of advertisements, contradictions in terms occur due to an abundance of the mentioned advertisements and most of these advertisements fail to reach their targets. Briefly, it is possible to say that the rate of response to visual advertisements has declined.

On a changing and differentiating axis, advertisers are also, inevitably, captured by the fragmentation of the media. In the present day, there is a shift from advertisements for large masses to smaller masses to niches and even to individuals. According to Kirby and Marsden (2006, p.xix), the positive return on investment (ROI) rate from TV commercial campaigns is 18%. The average rate of return to sales on each dollar spent on commercials is 54 cents. In the last ten years, TV advertising costs have increased by 256%. Only 14% of the people believe the given advertising information. 90% of people zap past TV commercials. Approximately 3000 advertisement messages are seen every day. The list against conventional advertising continues to expand further.

Mass communication is a type of communication without an exact subject, or with a hidden subject. Here, the hidden subject refers to the viewer or spectator. Besides, commercials with a common element are created with the assumption that the interest of most commercial viewers arises by sensuality. In the technological environment, that emphasizes personal characteristics. The major reasons for advertisers' to shift into new alternatives are required to be explained. The users of computers and mobile systems want to go into the process of searching, developing an opinion on the product, and go on shopping. This and similar conditions hint that being faster will further increase the chance of being successful.

Mobile instruments are among the most prominent instruments in these types of advertising activities towards fragmentation into a smaller scale. The point mobile phones have reached in the last ten years has been influential. In their becoming influential can be compared to the rapid development of PDAs. The reason is that mobile instruments constitute a piece of the developing process together with the new personal communication technology. People are gradually becoming more conscious on marketing, and the counter-opinions against conventional marketing campaigns are increasing day by day. Here, "No Logo: No Space, No Choice, and No Jobs" kinds of books, consumption groups and social networks have had a great impact. On the other hand, the growing and developing media channels also make it more difficult to attain the correct objectives. And people are inclining towards individual or self-suitable options rather than mass consumption. Moreover, people demand technologies that block advertisements in order to overcome being continuously attacked by aggressive marketing, and they are directed towards being protected from such attacks.

Ignoring the benefits for businesses and advertisements in the present environment where almost everybody has a cell phone would be irrational. Brands which initially reached consumers via text messages have now entered into a stage where efforts to use more creative and alternative means exist in this field. However, the intensified SMS advertisement use is now also causing consumers to resist against this type of advertising. Therefore, mobile instruments are now employed not only through SMS but also through various other technologies such as MMS, WAP, and Bluetooth.

Mobile marketing is a part of the utmost personal contact environment compared to other marketing environments. Companies using mobile channels are able to communicate with the targeted person at any time of the day anywhere. One of the major indicators of the point the mobile channel has reached is its use during the 2008 election campaigns in the US. Its results imply how powerful mobile advertising would be in the future, if used properly (http://marketman-onair.blogspot.com/2008/05/reklam-mecralari.html).

Having emerged as an alternative to conventional marketing and advertising channels in the last ten years, the field of use in "mobile advertising" has expanded quite a lot in time due to filling the gaps of other marketing channels with its features as hitting the target mass at the right time and location, and measuring interest towards the product. As the use of mobile devices expand in the globalizing world, the importance of mobile advertisements also increase and efforts are carried out better and more effectively use the mobile marketing techniques.

Mobile advertising is the communication of direct marketing messages on CRM, advertising and promotion to mobile phones through different scenarios and practices concerning brands to customers.

Mobile advertising is on the verge of entering the mobile phone market as a new source of income. Minds are blurred yet regarding how personal advertisements should be shown and which advertising models should be employed. Therefore, companies continue their mobile advertising practices through the trial and error method. Various intra-system applications such as mobile video, sponsored call, coupon systems, wap banners, location-based advertisements, mobile search and mobile Internet arouse interest.

In order to understand the actual potential of mobile advertisements in a multi-platform environment, we need to carry out serious measurements in this channel and proportionately compare mobile advertisements of instruments such as mobile phones with small screens and limited messages but with instant access ability to mass communicative instruments such as TV, and other portable instruments similar to computers which are portable but bulkier compared to mobile phones. This effort would also suggest the measurement of mobile phones among the mentioned three screens, and that of the trials on these instruments. Herein, the remarkable difference is the ability of mobile advertising to communicate content via multi-channels to the targeted market and the targeted individual himself or herself (Reedy, 2008).

It is known that 90% of global brands have been planned to be allocated a budget for mobile advertising in 2008. The mobile advertising market is estimated to achieve an 11 billion dollar turnover by 2011 (http://www.sanalajans.com/taxonomy/term/34).

According to Leppäniemi and Karjaluoto (2007), mobile advertising can be divided into five categories. The first category is defined as the web category, and the mobile Internet is used in this category (with banners and interstitials such as mobile calling and mobile portals). Most mobile portals are designed for mobile phones. The second category is the publication category. Mobile radio broadcasts, Mobile TV broadcasts, trials or recently launched mobile technologies such as DVB-H, DMB, DAB-IP, MediaFLO and ISDB-T are included in this category. The third category is broadcasts on a narrow field such as mobile casting (a podcast designed to be downloaded by a mobile phone in order to be instant or later played by personal computers, mobile phone, MP3 player, video player or laptop computers), and bluecasting which is sends messages via a Bluetooth transmitter of a Bluetooth device to any mobile phone. The fourth category is the systems defined as the physical scanning category which includes various methods and technologies on touching, showing and scanning (e.g. Hypertag, RFID, barcode, UpCode) and which can be used with a mobile phone or which can access the Internet via mobile phone from a target point. And the last category includes other advertising formats which do not suit any of the previous categories.

According to the E-marketer's report on "Mobile Advertising: After the Growing Pains", US mobile ad spending in 2007 was 878 million dollars and the figure is projected to reach 1.7 billion dollars at the end of 2008 and exceed 6.5 billion dollars by 2012. (http://www.marketingcharts.com/direct/mobile-advertising-spending-to-surpass-65-billion-in-2012-4097/?camp=newsletter&src=mc&type=textlink retrieved 4/5/2008).

According to the E-marketer, although various mobile communication systems and similar products were introduced in 2007 with a sharp influence on this sector, mobile advertising could not break into the mainstream during 2007.

Next generations will prefer small screens to larger screens; thus, suggesting demands for preferring smaller screens in the field of advertising. Mostly written message elements are rather prominent with respect to the currently available devices in mobile advertising.

Spending on Mobile Advertising

According to the E-marketer, text messages will continue to account for the vast majority of mobile ad spending in the near future, and even the share of text messages in mobile ad spending is expected to reach 88% in 2008. This figure is expected to decrease to nearly 60% in 2012 with 23% share for mobile search advertising and 8% for mobile display ads. Advertisers worldwide are expected to spend 19 billion dollars on the mobile platform in 2012 compared to 2.7 billion dollars in 2007 and 4.6 billion dollars in 2008 (See Figure 1).

It is estimated that around 74% of the mobile ad spending in 2012 will be for text messaging campaigns. According to Gauntt, who prepared the report of E- marketer, the US seems to be the most active region in web advertising as it is the largest interactive economy (See Figure 2).

Figure 1. Mobile advertising spending worldwide

Mobile Advertising Spending Worldwide, by Format, 2007-2012 (millions)

	2007	2008	2009	2010	2011	2012
Mobile message advertising*	$2,560	$4,200	$6,440	$9,260	$11,960	$14,173
Mobile display advertising**	$52	$142	$338	$629	$945	$1,203
Mobile search advertising***	$83	$244	$597	$1,290	$2,345	$3,773
Total	$2,695	$4,586	$7,375	$11,179	$15,250	$19,149

*Note: numbers may not add up to total due to rounding; *spending on placement in text messages, includes direct spending on message campaigns as well as spending on promotional coverage of end-user messaging costs; **spending on display banners, links or icons placed on WAP, mobile HTML sites or embedded in mobile applications such as maps or entertainment services (eg games or video); ***spending on sponsored display ads and text links that appear alongside mobile search results, as well as spending on audio ads played to mobile phone callers making a directory inquiry*
Source: eMarketer, March 2008

092628 www.eMarketer.com

Figure 2. US mobile advertising spending, 2007-2012

US Mobile Advertising Spending, 2007-2012 (millions)

	2007	2008	2009	2010	2011	2012
Mobile message advertising*	$810	$1,470	$2,380	$3,060	$3,830	$4,500
Mobile display advertising**	$34	$85	$186	$327	$453	$541
Mobile search advertising***	$35	$107	$242	$531	$910	$1,484
Total	**$878**	**$1,663**	**$2,808**	**$3,917**	**$5,194**	**$6,525**

*Note: numbers may not add up to total due to rounding; *spending on placement in text messages, includes direct spending on message campaigns as well as spending on promotional coverage of end-user messaging costs; **spending on display banners, links or icons placed on WAP, mobile HTML sites or embedded in mobile applications such as maps or entertainment services (eg games or video); ***spending on sponsored display ads and text links that appear alongside mobile search results, as well as spending on audio ads played to mobile phone callers making a directory inquiry*
Source: eMarketer, March 2008

092634 www.eMarketer.com

However, the total spending in the Asia-Pacific region is expected to be larger in 2012 with the adoption of the mobile phone as the primary interactive screen by huge middle classes in China and India.

Mobile advertising is the display of "call-to-action" message or banner on the mobile phone similar to those of the Internet and television ads. For this purpose, SMS, WAP, MMS, Mobile Video or a different application mix can be used.

According to Bulander et al, mobile advertising is divided into three as Broadcast, MANET (Mobile Advertising Network) and dedicated connection. Broadcast is subdivided as local and global, while dedicated connection is subdivided as divided push, pull, and smart push and pull (http://www.nielsen-mobile.com/documents/CriticalMass.pdf).

Studies on mobile advertising have shown that 14% of the mobile data users in the US ignore mobile advertisements, which implies a low level of recognition on personal advertisement reception. Mobile marketing can be successful if it is directed properly because it is also able to measure reactions at a much faster and on a more personal scale than all previous media instruments. It is an activity that could bring success so long as it does not negatively influence individuals' lives. Similarly, it should possess the characteristics of acting with very different meanings in recognition of the difference, and building an interaction through this difference. People generally tend to continue using elements that add value to their lives. Mobile marketing activities should also be structured so as to increase individuals' wants. As the basic approach is customer orientation, mobile marketing attempts are strongly expected to be efficient. Even on certain points of interest, as with the provision of free of charge communication, some customers may even begin to long for receiving mobile ads.

23% of the data users in the US have further expectations. Many mobile data users are open to marketing experiences which require supporting the general mobile media experience costs. 32% of the users indicate that they would accept mobile advertisements if it provided a discount on their mobile communication bills.

Mobile Message

Mobile message is currently divided into three basic messaging groups. These are SMS, MMS and IM. According to Warren, Lyall and Owen (2007), among the present mass mobile marketing solutions, the channel through which a brand can reach the customer by 100% is SMS messages. Mobile short codes can be added to any media and a wider communication with the customer can be attained. Branding mobile micro sites may enable the presentation of more different and complex messages through connections that emphasize SMS. For example, practices such as giving out coupons through the mobile channel, or mobile ticket applications are suitable fields for expansion. Although the MMS has a better branding talent, it currently has certain gaps and defects, a high cost and it requires some complex operations. The developments for MMS messages have not been completed yet. Instant Messaging (IM) is also regarded as a future potential.

One of the advantages of asynchronous interaction is the user's ability to take part in communication at any time and place they wish. Besides, they have the opportunity to re-visit the discussions. As the period between the conversation, messages and replies through this type of interaction is very long, it is named as "slow motion".

Mobile Portal

A mobile portal enables a joint, personalized access to data, expertise, and applications by using mobile devices. According to Tsalgatidou, Veijalainen and Pitoura (2000), mobile portals can help users in this respect by offering personalized and localized services, and by taking into account the user's preferences, language, previous behavior and so on. Warren, Lyall and Owen (2007) also uttered that mobile banner advertisements, interstitials and sponsored contents can be presented with the skills and equipment to be developed by the operators in the mobile content section. In order to see the actual effect of the sponsor supported content, a good list of equipments is required. Interstitials are the ads that appear on a separate browser window while uploading a web page. They mostly consist of presentations with flowing graphics or small applications then the traditional banner advertisements. According to research, the users click on interstitials more than banner advertisements. Meanwhile, some users complained that it slows down the page desired to be accessed (http://www.webopedia.com/TERM/i/interstitial.htm

retireved 7/6/2008). It can be realized on mobile system in different ways with advertising and similar communication permits to be obtained via portal before the search.

Mobile Off-Portal & Search

Due to the limitations in use and the walls bonded by the operators, mobile Internet sites have not received much interest so far. However, content brands are willing to include mobile research in addition to other research when creating their contents with the present technology that presents the features which could make this possible especially the means of real-time mobile Internet research have begun to develop. Here, also the influence of creating marketing opportunities through content brands is noted and the use of such sites is increasing. Even, they are tried to be developed as advertising models which could be included in major search engines such as Google and Yahoo. From 2009 onwards, proximity based marketing applications will become easily receivable by mobile phones (Warren, Lyall and Owen, 2007).

Mobile Search

The human being is always inquisitive since his or her birth. From researching, the human being benefits from the structure with comprehensive information circulating in the digital environment. The information in this structure shows dramatic differences in terms of the approach to the subject and evaluation of the subjects, and literally requires running your head against a brick wall while working in an environment with millions of pieces of information on a certain subject. In parallel to the development of the digital environment, first Internet and then the search engines have stepped in. The mobile versions of these search engines have also been developed for ease of reaching specific information. Yahoo One Search 2.0 service of Yahoo leads the groups that deliver such services. (http://mobile.yahoo.com/onesearch). Web2go provides a better Web browsing experience on your mobile phone by making it easier to view and navigate (http://designtaxi.com/news.jsp?id=22245&monthview=1&month=11&year=2008). Studies on the configuration of the search engines for providing the best service are ongoing. The purpose of the categorization and grouping of search results on a screen, and the development of the related functions is to prevent the chaos, and to filter and take beneficial information from the environment which is also named as digital trash. The users of the search engine can also send text messages and obtain results from the system.

The interview with Tsirulnik (2008) indicates the view of the sector regarding mobile search. Accordingly, it is seen that the sector does not consider its customers as fully prepared for mobile search yet and even they are suspicious on how much of the idea is adopted by customers, how much they can search and to what extent they have understood the means provided in this regard. Meanwhile, the utilization volume of the call function is not dense enough for website optimizations. As not all of the mobile phones that are available in the market belong to different generations have the same capabilities, an increase in the number of users with the means such as GPS WAP and innovative thinking should also increase. 3G and higher performance mobile phones will drive people and organizations to "mobile search" because it will be possible to make all current Internet searches via mobile phones and similar mobile instruments with the same, or maybe even with a higher speed. Furthermore, time is required to establish information obtained through contingencies in this field. It requires reaching an experienced critical mass in the pursuit of speed. The experts predict the required situation will develop in the market in 10 years.

Mobile Broadcast

The number of mobile phones ready for mobile TV, video and radio broadcasts is gradually increasing, although currently the number of such phones is limited. Actually almost all of these features mentioned above are used in Japan. Meanwhile, the recording and storage capacities of mobile instruments are also gradually increasing. Consequently, mobile TV and radio broadcasting will turn into an important income generator model. These channels present unique opportunities to their clients in terms of traditional advertising and product promotions. As the customers are not willing to pay for the content, the owners of these media types are rather directed towards advertising and a sponsorship type of business models. In addition to the higher capacities of hitting the target compared to the conventional broadcasting, their costs are also expected to be lower (Warren, Lyall and Owen, 2007).

Mobile Content

Mobile content is any type of media which is employed on mobile phones, like ringtones, graphics, discount offers, games, and movies (http://en.wikipedia.org/wiki/Mobile_content retrieved 4/26/2009) Games, images, mobishows, music and videos are the most used contents. These are very important for marketers that want to reach and gain their customers. Mobile phones are the instruments that can be easily personalized by individuals with their games, sounds or entertainment aspects. Sponsorships and the issues such as advertisements in games should be supported in terms of rapidly exhilarating the market. The Internet, mobile social networks, and fields such as videos and blogs that are made by the users also create cost-effective opportunities. (Warren, Lyall and Owen, 2007)

Mobile CRM

Mobile CRM is one of the main opportunities, which has not been used widely for brands. According to Eklund (2002), mobile CRM is a solution that extends the reach of customer relationship management applications to any user remotely whenever and wherever needed. Through dense brand interaction, mobile CRM is expected to create a difference and high customer spending by creating a difference in terms of service towards obtaining loyalty. Examples of mobile CRM are the SMS customer service messages sent by banks, or the customer communication messages by home delivery companies or airlines (Warren, Lyall & Owen, 2007). However not all CRM applications are ripe for mobilization (Hildreth, 2006). This gives a hint for CRM and mobile firms to create new kinds of improvements for the new types of customers.

Mobile Locals

Location Based Marketing is also named as proximity marketing. Proximity marketing is a method of distribution where advertising content is presented through associating with a certain place. Here, it is possible to transfer the content through obtaining the approval of people at a certain location. Also the necessary equipment should be available (http://en.wikipedia.org/wiki/Proximity_marketing). According to Warren, Lyall and Owen (2007), proximity marketing experiences are presented during the presentation of certain up-to-date advertisements or promotions. A similar study was carried out by Yamamoto, Candan and Aydın (2008). Such broadcasts are transferred to customers particularly through Wi-Fi or Bluetooth boards, which are installed at a certain place. The reason for such attempts is the tremendously low cost and the presentation of the promotions to the clients at exactly the right time. With promotions presented through this method, customers are informed faster and they are made available to place their offers in a short time. At present, such activities with development potential for companies both seem technically complex and they can also be perceived as spams. With further development of technical features of mobile phones, the mobile phone will also become a mobile payment channel in addition to being an instrument of promotions and so, they will be more attractive for users.

Mobile Handsets

Mobile phone branding is the creation of alternative channels by a group of alternative channels to achieve their objectives. In addition to physical branding, pre-installed branded materials, which can be presented by mobile phone sellers in groups, can also be included. Advertising or promotions can also be sent instead of the screensaver depending on the location of the telephone which remains idle. Advertising-based free of charge calls are already realized by MVNOs. They can also be targeted as similar prepaid sections (Warren, Lyall and Owen, 2007).

Mobile Terminals

Mobile terminals also appear as prospective target platforms in addition to mobile phones and PDA's. These terminals are spread on a wide area. These terminals can also be defined as the systems that can present multimedia content and high accessibility to portable devices. However, access to the individual at a high level is of utmost importance especially with regard to issues such as privacy. Also, mobile advertisements that are separable from spam messages may be influential on individuals. Moreover, the limited user interfaces of mobile devices appear as another troublesome issue. Bulander et al (2007) presented a mobile advertising solution approach in their study. They named this solution as MoMa. This project was supported by the German Federal Ministry of Economics and Labor (BMWA). MoMa is a highly personalized and content-sensitive system that can guarantee data protection in mobile advertising. It emphasizes the requirement of distinguishing public and personal information in this context.

REFERENCES

Aaker, D. S., & Bruzzone, D. E. (1985). Causes of Irritation in Advertising. *Journal of Marketing, 49*(2), 47–57. doi:10.2307/1251564

Abramovich, G. (2008). More Targeted and Creative Mobile Campaigns. Mobile Marketer's Mobile Outlook, January 9, 2008.

Barnes, S., & Huff, S. (2003). Rising Sun: iMode and the Wireless Internet. *Communications of the ACM, 46*(11), 79–84. doi:10.1145/948383.948384

Buchanan., et al. (2001). Improving Mobile Internet Usability. WWW10, Hong Kong, China. Retrieved May 1-5, 2001 from http://www10.org/cdrom/papers/pdf/p230.pdf 7/30/2008.

Bulander, R., Decker, M., Schiefer, G., & Kölmel, B. (2007). Advertising Via Mobile Terminals – Delivering Context Sensitive and Personalized Advertising While Guaranteeing Privacy. Retrieved from http://www.springerlink.com/content/u0l885871m062767/

Deleon, B. (2004). Marketing on the Go. Ft. Lauderdale, FL: DGXMedia.

DMA. (2005). Mobile Marketing Best Practice Guidelines. London: Direct Marketing Association (UK) Ltd.

Eklund, R. (2002). Mobile CRM Comes of Age. Retrieved 4/26/2009 from http://www.destinationcrm.com/Articles/ReadArticle.aspx?ArticleID=45397

Hildreth, S. (2006). Mobile CRM Makes Its Move. Retrieved 4/26/2009 from http://searchcrm.techtarget.com/news/article/0,289142,sid11_gci1193465,00.html.

http://designtaxi.com/news.jsp?id=22245&monthview=1&month=11&year=2008 (n.d.).

http://en.wikipedia.org/wiki/Mobile_advertising (n.d.). Retrieved 8/28/2008.

http://en.wikipedia.org/wiki/Mobile_campaign (n.d.).

http://en.wikipedia.org/wiki/Mobile_content (n.d.). Retrieved 4/26/2009.

http://en.wikipedia.org/wiki/Proximity_marketing (n.d.).

http://marketman-onair.blogspot.com/2008/05/reklam-mecralari.html (n.d.).

http://mobile.yahoo.com/onesearch (n.d.).

http://wataniya.yahoo.com/glossary.html (n.d.).

http://www.marketingcharts.com/direct/mobile-advertising-spending-to-surpass-65-billion-in-2012-4097/?camp=newsletter&src=mc&type=textlink (n.d.).

http://www.nielsenmobile.com/documents/CriticalMass.pdf (n.d.).

http://www.thebesengroup.com/downloads/Mobile.Data.Service.Portfolio.pdf (n.d.).

http://www.thinkdutchess.com/telecommunications/glossary-of-terms.html (n.d.).

http://www.webopedia.com/TERM/i/interstitial.htm (n.d.). Retrieved 7/6/2008

Kirby, J., & Marsden, P. (Eds.). (2006). Connected Marketing. Amsterdam: Elsevier.

Kleijnen, M., De Ruyter, K., & Wetzels, M. (2007). An Assessment of Value Creation in Mobile Service Delivery and the Moderating Role of Time Consciousness. *Journal of Retailing*, *83*(1), 33–46. doi:10.1016/j.jretai.2006.10.004

Klein, N. (2002). No Logo: No Space, No Choice, No Jobs. Basingstoke, UK: Picador.

Leppäniemi, M., & Karjaluoto, H. (2007). Mobile Marketing: From Marketing Strategy to Mobile Marketing Campaign Implementation. In Proceedings of the 6th Annual Global Mobility Roundtable Conference, Los Angeles, June 1-2. Retrieved 4/26/2009 http://www.marshall.usc.edu/assets/006/5568.pdf

Necat, B. (2007). Distance Learning for Mobile Internet Users. Turkish Online Journal of Distance Education-TOJDE, 8(2), 3. Retrieved from http://tojde.anadolu.edu.tr/tojde26/pdf/article_3.pdf

Off the Fence: Will Mobile Ever Become a Mainstream Ad Channel? (2008, August 12). Mediaweek. Retrieved September 28, 2009 from http://www.brandrepublic.com/MediaWeek/Features/Analysis/838684/Off-Fence-Will-mobile-ever-become-mainstream-ad-channel

Pihlström, M. (2008). Perceived Value of Mobile Service Use and its Consequences. Swedish School of Economics and Business Administration, Helsinki Finland. Retrieved 7/7/2008, http://dhanken.shh.fi/dspace/bitstream/10227/269/1/176-978-951-555-977-7.pdf

Postman, N. (1993). Technopoly. New York: Vintage Books.

Reedy, S. (2008). The Three Screen Advantage. Retrieved from http://telephonyonline.com/wireless/news/telecom_threescreen_advantage/index.html

Saarikoski, V. (2006). The Odyssey of the Mobile Internet. Retrieved from http://www.tieke.fi/mp/db/file_library/x/IMG/20156/file/Saarikoskivaitoskirja.pdf 7/30/2008

StrategyAnalytics. (2008). Understanding the Mobile Ecosystem. White Paper. Prepared for Adobe Systems Incorporated. Retrieved 4/26/2009 http://www.adobe.com/devnet/devices/articles/mobile_ecosystem.pdf

Tsalgatidou, A., Veijalainen, J., & Pitoura, E. (2000). Challenges in Mobile Electronic Commerce. In Proceedings of IeC 2000 3rd Int. Conf. on Innovation through E-Commerce. Manchester UK, Nov. 14th –16th 2000. Retrieved 4/26/2009, http://cgi.di.uoa.gr/~afrodite/IeC_Manchester.PDF

Tsirulnik, G. (2008). Mobile search: Not there yet. Retrieved 5/23/2009 from http://www.mobilemarketer.com/cms/news/search/1179.html

Warren, R., Hakes, K., & Simmons, A. (2009). 2009 UK Mobile Marketing. Retrieved 4/25/2009 from www.firstpartner.net

Warren, R., Lyall, K., & Owen, C. (2007). 2007 UK Mobile Marketing. Retrieved from www.firstpartner.net

Yahoo! oneSearch Launches as the Mobile Search Service. (n.d.). Retrieved 4/26/2009 http://designtaxi.com/news.jsp?id=22245&monthview=1&month=11&year=2008

Yamamoto, G. T. (2003). Bütünleşik Pazarlama (Integrated Marketing). Istanbul: Literatur Yay.

Yamamoto, G.T., Aydın, K. & Candan, B. (2008). Mobilizing World: New Challenges for Advertising. E-Business Review, 8.

KEY TERMS AND DEFINITIONS

Mobile Advertising: Mobile advertising is a form of advertising via mobile (wireless) phones or other mobile devices. It is a subset of mobile marketing (http://en.wikipedia.org/wiki/Mobile_advertising retrieved 8/28/2008).

Mobile Campaign: A campaign, usually marketing, advertising, or public relations-related, through which organizations are able to contact and interact with their audience through SMS (text messaging). This form of campaigning allows organizations to reach out and establish relationships with an audience in a more individualized, intimate way (http://en.wikipedia.org/wiki/Mobile campaign 8/28/2008).

Mobile Internet: Refers to gaining access to the Internet using a lightweight, handheld device, such as a PDA or cell phone. (http://www.thinkdutchess.com/telecommunications/glossary-of-terms.html 8/28/2008).

Mobile Marketing: It is wireless marketing or mob-marketing which describes marketing on or with a mobile device such as a mobile phone (http://en.wikipedia.org/wiki/Mobile_Marketing 8/28/2008).

Mobile Search: An evolving branch of information retrieval services which uses mobile platforms and mobile handsets or other mobile devices. These services enable users to find mobile content interactively on mobile websites. Mobile content shows a media shift toward mobile multimedia (http://en.wikipedia.org/wiki/Mobile_search 4/26/2009).

SMS Advertising: Advertising and messaging to customers via SMS.

WAP: Wireless Application Protocol is a special way of formatting content so that it can appear on small screens such as micro browsers like those on mobile phones (Deleon, 2004, p.151). WAP is a technology that enables mobile phone users to look at specially converted web pages, order goods from the Internet and use email from their phone screens. It has a set of standards that allows Web access on mobile devices (wataniya.yahoo.com/glossary.html 8/28/2008).

Chapter 5
Mobile Customer

Organizations and people, whenever they need to, aim to fulfill their requirements with suitable products and services. The suppliers of these requirements perform a certain marketing process until the requirement is fulfilled. In this process, first the attributes of the product or service are determined. Various studies are made during the identification of these attributes. These studies are evaluated from the customer's perspective for conformity to taste, functionality, suitability to place and culture, convenience for budget and time. The purchasing process is concluded upon the final decision of the people or organizations that see the suitability of the product to fulfill their requirements.

In this context, the mobile customer tries to cover his or her requirements with the knowledge based on information obtained from the virtual environment, and reaches facts via virtual data, and the customer creates a new structure with different features.

This section aims to detail the changing customer structure, the individualization and the experienced changes and considers the transformation of individuals into mobile customers. Generations who depended on products of high-technology development, which show themselves with different aspects, have also brought differences to marketing. They have both changed the production and service approach, and began to benefit from different types of products and services. In addition to other environmental conditions that affect life, human generations who change due to the effects of the means of the age should also be examined in this context.

INTERACTIONS AND MOBILE TRENDS IN THE WORLD

The latest developments in the world and the efforts towards the development of east and west integration create serious influences on customers and cause changes in the perception of the customer. According

DOI: 10.4018/978-1-60566-916-8.ch005

to Koç (2008), the steam boat lies in the basics of developing east-west integration. Owing to the steam boat, long distances became accessible in shorter period of time. The east-west integration essentially upholds two basic changes. One of them is related with crossing the borders and it is based on the facilitation of transport with the development of transportation vehicles. The other one is the change of the mindset which was determined in the interdependent social life. So far, integration has reached to such an extent that it has caused globalization. The world is now perceived as a global totality, and it is also considered as a global place from the point of marketing. As geo-political borders have been eliminated, production has also lost its nationality (Candan, Aydın, Yamamoto, 2008). On the other hand, the world economy has grown rapidly and the reflexes suitable for this economy have arisen and developed. Small players with correct bounces gained strength, while some major players went into shrinkage.

With the development of the means of communication in addition to transportation, the consumer became tremendously strong in terms of gaining knowledge. The world has turned into a giant market. Life styles are shifting towards the formation of the homogenous consumer. The involvement of mass communication instruments especially supports these homogenous formations, influences purchasing which is an important feature of the customer, and paces up consumption. In this context, Koç (2008) compares the world to an auditorium and explains that 6.5 billion people are sitting under a giant dome and share an environment surrounded with immense message rain. Actually, what make this message rain heavier are communication technologies themselves because serious developments have occurred since the invention of the telegram and although 100 years of work was required to connect one billion users to each other after the invention of telephone, the development of computers and mobile communication instruments in the second stage added 10 billion users to the milieu of communication in only 10 years.

Human is a being that desires a social environment and accumulates values in that environment. In the mean time, the human develops a consciousness in terms of protecting the individual values and interests. Therefore, the human being always lives in a dilemma. However, according to Atasü (2008), modern human is in a stronger dilemma. S/he wishes to both be independent and unique, and belong to a certain community. This is something that should especially be taken into consideration and evaluated by the marketers.

Every new occasion causes to certain new opportunities and threats for the organizations that fulfill the requirements and demands. They are influential in the customer approach and perception that is created as a result of some certain marketing efforts, and therefore, the activities require revision. Besides, a revision is also required to assure the access and address to the customer units with innovative approach. During the evaluation of mobile customers and their features, especially the discrepancies of ages and generations should be taken into consideration. The generations, which are described as i-v-millenium and developed in the last 10 years basing on innovation, as well as the cultural changes brought along by the mobile, should be examined. For example in the US, the time of speaking on mobile instruments increased from 30 minutes to over 260 minutes in the last 25 years; meanwhile, the GSM penetration ratio in Turkey was 90% as of June 2008. Also the number of prepaid callers in the same period is 80%. In 2007, the speaking time in Turkey (MoU) was measured as 78 minutes. Mobile income of Turkey in 2007 was stated as TRY 12.8 billion (TRY 9.6 billion in 2006), and mobile investment as TRY 1.4 million (TRY 1.09 million in 2006) (Aykara, 2008). Therefore the customer, customer profiles, tendencies, innovating and diversifying customer generations, and approaches to the product/service have maintained their actuality as the important subjects of mobile marketing.

TOWARDS THE MOBILE CUSTOMER

The needs towards the procurement of news, information and communication at any part of the world have become easily achievable because of the information technologies. For example, a TV serial creates an immediate effect at any part of the world, not only in the country where it is produced. In this environment, the broadcasters and marketers have now become highly professional on making use of the TV, radio and similar media instruments in order to create an overseas effect. The products included in the communication system are distributed to various geographical regions abroad such as the US, Japan, Korea, Finland or Turkey, regardless of where they are produced. YouTube, Facebook and similar social communication websites have enabled the access to such features at almost personal level. A video clip which is shot in a small town of Canada or the imitation of a famous band by a Chinese group in a province of China can easily become one of the most watched videos of the world. Therefore, the affects in the communication environment are quickly distributed via social communication sites and they influence the users' lives, and turn into a new environment through which the communication is carried out.

Photographs, videos, short films and symbols are distributed to the entire world via similar channels. Everybody exposed to them are automatically influenced by the elements such as figures, symbols and images. The elements which are presented in this context take their places in the memories of individuals in a short time and spread in the entire world. This is in turn leading to a symbols-based thinking. And finally, the behaviors and preferences arising from this thinking are gradually becoming more similar to each other.

Apart from them, the utterly new characters in an utterly new environment take their places in the new structure with the desire for differentiation of the players in the environment. Even, many individuals make use of the websites such as secondlife or social networks with the new online characters in order to perform some of their actions, which they desire but fail to do in the real life (Yamamoto, 2008).

The world experiences the communication bombing arising from the North America and Western Europe, while the urban sections of underdeveloped or developing countries are eminently influenced from this bombing and the western culture. Due to these environments that have been used by the west in the last 20 years with the desire to create a dominant type; a western style of living is assimilated in the lives of the people in the entire world. The idea of changing the family structure is supported with the structures such as nuclear family, etc, and serious efforts are spent in order to create the life styles similar to that of the West. On the other hand, certain production of the eastern culture are moved and spread on internet with the elements such as "anime" and "manga"; and they are tried to be adopted to young people through the fashionable virtual trends. This results in the major interactions caused by the globalization.

With the data obtained from the communication media, the individuals in the developing countries see how the people in the other countries live. They begin to imitate and turn the virtual values into reality by seeing how the people in other countries use the space in their homes, how they communicate, what they communicate about and how they spend their holidays. Birthday celebrations and the brands which are used in order to determine the life styles are also among the examples which are easily revealed.

The people of the developed countries are one or more steps beyond the citizens of other countries with the advantages of visiting and interacting with anywhere they like by the increased means of travel. As the means and circumstances advance, certain effects of convergence appear on people.

Regardless of how closer the east and the west became and how certain styles and methods have changed; certain individual and cultural differences continue at various levels of society. However, virtual forums or Internet communication in the world carry certain ideas and opinions directly to the communication environment with its features passing beyond borders in a different habit and emphasizing individual approaches more than ever. This milieu causes changes of certain rules and structures. The mobile customer is the customer who can make use of this habit and the unlimited environment. While approaching the mobile customer, it should not be forgotten that an entirely different customer group which requires consideration with evaluated aspects of individuality has emerged.

Nonetheless of the source of news, people have begun to demand the best of everything with the best conditions: best house, automobile, fishing rod, mobile phone, etc. In this environment, a more self-centered and individualized prototype customer image is encountered immensely. There is a course shifting towards different social environments compared to the past which identifies individualism with being privileged, and where the customer tries to obtain the maximum benefit from technological developments. For example, mobile phones have moved people away from the phenomena of meeting face to face, but made them virtually closer. It is apparent that this is a different approach.

At present, through the agency of mobile instruments, the new consumer mass receives messages and replies to the messages of interest on the mobile environment literally everywhere: at home, office, car, and on holiday. The work performance talents of marketers in this mass should also be different. People, who can be accessed even when they are called in an irrelevant time and place, have begun to call and text each other more frequently as mobile communication has gradually become more economic. Marketers consider this situation as a benediction as it creates the opportunity to individually receive marketing messages anywhere. Therefore, mobile communication is becoming more and more interesting. An individual, who has a mobile phone, always keeps it near and almost always has it turned on. That's why, mobile marketers and their more specialized counterparts: the brand and content providers have gained the opportunity to access the target mass at any time and everywhere. The mobile communication environment with its developing infrastructure can be used not only to exchange messages but also for shopping in a very short time through payment. Thus, it turns into an unlimited platform from the marketers' view. On this platform, it is possible for individuals to merge their powers and create multi-layered synergy and effectively use it.

Mobile, Individualism and Change

Technological opportunities have not occurred suddenly and automatically. They have been created with the transfer of certain knowledge between generations and with the processing of developing knowledge by creative minds. The most important materials which provide transfer and repeated reminders of this knowledge are the written materials. Books cover the widest area among all written materials. Addressing the book as one of the important phenomena which has affected humanity is important in terms of more clearly expressing the interactions and the changes in life. The "book" has created social factors beginning from the thirteenth century, and enabled the creation of the individuality concept by supporting the individualization of the human being.

Mobile systems that have been developed in our age are actually one-to-one influential in the individuality of life styles. Mobile communication and instruments aim to use individuality obtained via the book in the most efficient manner. Chokvasin (2007) has derived a socialized individualism analyses of Habermas and Heiddeger's questioning of everything related with technology, and expresses that mobile phones enable individuals in a communicative universe to confront their new technological stands. This confrontation leads individuals to behaviors which are structured by data selected from a chaotic environment based on different perspectives.

Mobile instruments support the formation of a different lifestyle by maximizing individuality with their features. Most people regard the mobile phone as their inseparable parts; some even feel naked without their mobile phones. The mobile phone did so well for people that it began to take place even in the core of lyrics and movies, becoming an integrated part of humanity. For example, the band "Blue" has a song with the title "Without You". The chorus of the song is as follows (www.musicsonglyrics.com/B/bluelyrics/bluewithoutyoulyrics.htm 8/29/2008):

"I couldn't live without my cell phone,
I couldn't live without my 4 x 4,
I couldn't live without my credit card,
I couldn't live without you."

These words can be actually considered as a comment or summary of the common form people have been attracted into. In movies, mobile phones have already taken their places among the most commonly used items after automobiles and church entities. Furthermore, mobile phones or similar mobile instruments are not considered just as an innovation but as one of the critical instruments used.

Despite the rumors which indicate that text messaging would physiologically elongate the activated thumb, mobile instruments are used for many social activity-related functions. Besides, they help eliminate queues in social interaction and registry transactions (at hospitals or for payments to tax offices, or water, electricity and gas), and they have taken part in almost every aspect of life. Mobile communication and technology, which has been so much interlaced with life, cannot be considered isolated from marketing. From the point of marketing, it is not only important from the corporate perspective, but also in terms of converting the new user to demand-oriented marketing that guides the buyer or seller. There is a market which develops and changes by producing new terminologies that turn into converged and mixed elements while looking for the answers to questions such as who is whose customer, who is the buyer, and who is the seller.

New occupations have arisen on account of mobile marketing. The mobile telephone has also touched the language we use. It even influences the use of languages with its terminology oriented towards creating images through "abbreviation". By removing the vowels in the written language, a differently created situation arises and is gradually accepted by society although it is not liked all the time. Most of the things which were previously learned as spelling rules are not used anymore. In this change, in addition to the abbreviation or symbolization on Internet, the text messaging feature of the mobile phone provided a great contribution.

Such changes in language may allow for the liberalization of the language and may influence people in terms of expressing their ideas in a different way, and may use the language without caring much for the rules while making use of the technological devices. The different values to be attached in an important way by freer and more liberal people create a social problem, suggesting the necessity of understanding and considering the new customer structure and values. Under the new customer structure, the present classical structures may be insufficient because, arguments (which are proved with convincing evidences) which govern the classical marketing mix with 4 Ps; Product, Price, Place, Promotion are gradually changing and losing their validity regarding their coherence with current customer behaviors. According to Shiffman (2008, p. xii), in the classical paradigm, organizations create products, define special features and properties for them, put a price, select the places to present them and provide service, and aggressively market them to the customer with various methods such as advertising, public relations and direct mail. This marketing approach can be framed with the creation of a limited number of focuses.

With globalization, activities in the world creating the opportunity of an unlimited number of focuses in marketing have continued as connected to a speed spiral. Humankind is affected by technological developments that has caused new world formation also in marketing, just like it did in all other issues. The consumer of the new world changes into individuals who either can shop, have hardly spare time for shopping, or think they have plenty of time. In an environment with very different tastes and tendencies as well as mass and service delivery of dense information, the creation of individual customer tastes has also been possible. This has let marketers act in different approaches specific to individuals in both the classical and modern approach during this transformation period. Actually, each individual in the present state becomes a piece of the marketing element by either trying the products, recommending them or being influenced by them.

The new customer group with a new approach benefits by the environment of information technologies, and respects the existing facilities in this environment. These people embrace and easily accept everything that is technologically developed. These customers are also in a more dynamic structure and pay great attention to mutual interaction (Craig, 1999). When consumers, who currently know how to use computers and benefit by Internet services, consider to buy a product or service, the first thing they do is to collect information on the related product or service by searching on the search engines via Internet or their mobile phones. In addition to the price and product features, it is also possible to examine the ideas of other users on this product or service. The manufacturer who wishes to influence the consumer is required to be present on the Internet, ranking on the top of the list in search engines, and provide an environment where the users of the product or service can clearly state their ideas. At present, such approaches create an unexpected level or relationship and interaction between the marketers and customers (Shiffman, 2008, p. 137).

This level of relationship causes producers to directly enter in the market, preventing marketers from acting freely as they previously did. Many organizations also began to make use of technological resources with a guerilla tactic, adopting aggressive policies. In order to be good at marketing during this swirling fast period of change, it is required to rapidly accept change, and quickly evaluate not only what is going on in the close vicinity but also in the remote marketplace, and in addition to the marketing information systems, data-mining is also required to be correctly employed. Marketing plans are inevitably required to be adapted to the new world and new technologies.

Due to the technologies of the present time, consumers are able to easily find any kind of requested or required product, service or price via Internet, and compare both different and similar products. In order to do that, the thing the consumer is required to do is, simply, to write on one of the search engines, reach the product data according to the obtained results with single click or a few clicks in only seconds, and make a decision by comparing the prices. Curious consumers can also easily learn about complaints and similar comments on the product or service from these web sites. The solution to these types of complaints can also bring along serious opportunities for organizations. Unless organizations can resolve their complaints regarding the products and services, they would accumulate in the technological communication environment and create unfavorable effects against the product or service and become negative word of mouth marketing elements.

In the present marketing environment, the consumer is not required to go to the place where the product or service is presented. Thanks to Internet and mobile instruments, consumers can place orders from their chairs, and immediately remove all the negativities of traditional shopping (loss of time, traffic, parking lot problems, queues, and product search efforts).

Nevertheless, it has also paved new ways for product and service delivering businesses. Through the agency of Internet and mobile devices, businesses have the opportunity to promote and sell their products and services in the overall world in the most economical way, and directly reach target consumers via web sites and WAP. Sales through Internet or mobile devices sometimes provide incredible advantages to corporations and individuals compared to other sales while eliminating the necessity of paying thousands of dollars to open a store at the best place, and to invest in the store decoration or to sell to big department stores. Actually, these issues are gradually becoming less important for consumers. Therefore, costs decrease during marketing activities via Internet and mobile, which is in turn, reflected on the sales prices of the products.

With regard to the sales of devices, instruments and services which include mobile technologies on a global aspect, it is noted that the sales trend rapidly increases when this technology is used, yet the increase differs as per geographical societies (countries). For example, although there are more than 50 million mobile phone users in Turkey, the user mass has not been fully saturated yet. The market has reached saturation in countries with high technological development such as the US and Japan. In these countries new technologies or differences are in demand which provide some certain additional facilities from the consumer.

According to Gartner's report, mobile phone sales to end users in Japan decreased by 10.1% to 13.2 million in the first quarter of 2008. The reason is that different telephone features that would influence growth did not in the first quarter of 2008. However, in the first quarter of 2007, digital broadcast reception and digital broadcasting with integrated services – terrestrial (ISDB-T) tuners were added to the music playback function of mobile phones, causing consumers to replace their devices. According to Mitsuyama from Gartner Inc.: "The proportion of WCDMA devices sold in Japan decreased from 63.2 per cent in the first quarter of 2007 to less than 30 per cent in the first quarter of 2008. Conversely, the proportion of high-speed downlink packet access (HSDPA) devices rose sharply from 1.9 per cent to more than 35 per cent in the same period." (http://www.gartner.com/it/page.jsp?id=680207 8/28/2008).

MOBILE CUSTOMER AND CHARACTERISTICS

The mobile customer is the person who is diverted towards shopping in the gradually mobilizing world with the influence of information spread via mobile instruments. A solid product is not necessarily required in this mode of shopping. The service can be a game, or other downloadable elements. According to Öymen (2008), this group is defined as information glutton consumers. They have certain very new characteristics which do not rely on the elements such as brand loyalty, habits and trust. These consumers stand against the institutions with their environment and benefit from the power of social sharing networks on influencing consumption tendencies. It is apparent that as one of the most practical communication instruments: the mobile phone – its habit of use will greatly influence the consumer. Brand loyalty is a change brought by interactions in the existing order. Therefore, finding out different ways is required to develop loyalty. Trust is an issue that all organizations and institutions are required to establish for the sake of sustainability. And for the mobile customer, trust is among the most essential issues.

The most evident characteristics of a mobile customer are:

- They like using technology.
- They obtain information from the technological environment.
- They do not care much about the elements such as brand loyalty, habit and trust.
- They appreciate the social sharing idea more than the corporate organization.
- They are willing to easily reach whatever they need.
- They adopt attitudes which vary according to age and social status.

It is possible to separate mobile consumers in very different segments. Separation according to demographic features is an important basis of segmentation. Apart from that, these consumers can also be classified according to their location, educational background, profession, age, environmental means, and fashion. Different segments may be created after the data related with the mobile phone are collected according to demographic properties, technological tendencies and the market share.

According to Bayne (2002), there is not a typical, single-minded mobile user profile. Mobile devices with different features are used due to many different reasons according to place and situation. Many cable sites or broadcasts are not currently suitable for wireless functioning. According to Bourke (2006), mobile marketers are required to evaluate and dwell on personal requirements and demands of mobile phone users. Before attempting any communication campaigns, detailed plans should be made on how to make the campaign, how to reflect it on the customer and how to establish the brand-price relationship. For example, according to the Yougov report from 2006 MobileLife (MobileLife, 2006), mobile phone users in the UK are as follows:

- **Mobile Generation:** Single, style-conscious, ages between 18 and 24, students and young adults in their first employment.
- **Phonetics:** Single, ages between 18 and 34, they consider mobile devices as their most important electronic devices.
- **Practical Parents:** Cost sensitive, families under the age of 34, who select mobile devices according to price.
- **Thumb and fingers:** Married, middle-aged or retired people with children or grandchildren.
- **Clever connectors:** Ages between 25 and 44, rich families and professionals who use mobile phone in organizing their businesses and social lives.
- **Silver selfish:** Rich married couples who reached the retirement period.

These are the segments identified in the Yuogov report and they are considered meaningful data for mobile commerce in the UK. In the researches that were carried out five years ago on mobile, it was expressed that older but well-trained consumers were more prudent on mobile advertisements (Kaasinen, 2003). According to the example, it can be inferred that expectations of the "clever-connected" group will be affected.

The most important communication tool before the introduction of mobile phones was the land phone. Although land phone utilization rates have not changed much, usage frequencies other than connection to the Internet may decrease. The number of people shifting to the mobile phone instead of the land phone is gradually increasing. Both cost and the ease of use affect this preference. For example, in a questionnaire applied to 1,100 university students in the age range of 18-34, 51% of the participants replied to the question "What is your frequency of using land phone in daytime?" as they did not use the land phone at all while the rate of those who just preferred calling via land phone is 2% (Gençlerin Tercihi Mobil Yaşam, 2008). This questionnaire is a small example that suggests usage tendencies while indicating that mobile phones will be used more frequently by the latter generations. In short, mobile phone has stopped being a substitute of the land phone, and turned into a product which affects life entirely on its own.

Technological tendencies can also be considered as important segmentation instruments in this context. As information regarding the use of communication technologies increases and expands, the tendency towards mobile products also increases. The saturation points of the market should also be seriously considered and taken into account for evaluation. A new hunger appears with every new technological generation while the saturation for former technologies is apparently obtained.

In the decisions during production stages, the mobile customer has even gone beyond being the king and became internalized. This internalization: the customer, the customer's character, requirements, and preferences are interrelated with the offered product and service. Although a relatively short time has passed, a large number of customers have reached a digital interpretation level which enables them to use high technology comfortably.

Characteristics of the Mobile Customer

Outlining the general characteristics of mobile customers is an important problem in terms of explaining what draws the interest of mobile uses, how they can be attracted and how they can be served based on the users' requirements. Mobile phone manufacturers, the service providers and mobile marketers are in the pursuit of finding the answer to these problems. Actually, the characteristics created by mobilization on the human can be considered almost more difficult than breaking the Enigma code. Mobile telephone manufacturers are arousing product differentiation with a tactic which has virtually not been noted for any product so far. During the continuing mobilization process, technology is rapidly innovating itself and every innovation is used as a marketing instrument.

When considering mobile customer characteristics, the evaluation of the tendencies towards the means offered by the mobile phone or similar devices is expressed only by touching. Here, touching is related with touching buttons, touching life, and meeting and adopting new values.

Mobile telephone and similar devices gradually make human life freer. According to Öğüt (2008), mobile devices facilitate the person's passage from the "field of obligation" to the "field of freedom" by introducing a development in parallel to an ethical understanding that puts forth the individual and universality that passes beyond old thoughts, and that aims to move forward the humanity which is led by freedom. This is contrary to the conservative ethic which is tightly attached to strict social sanctions and past values based on obligations and duty. It indicates that the mobile will support the changing of the customer character towards liberalization. Some problems will also appear as the area of personal freedom expands. Of course, in order to resolve those problems, asking the question about the extent of freedom might be required once more.

Modern society created an earthly ambit of morality instead of the dogmatic values that shaped the Middle Age. All innovative trends introduce a restructuring based on the denial of its roots. However, societies can never cross out their entire histories. To the contrary, they tend to bring along their roots. Moreover the relative ethic - which defends that the good and the bad, the right and wrong, what should be done and what shouldn't be done change from one person to another, one culture to another, and one period to another, and prepares the reply to questions of according to who and what - is in continuity. It is claimed that the allowed or forbidden activities are shaped according to the character of the person, and the connections established with the roots of the activity.

Mobile customers know different ways of acquiring information, and are located in a "society-based" relational cycle. The social network web sites that crease a society-based relational cycle and information sharing storages such as blogs, podcasts, and wiki can be shown as examples in this regard (Roe, 2008). Those who can use these ways well will create positive values such as benefiting from all means of the Internet. Mobile telephones which comprise not only the Internet, but also many different functions and continuously new technologies serve their users.

Using mobile phones as also with the MP3 and radio is considered quite a simple function. During the time of the walkman twenty years ago and discman ten years ago, it was not imagined that people would become so internalized with earphones on their ears. The measurement and evaluation of the type of music or radio channel listened to by these users and the frequency of listening to music may facilitate the marketers to act according to behavior. As mobilization has turned into a piece of life in every stage, it can also be considered as an indicator of contemporary life. For example in Poland, bus tickets can be obtained via mobile and people can get on the bus by using their mobile phones instead of tickets. For example, the wireless system has been put into use at most airports in May 2008. If the cost decreases, then everyone will be able to check and send e-mails via WAP. Issues such as momentary message exchange have already become a piece of normal life. However, issues such as the frequency and features of these messages have not been determined yet.

People are spending efforts to differentiate their mobile phones with various accessories such as wall papers or ringtones, and to personalize their accessories, melodies and screen images with coaters. Personal tastes and requests can be identified on the product and people can accept it only according to their self-preferences.

The mobile device is also considered as an assurance for emergencies. These instruments are provided with the feature of informing emergencies even if they are out of battery or coverage; thus, providing security for possessors.

It seems that accessibility, which may also cause disadvantages in addition to advantages, will influence human life for a long time.

MOBILE CULTURE

Despite different uses in the previous period, mobile phones are currently used during the daily lives of many people in their businesses and existences; it coordinates the lives of people and so, presents a brand new life style creating a different culture. Mobile devices have caused a shift from face-to-face communication to talking and chatting via virtual communication and towards writing messages or sending pictures. Mobile phones organize a new life culture which redesigns features such as wishing for a quick recovery or messaging for birthday organizations, expressing mutual love and respect, directing commercial changes, as well as other features which allow for the remote performance of many actions.

Having eliminated the place and time concept with accessibility to people at any moment, anytime and anywhere, mobile phones can enter restaurants, buses, parks and even public toilets, and others can secretly access to talks with these instruments and that appears as an important problem beyond what the manufacturer and the user can do, bringing a brand new meaning and definition to the subject of

Table 1. Generations

Generations
Age and life phase (ontological factors)
Era and Technology
Events and experiences (historical factors)

communication (Ling, 2004, p. 4). This situation, which means the continuous monitoring of personal areas, may cause some distressing circumstances (see private sphere).

According to Shiffman (2008), Web 2.0 creates a ready-made environment for in-depth contact of the customer and potential clients with brands of products and services in this culture.

In this context, it is possible to talk about a culture which is informative or easily informant, individualistic, society-based referenced, more freely thinkable despite the traditional understanding along the lines of what is right and wrong which is more clearly observable in the work environment, but regardless of whatever happens, more open to communication and which can use all the audio-visual elements of communication, and therefore be more creative.

Different Generations and Mobilizing Marketing

In the world, certain conflicts, disagreements and different behaviors have occurred between the old generations and the following generations throughout history. People can be separated into different generations in terms of various factors such as age and life style (ontological), time and technology (sociological), events and experiences (historical) (see Table 10).

According to that, it is very important to consider the individuals of the mobilizing consumer group as a generation. Mainly X and Y generations, the generations such as i, v, millennium are different from each other both in terms of the users of the devices and in terms of benefiting from the functions.

X and Y Generations

People who were born in and after 1960 are defined as X Generation in the US (Gen Xers or Generation 13), E Generation or the 68 Generation on Continental Europe, "Génération Bof", which means "I don't care or whatever" generation, the "burn generation" in Iran, the "crisis generation" in South America, "glasnost-perestroika" and sometimes the stockers and watchmen generation in former communist states, and "shin jin rui" "the new human being" generation in Japan (http://www.answers.com/topic/generation-x retrieved 9/14/2008). The lifestyles of these generations have shown significant changes compared to previous generations, and the shopping habits and tendencies developed in a different way. Strauss and Howe (1992) are concerned with certain factors which they believe shapes Generation X. These factors are listed as follows (http://www.answers.com/topic/generation-x retrieved on 9/20/2008):

- Increase in divorces
- Increase in the number of working women
- Zeroing of the population increase rate
- Change in the characters of horror movies

X Generation accounts for approximately 17% of the population in the US, 22% of the population in the UK (http://www.statistics.gov.uk/census2001/pyramids/pages/64.asp), and 54.5% in Turkey (according to 2007 figures) (http://tuikrapor.tuik.gov.tr/reports/rwservlet?adnks=&report=turkiye_yasgr.RDF&desformat=html&ENVID=adnksEnv). The X Generation is considered as an independent and skeptical group grown under the shadow of the Baby Boomers band, known as the post war children. The oldest among the members of this group are around 30-40 years old now, and they have become full consumers, with a certain income, most have families and own a house.

The people who were born in 1994 and afterwards are known as Generation Y. They account for approximately 25% of the population in the US, 25% in the UK and 18% in Turkey. This group is generally considered as idealist, optimist and patriotic. Members of generation Y consume the media by dividing it into smallest points. Of course, there are many common points and many things mutually experienced and shared by members of Generation X and Generation Y. For example, members of both groups were influenced by the crisis and recession, they lived with TV and they were even born after TV, they merged into Internet and grew up with other personal technologies. In consequence, both of these groups consume products and services more differently compared to former generations, and especially use the products and services that are created in the time of their generations in a different way compared to the next generation. The use of traditional marketing channels between X and Y generations are gradually becoming more difficult. Flesichner (2007) considers online marketing and direct mailing opportunities as one of the strongest ways of marketing to Generation X and Generation Y. Besides, in order to understand the new applications, it is required to compare them with basic marketing practices and carry out them out with the attitude based on these differences. According to Sheahan (2006), the Generation Y group is the group most connected to technology in history. This generation has turned into a group attached to the "system" and market.

The Millennium Generation

According to the Millennial Fact Sheet issued by Deloitte in 2005, Generation Y is considered as follows. Although not a certain birth year is mentioned, those who were born between 1982 and 1993 are considered in that group. They are the children of those who were born during the baby boom. They are the Generation X's siblings. 38% of them define themselves as 'non-white'. This is the generation of new age heroes. They are also named as Generation Y, Generation Why, Generation Next, Nexers, Milleniums, Digital Generation, Echo Boomers, Boomlets, Baby Busters, I Generation, Net Generation, Netizens, Gaming Generation (along with the younger members of Generation X). This group is also named as "Generation Now" as they use and reflect technology in a fast and satisfactory way (http://en.wikipedia.org/wiki/Internet_generation). Also, similarly, they are named as the computer generation; Generation D (Digital), Generation M (Millenium or Multi Task), Google Generation or Generation Q (Quiet).

The people in Generation Y have a good education, integrity is appreciated and the families create their role models. In this period, fathers contribute more to child care and bringing up. There is a continuous change in the modern organizations where they are located. Their characteristics include being technology-oriented, in 7/24 connection, self-confident, optimistic, independent, hopeful, adamant, determined, success-oriented, life style-centered, global urban and society-oriented, service oriented, entrepreneur and like changes.

Table 2. Business approaches in the new generation (Source: Deloitte (2005). Millennial Factsheet Who are the Millenials? a.k.a. Generation Y retrieved from http://www.deloitte.com/dtt/cda/doc/content/us_consulting_millennialfactsheet_080606.pdf)

Approaches	OUT	IN
Company Philosophy	Slow and Clumsy	Well organized and effective
	"Suits everyone"	Specialized and targeted
	Passive learning	Interactive and just in time information
Personal Behavior	Speak on behalf of the company	Say whatever the consequences are
	False pretenses	Brutal honesty
	Feckless but political oriented	Cares for respect
	Plays games	Shows reliable behavior
	Intolerant	Open-minded, integral
Time	Next week or next year	Now
Reason	Do this because: I said so We always do it like that here You have debts to pay	Do this because it is rational
Concern for the future	Company stairs; retirement	Bright CV; Personal retirement account

One of the features of Generation Y is that they did not spend a part of their lives without computers. It is possible to say that they have the mind-set of a reverse-directional flow of information. They act with the philosophy that the younger you are, the more information you can obtain, and the entire information is only one-click away from them, and people compete in creating information-looseness for themselves. As the entire world is only one click away, the idea of acquiring knowledge has changed. It is natural for employees with the habit of using readily-given knowledge to show different characteristics also in business life. The members of this group are good at team work with their friends. They are cooperative, skillful and innovative. They like struggling with a different energy, and they are continuously in search of differences. They want to do things that are worth for their effort. They want to be heroes, they are impatient, comfortable in speed and change. They are flexible and feel better when they explore new places. They get on well with mentors. They are value-oriented. They expect respect.

This generation is also named as the I-generation after the booming of the iPod. Because the people in this group do not only use iPods, MP3 players etc., but they also make use of services such as Internet forums, email, Wikipedia, search engines, MySpace, Facebook, YouTube and densely use Internet in this context.

At present, people between the ages of 15 and 25 prefer to simultaneously consume certain things in different fields. These consumers, who are defined as multi-tasking, are able to use more than one medium and field at the same time. For example, while sending a SMS, they can also watch movies and listen to music from an iPod.

The Cell Generation

This generation, which is brought up by families who are around 31 years old at present and who become parents for the first time, is defined as the Cell Generation or Generation Z (http://en.wikipedia.org/wiki/Generation_Z retrieved 4/5/08). They are considered as a rapidly maturing and rapidly growing group educated at very early ages and are exposed to marketing and its elements when they are very young. This group quickly understands the Internet and it is a technologically brought up generation. The people in this group are also grown with a multi-tasking ability. They pay attention to speed rather than correctness when being directed from one issue to another and from one work to another. Although they are not included in the labor force at present, 10% of this group will be included in the labor force in 10 years time. In 10 years time, the number of new recruitments is expected to be less than the number of retirements. Therefore, this generation will appear with an aging population and it should be considered as a different generation in an environment with a global request for labor force due to the problems caused by the diminishing skillful labor force, and thus marketing activities should take this into account.

Generation V

According to Havenstein (2007), with Generations X and Y, marketers are obliged to also consider generation V, which is in a new and mixed structure. Generation V possesses certain stereotypical characteristics of the generations that follow the baby boom. Although some people define it as the latest generation, the marketers of Gartner Inc. suggest that individuals should not be defined according to their birthdays but according to their online activities.

As actual behavior, Generation V can be considered as a group which prefers online experience instead of physical experiences. Here, attention should be paid to online differentiation. Because the traditional demographic data such as the age, name, sex and address are not sufficient for the correct routing of marketing messages by companies which desire to sell products and services to these customers.

Here, one more step should be taken and customers should be evaluated not only in terms of online activities, but also mobile activities. As mentioned earlier, marketing products and services to customers with traditional methods is not sufficient. It is true that this group has a multi-layered demographic structure. The V group is also included among those who are familiar with the virtual world, who like video games and social networks, and who are shop from web sites such as Amazon.com and e-bay.com. This social group, which is known as Generation V, is the group of people who can simply adapt to what is in force or what is valid. Apart from actual status definitions, people in this group should be considered as a different social status which appears with the people's own actions included, and which is not considered as functional in the physical world.

This generation is able to work with marketers anonymously, and they can establish communication in mutual interaction. According to Sarner (2007), in most online behaviors, this type of customer establishes their online characters rather than their actual personalities. In this context, those who are willing to work with generation V are required to meet certain online characters and create new tactics during the purchasing processes of these characters in the virtual world. Within the context of relationships to be developed with generation V, which has a volatile structure, in addition to the describable generations of X and Y, the characters that appear in the virtual environment and the effects of these characters on the customer's experiences should be taken into consideration; and it should be known that disclosing the fact behind the purchase of product and service has become crucial for marketers. (http://www.worldsinmotion.biz/2007/11/gartner_recommendations_on_gen.php retrieved 4/5/2008). Of course in time, more data will be revealed regarding the taste, shopping behavior and shopping approach of the generations and information collected through data mining will be better evaluated, and the actual desires and requirements of the customer will be covered. Apart from that, Gartner claims that from 2015 onwards, money spent in marketing and sales for multi online characters shall be higher compared to that for customers who do not shop via Internet. The difficulty in their identification and gap in the market may develop by targeting these groups which cannot be identified as customers. Therefore, it is stated that generation V is an important area in the next 10 years for all purchases.

Gartner recommends the following for companies which are willing to market to Generation V.

- To organize products and services according to multi-online characters, and even creating their mobile characters
- To sell to the adopted character rather than the person
- To create virtual environments which can enable the discovery of the customer during shopping in a more efficient way
- To shift investments from known customers to unknown customers,
- To create a worker profile suitable for this new type of customer and virtual environment
- To concentrate on those who affect society

Generation V with its online culture, consists of individuals who can only be virtually defined. These individuals realize their relations and use services and communications primarily through the electronic environment (http://en.wikipedia.org/wiki/Generation_V retrieved 4/5/2008). As can be understood, generation V acts outside of the factors such as age, gender, social status or geographical location, and prefers a digital environment in finding, acquiring and sharing information.

There are four different customer structures according to the connection level in society that creates generation V. They are considered as creative, participant, opportunist and observer. The characteristics of these segments, which are the classification of Gartner are defined as follows.

Those who are in the creative group account for 3% and they have the original content developed skill. They play an important role in the presentation of products and services. The participants' share varies between 3 to 10%. They do not start a relationship but they join it. Participants may recommend products or services to customers during the product or service purchasing period. Those in the group of 10-20% evaluate opportunities related with purchasing decisions. This group provides a kind of contribution to marketers by providing detailed information to people who can reach a technological environment. Opportunists can add value to the conversation with regard to purchasing. The audience group accounts for the remaining 80%. They can indirectly contribute to society they are included in, yet they do not create a focus point. They are observers in general, and all users in the communication environment are included in the system with this type of relationship.

Actually, generation V is a customer group defined through associating their general behavior, attitude and interests with online environment. Generation V has a feature which is of a different manner than "the generation of culture symbols". The technology-oriented understanding is gradually expanding and it is going under certain changes which would enable the growth of technology-oriented generations. They create a segment for many generations that have sometimes a supporting and sometimes a preventive approach. Here, the virtual environment is considered as a game field organized in a worthy order according to capability, motivation and effort and according to virtual environment, age, gender, class and income issues (Sarner, 2007).

At present, someone who might be considered a child yet possesses the role of a consultant who provides opinions on how to use computer software. An ordinary office staff may appear as a player who plays in the latest steps of a computer game which is very difficult to access. In order to obtain strong social incentives such as gaining prestige, being effective, and obtaining individual development, wider masses spend more time in the virtual world. Here, new processes and abilities are required to be evaluated according to their personal characteristics rather than demographic characteristics (Gartner Recommendations on 'Generation Virtual', 2007).

The new generation added to the labor force seeks a higher work-life balance and if the performed work does not meet their expectations, then they immediately begin to look for another job. Therefore, these people should be motivated by a manager to reinforce their perception that their job is not slavery. In addition to that, a suitable style of entertainment is also required to be found in order to bring entertainment to business life. However, pranks and practical jokes related with colleagues are things that sometimes alienate people and which should be avoided by any means.

PREFERENCES & EXPECTATIONS

Different generations are equipped with different thoughts and they have different opinions and behaviors. It is clear that created preferences will also be different when the related characteristics are considered. Actually, new generations are based on what is produced by previous generations. New generations expect more practical and easier solutions from the producer and marketer, while others prefer to take part in the process and put forth issues such as benefitting from products which are created according to their expectations.

One of them is after fads, while the other group spends efforts to overcome the difficulties of compliance to groups or to the social environment. In order to maintain an equal distance between both of these groups with different views of life, marketing essentially requires approaches with different perspectives. The new generation aims to be successful no matter what happens. They are in the pursuit of the best and easily accessible one of everything. All inter-generational conflicts arise from differences of priorities. This conflict is actually an indicator of development. To rule the future, awareness on the dynamics of the future is required. Furthermore, these dynamics should also be considered as a highly effective instrument, and as a value defined as the forecast in the configuration and orientation of marketing plans. Technology facilitates obtaining required data for classification according to personal characteristics, and also facilitates the performance of demographic classification. Real data obtained from the virtual environment are provided with a clearer and more concrete form.

These means also cause differences in terms of the customer's expectations. It is quite natural to observe wide differences among expectations of the new generations as the Generation V group is directed to a more active attitude while the Generation X group indicates a more passive attitude with its personal self-perspective.

The expectations of Generation Y are as follows (http://www.growingupdigital.com/Fngconsum.html retrieved 10/3/2008):

- Availability of options
- Self adaptation of products and services
- Changing mind
- Trying before buying.

Due to this volatile structure, retaining the Generation Y customer is not easy. Communication and information developed in the field of technology with the claim of being an information society causes other changes in addition to digital approach with its impact on the expectations of both X, Y and other generations. The approximation of advertising and sponsorship relationships and what is pumped into the environment by systems that facilitate the configuration of products, services and even characters provides the opportunity to both entertain and develop their hobbies which is important. The request to be held subject to a special treatment as a customer is expected to become a new norm. The new location and subject-dependent marketing effort result in the requisite for paying great attention to the expectations of organizations and individuals which also necessitates marketers to pursue new expansions. The taste-satisfactory marketing efforts may be another important expectation. The difficulties on direction towards not only the requirement but also tastes and introducing the most suitable offer for the customer will become the most important problems for marketers.

BEHAVIORS

Innovation is the replacement and improvement of anything old, harmful or unsatisfactory with the new, beneficial and satisfactory. Therefore, in order to conceive contemporary lives, regardless of the generation, it is required to deal with and conform to the innovations that arise in daily life. Benjamin (1999) philosophically considers innovation as a component of the flow of life, explaining the reason for people's sharing the newest things with each other as probably to gain victory against death, which is not something new. However, every innovation needs some time for internalization.

The short lifespan of the "virtual" environment suggests that information products will also have a short lifespan (Böhm, 2001). In our age, in addition to the influence caused by the short lifespan, information created in the virtual environment also faces the risk of losing its importance until its value is recognized by the human brain, which characteristically perceives it to be late.

Then, to maintain harmony between the demands of the contemporary technology and human behavior requires questioning. How changes in the new generations affect ideas and approaches and how this influence would reflect on society should be dwelled on, and the ways to obtain harmony should be revealed. In this context, foreseeing consumer behaviors between production societies which guide marketing activities and the technology-consuming societies or information societies is a great concern.

Augé (1995) defines the world as "non-places" as having no identity, no history and no urban relationships, which is created on individual terms for a temporary, short time and without an established place. With the development of the transportation, tourism, entertainment and commerce sectors which eliminate the concepts of time, place, volume and distance, and with the restructuring of communication through wired or wireless instruments, the peculiar assets of individuals (wisdom, information, capital etc.) and the results of communication with an individual with another image should be evaluated and actions should be taken according to them. Meanwhile, borders between the real world and the virtual world are also removed as they interlace with the virtual world through brand new definitions of the time, volume and distance concepts, and at these times, the concepts of the real world and the virtual world are mixed in each other.

In this context, how should the personality differences of people who benefit from innovations at different levels and by different means be neutralized or considered and how would the pains of innovation in behaviors be removed are also among the questions to be taken into account. Atasü (2008) suggests that individualism is much more related with collectivism than is actually believed.

There is the risk of over-fragility caused by individualism, which has been fortified with the Internet and then the mobile environment, and the risk of unbalancing the details, damaging the sense of society, and "individualization" and personalization of everything as a result of the mentioned over-fragility. The threat of shifting societies towards more egoist norms through these approaches should also be evaluated. "Would such members of society choose individualism or collectivism for their lives? How will the behaviors based on these two approaches be explained?" These are among the questions for consideration.

From the past to the present, certain theories exerting to explain the behaviors of the customer have tried to describe the expansions adopted in this regard, and they mention the results concluded from abstract to concrete. According to the enhanced view of the Theory of Reasoned Action (Ajzen 2001, Bentler and Speckart 1979, Eagly and Chaiken 1993), the past-time behavior of an individual explains his or her present-time behavior. It means customers prefer shopping at places where they previously shopped with the expectation of obtaining the same benefit. Corstjens and Rajiv (2000) explain this approach as the request for minimizing the cost of thinking and psychological affection related to their previous likings. This is defined as the inertia effect, and it automatically occurs without conscious thinking (Vogel, Evanschitzky & Ramaseshan, 2008). The continuity of this tendency is required to be maintained in mobile shopping, and the trust mechanism should be correctly operated on the customer. Mobilization is also related with the interest-awakening feature of the innovative processes. How problems about closing changes without any damages are resolved and absorbed are not clearly known because mobile marketing appeared as a trend realized in the last 2 or 3 years as a result of the means provided in the last 20 years. It is not possible to say that the "Theory of Reasoned Action" is sufficient for mobile marketers. However, as individuals perform certain actions on digital media, the behaviors related to them have gained meaning.

McKenna and Bargh (2000) highlight four main differences among real life acts and the selfness, identities, social interactions and relations of individuals on the Internet. The anonymousness of the person or hiding the name to be free of relationship development based on physical environment and appearance, and acting controlled on arranging time and speed in the interaction is common of individuals on the Internet.

When these differences are considered from the mobile marketing view, the individual has an environment sufficient to hide his own number when building communication. Together with that, mobile devices enable communication to be performed within the given permits. People or organizations act in an environment where they can more easily express themselves by disclosing themselves and their identities, and even this environment is considered as more reliable than the Internet. The interaction is momentary and immediate while certain infrastructural problems arise with regard to the physical environment. From the point of mobile marketing development, the only environmental problems are the use or ease of use of devices and the number of users, the insufficiency of facilities which provide access or uncovered areas, and the high prices of 3G and similar ones. Nevertheless, the use of mobile Internet presents certain problems which resemble those of the Internet as the mobile connects to the

Internet. However, such problems should be considered as easily solvable after the full-establishment of the 3G system.

According to De Souza e Silva (2006), mobile phones also cause changes on the physical environment as much as changes in the social environment. In cyber places, it is possible to perceive someone else or be together with someone else without requiring any physical action or moving any physical objects. So, cyber places create hybrid domains and enable circulation of the individual's information in cyber places (Urry, 2002). People can easily talk on mobile phones while walking which means they can still be in a social environment while moving in a physical environment (Klein cited in De Souza e Silva, 2006: 29). This absent presence transforms the physical environment into non-place.

Mobile telephones also cause the formation of hybrid domains similar to the Internet, but they are more minimized and narrower. Even, with their digital interfaces, mobile phones enable reaching people at anytime in hybrid domains. The portable interfaces in the mobile environment are merged in the physical environment, thus blurring the environment where real life and digital environment co-exist. With mobile phones, people can carry the Internet wherever they go, feeling as though they are everywhere at the same time (http://www.modernexistence.net/mobile-phones-connecting-and-disconnecting-people-and-places/ retrieved 10/2/2008). However, as there are certain specializations developed on monitoring the environment, they can also be used threatening individuality. Being aware of these two conditions, the issue of how users would make use of their preferences will be shaped by their experiences.

With the development of web surfing functions of mobile phones, the idea of using the mobile environment will become more effective for marketers. It is apparent that the behavior in cyber places would be different than those in the physical environment.

Mobile phone users generally come and go between the emotions of devotion and detachment, and it also reflects on their behaviors. For example, a received phone call during a dinner breaks takes the individual off from the current social environment for a moment and takes him or her into another environment; therefore, the communication environment may influence the actual environment of the individual and turn it into a different state at the same time (Doyle, 2007). The question raised by Doyle for the mobile phone as an instrument susceptible to discussing from many perspectives is: "Is it building communication or just being in a placebo?" Nevertheless, the question raised is related with the reconsideration of the utmost fundamental issues such as "Who will communicate with whom at which domain?" According to his arguments and the reflection on the individual's behaviors against individuals, and the social behaviors, Doyle draws the attention that this situation has a multimodal character. He hints warnings on the results of social imports which might be caused by the virtual interlacing of the private domain and social domain at the same time. Due to observed inconveniences, using the mobile phone in the classroom is prohibited in most states of the USA and in certain countries. According to Meyrowitz (1985, p. 315), breaking off the unnecessary connection in social relationships is gradually becoming more difficult. Everybody has begun to interfere with everybody at any time whether it is necessary or unnecessary.

According to Harris Interactive, different generations have already separated and identified themselves as different from other generations in terms of behavior (http://www.harrisinteractive.com/news/allnewsbydate.asp?NewsID=1328retrieved 10/4/2008).

In studies held in order to understand how different generations are perceived by the society, it is noted that these generations have more than one of the three strengths: being admired, being productive and having a positive effect.

The generation before Generation X is foreseen as that which creates positive influence in society by 35%, and it is followed by Generation X with 25%.

The pre-world war II generation, which is defined as the Silent Generation, is described as the most widely admired generation with 33%. The silent generation is also considered as the most generous generation with 40%. Baby Boomers are second to Generation X with 33% in terms of generosity.

The group described as the most productive is Baby Boomers with 45%. They are followed by Generation X with 32%. The most self-indulgent group is Generation V with 53%. Generation X follows with 25%.

A 41% plurality sees Generation X as the most innovative. It is followed by Baby Boomers (25%) and Generation Y (22%).

The most socially conscious groups are Baby Boomers (34%) and Generation X (26%). 32% of Generation Y would prefer being renamed as the "Internet Generation". 25% of Generation X would prefer being renamed as "Generation Tech". Those in the Baby Boomers group are happy with their name.

Customers who accept to receive an advertising message from a mobile device are young, pioneering and vanguard people.

Katz (2006) exemplifies configuration of the effects created by the mobile phone in awakening desire for relationships in the perception of human values, and scrutinizes decisions regarding the identification of mobile phones with personal images. These images cause the gradual approximation between public domain and personal domain. The symbol chosen by each individual may differently reflect the situation and different images are preferred depending on the culture, person or information environment. Here, one of the important points is to know the character located in mass circulation affecting the individual, and act accordingly or to stimulate buying desire with conforming transmissions.

The Creativity Mechanism

Design-making is leading one of the most important characteristics which separate human kind from other species. Designing is the foundation of cultural issues-related sharing. Design-making also lies in the foundation of social construction; therefore, design is an indispensable value for social structure and it requires creativity.

According to Luecke (2008), creativity is not a state of mind or a personal nervous system, but it is the development and expression of new ideas for solving problems and satisfying needs.

The creative innovation or creative inspiration is included in a momentary design and even in the moment itself. It reaches from an ambiguity to clarity. Otherwise, what is done would not be anything else than repeating the evident (both in essence and in form). The past is assimilated in the present. Now it also includes the elements of the future. Then, the past, the present and the future remain integral in every aspect of life (Behramoğlu, 2008). On the other hand, each period has its own internal dynamism.

Lyons (2007) defines innovation as a "fresh" idea which creates value. According to Luecke (2008), who carries it one step forward, innovation is the synthesis, combination or solidification of information in order to create an original, new and valuable product, production process or service. In order to do that, an innovative mindset is required.

According to Altun, the spirit of time has support from the most suitable environment for innovation. Technological demands create new balances on what has not been done, trying what has not been tried and creating what has not existed by an innovative organization and being fast. Thinking differently and developing creative, innovative business ideas accordingly has entered on a demand basis; thus, featuring a value (2007). However, according to Schiffman, businessmen may not dream for others or suggest innovative or creative opinions for others and make plans to help people (2008). New technologies have led from the former narrow-minded customer behaviors of read-learn-watch culture to the participation and cooperation of the customer. A shared, social, global, open and interacted customer relationship, which is produced or developed as requested by the user, has emerged (Shiffman, 2008, p. 2).

On the other hand, the media world changes with concepts such as personal publishing, digital video, podcasting and social networks, and the customers have become able to reach information when they want to obtain information. This creates an arbitrary or on-demand information access system, yet it is in need of creativity on developing technology to filter undesired messages (http://www.marketingpower.com/ResourceLibrary/Pages/Webcasts/Webcast082808.aspx 8/28/2008).

As creativity is required to be established on the production of fast consumption-based technological products to their extension to the end consumer, creating innovation turns into a piece of the marketing business. Marketers are required to redefine their own works so as to race against speed and time, and to be able to develop content within the framework of their customers' requests.

REFERENCES

Ajzen, I., & Fishbein, M. (2000). Attitudes and the Attitude-Behavior Relation: Reasoned and Automatic Processes. In W. Stroebe, M. Hewstone (Ed.) European Review of Social Psychology. Chichester, UK: Wiley.

Altun, Ş. (2007). İnovasyon Türklerin Genlerindedir (Innovation is inside of the Turkish Gens). Retrieved from http://www.bilgicagi.com/index.php?option=com_content&task=view&id=112&Itemid=139&limit=1&limitstart=3

Atasü, E. (2008). Düşünce Sefaletinin Kıskacında. Ankara, Turkey: Imge Yay.

Augé, M. (1995). Non-Places: Introduction to an Anthropology of Supermodernity, (trans. J. Howe). London: Verso.

Aykara G (2008). Telecomünikasyon Kurumu. Basın ve Halkla İlişkiler Müşaviri 9/9/2008 dated e-mail correspondance.

Bauer, H. H., Reichardt, T., Barnes, S. J., & Neumann, M. M. (2005). Driving Consumer Acceptance of Mobile Marketing: A Theoretical Framework and Empirical Study. *Journal of Electronic Commerce Research*, *6*(3), 181–192.

Bayne, K. M. (2002). Marketing Without Wires. New York: John Wiley & Sons.

Behramoğlu, A. (2008). Gelenek Gelecek ve Şimdi Üzerine Notlar. Cumhuriyet Pazar Eki, S. 1154, 4/5/08, s.11.

Benjamin, W. (1973). Charles Baudelaire: A Lyric Poet in the Era of High Capitalism. (trans. H. Zohn). London: Verso.

Benjamin, W. (1999). The Arcades Project. (trans. H. Eiland & K. McLaughlin). Cambridge, MA: The Belknap Press of Harvard University Press.

Bentler, P. M., & Speckart, G. (1979). Models of Attitude-Behavior Relations. *Psychological Review*, *86*, 452–464. doi:10.1037/0033-295X.86.5.452

Böhm, S. G. (2001). Fetishism in Consulting, or, The Dancing Techno-Knowledge Commodity. Retrieved 10/2/2008 from http://www.mngt.waikato.ac.nz/ejrot/cmsconference/2001/Papers/Management%20 Knowledge/Bohm.pdf

Bourke, C. (2006). How to Develop a Mobile Marketing Strategy. Retrieved 6/26/2008 from http://www.aerodeon.com/whitepapers/Aerodeon_MobileStrategy_v100.pdf

Candan, B., Aydın, K., & Yamamoto, G. T. (2008). A Research on Measuring Consumer Ethnocentrism of Young Turkish Customers Purchasing Behaviors. Serbian Journal of Management, 3(1). Retrieved 4/27/2009 from http://www.sjm06.com/3_1_2008.htm

Cara, F. (2008). Retrieved 7/4/2008 from http://www.smallsurfaces.com/2008/06/the-evolution-of-the-mobile-ecosystem/

Chokvasin, T. (2007). Mobile Phone and Autonomy. In S. Hongladarom & C. Ess (Eds.) Information Technology Ethics. Hershey, PA: Idea Group Reference.

Corstjens, M., & Rajiv, L. (2000). Building Store Loyalty through Store Brands. *JMR, Journal of Marketing Research*, *37*(3), 281–291. doi:10.1509/jmkr.37.3.281.18781

Craig, C. (1999). The Emerging High-Tech Consumer: A Market Profile and Marketing Strategy Implications. Academy of Marketing Science Journal, 27(3 Summer), 378.

De Souza e Silva, A. (2006). Retrieved from http://www.modernexistence.net/mobile-phones-connecting-and-disconnecting-people-and-places/

De Souza e Silva, A. (2006). Interfaces of Hybrid Spaces. In A. Kavoori & N. Archeneaux (Eds.), The Cellphone Reader (pp.19-45). New York: Peter Lang.

Deloitte (2005). Millenial Factsheet: Who are the Millenials? a.k.a. Generation Y. Retrieved 5/1/2009 http://www.deloitte.com/dtt/cda/doc/content/us_consulting_millennialfactsheet_080606.pdf

Doyle, (2007, May 6th). Call Waiting, Reality Dying. The Sunday Age. Retrieved from http://www.modernexistence.net/

Eagly, A. H., & Chaiken, S. (1993). The Psychology ofAttitudes. Fort Worth, TX: Harcourt Brace.

Fishbein, M., & Ajzen, I. (1975). Belief, Attitude, Intention and Behaviour: An Introduction to Theory and Research. Reading, MA: Addison-Wesley.

Flesichner, M. (2007). Marketing to Generation X and Y. Retrieved 5/1/2009 from http://www.evancarmichael.com/SEO/1282/Marketing-to-Generation-X-and-Y.html

Gençlerin Tercihi Mobil Yaşam. (2008). Cumhuriyet, 08/19/08, (p. 12).

Haghirian, P., & Madlberger, M. (2005). Consumer Attitude toward Advertising via Mobile Devices – an Emprical Investigation Among Austrian Users. Retrieved from http://is2.lse.ac.uk/asp/aspecis/20050038.pdf

Havenstein, H. (2007, November 18). Meet the Virtual Generation. Computerworld Sunday. Retrieved 4/5/2008 from http://www.pcworld.com/article/id,139748-c,researchreports/article.html

Howe, N., & Strauss, W. (2000). Millenials Rising. The Next Great Generation. New York: Vintage Books.

http://en.wikipedia.org/wiki/Anime (n.d.).

http://en.wikipedia.org/wiki/Generation_Z (n.d.).

http://en.wikipedia.org/wiki/Internet_generation (n.d.).

http://en.wikipedia.org/wiki/Theory_of_reasoned_action (n.d.).

http://en.wiktionary.org/wiki/individualization (n.d.).

http://tuikrapor.tuik.gov.tr/reports/rwservlet?adnks=&report=turkiye_yasgr.RDF&desformat=html&ENVID=adnksEnv (n.d.).

http://whatis.techtarget.com/definition/0,sid9_gci1189222,00.html (n.d.).

http://whatis.techtarget.com/definition/0,sid9_gci1235244,00.html (n.d.).

http://www.answers.com/topic/generation-x (n.d.). Retrieved 9/14/2008.

http://www.dtic.mil/doctrine/jel/doddict/data/c/01473.html (n.d.).

http://www.evancarmichael.com/SEO/1282/Marketing-to-Generation-X-and-Y.html (n.d.).

http://www.gartner.com/it/page.jsp?id=680207 (n.d.).

http://www.growingupdigital.com/Fngconsum.html (n.d.).

http://www.harrisinteractive.com/news/allnewsbydate.asp?NewsID=1328 (n.d.).

http://www.marketingcharts.com/interactive/generation-v-defies-traditional-demographics-5495/gartner-generation-virtual-engagement-levels-june-2008jpg/ (n.d.).

http://www.marketingcharts.com/television/how-consumers-prefer-to-receive-advertising-a-new-segmentation-5567/mri-ad-medium-preference-segmentationjpg/ (n.d.). Retrieved 10/4/2008.

http://www.marketingpower.com/ResourceLibrary/Pages/Webcasts/Webcast082808.aspx (n.d.).

http://www.musicsonglyrics.com/B/bluelyrics/bluewithoutyoulyrics.htm (n.d.).

http://www.nagc.org/index.aspx (n.d.).

http://www.pcworld.com/article/id,139748-c,researchreports/article.html 4/5/2008 (n.d.).

http://www.promarinternational.com/pdfs/IndustryStrategicStudies/ConsumerInsights/Gen%20Y%20snacks%20management%20summary.pdf (n.d.). Retrieved 10/4/2008.

http://www.quantum3.co.za/CI%20Glossary.htm#C (n.d.).

http://www.scribd.com/doc/270170/PERSONAL-MEDIATED-COMMUNICATION- (n.d.).

http://www.theage.com.au/news/opinion/call-waiting-reality-dying/2007/05/05/1177788462594.html (n.d.). Retrieved 10/3/2008.

http://www.urbandictionary.com/define.php?term=generation+z (n.d.). Retrieved5/1/2009.

http://www.worldsinmotion.biz/2007/11/gartner_recommendations_on_gen.php (n.d.).

Jochim, J. (1997). Generation X defies definition. Retrieved 10/28/2008 from http://www.jour.unr.edu/outpost/specials/genx.overvw1.html

Kaasinen, E. (2003). User Needs for Location-Aware Mobile Services. *Personal and Ubiquitous Computing*, *7*(1), 70–79. doi:10.1007/s00779-002-0214-7

Katz, J. E. (2006). Magic in the Air: Mobile Communication and the Transformation of Social Life. New Brunswick, NJ: Transaction Publishers.

Klein, N. (2002). De Souza e Silva's Personal Communication. November 8, 2002.

Koç, A. (2008). Integration, Globalization & Change. A Speech Given at Okan University (05/08/2008).

Koray, S. (2008). Küreselleşmenin İçyüzü. Bilim ve Ütopya, 170(Ağustos 2008), 4-13.

Leppäniemi, M., & Karjaluoto, H. (2007). Mobile Marketing: From Marketing Strategy to Mobile Marketing Campaign Implementation. In Proceedings of the 6th Annual Global Mobility Roundtable Conference, Los Angeles, June 1-2.

Ling, R. (2004). The Mobile Connection. Amsterdam: Elsevier.

Luecke, R. (2008). İş Dünyasında Yenilik ve Yaratıcılık (Managing Creativity and Innovation). Trans. T. Parlak. İstanbul: İş Bankası Kültür Yay.

Lyons, R. K. (2007, October 13). A special Report on Innovation. The Economist.

McKenna, K. Y. A., & Bargh, J. A. (2000). Plan 9 From Cyberspace: The Implications of the Internet for Personality and Social Psychology. *Personality and Social Psychology Review*, *4*(1), 57–75. doi:10.1207/S15327957PSPR0401_6

Meyrowitz, J. (1985). No sense of place: The Impact of electronic media on social behavior. New York: Oxford University Press.

Öğüt, H. (2008). Ahlakın Göstereni ve Gösterileni Kadın (mı)? [(Is) the woman that indicant or indicated of ethics?)]. *Varlık*, *76*(1211), 3–8.

Öymen, E. E. (2008). Bilgi Obezi Tüketici (Knowledge Obese Customer). Cumhuriyet Bilim ve Teknoloji Eki, S. 1112 (July, 11 2008).

Roe, R. (2008). Efficient Governments, Dictatorships and Resilience. Retrieved 9/13/2008 from http://managementmythbusters.blogspot.com/

Sarner, A. (2007). Gartner Recommendations On 'Generation Virtual'. Retrieved from http://www.worldsinmotion.biz/2007/11/gartner_recommendations_on_gen.php

Sheahan, P. (2006). The Connected Generation. Retrieved 11/1/2008 www.petersheahan.com.au/cpa/exe/exe_file_get.asp?fileName=TheConnectGeneration(2).pdf

Shiffman, D. (2008). The Age of Engage. Ladera Ranch, CA: Hunt Street Press.

Strauss, W., & Howe, N. (1992). Generations: The History of America's Future, 1584 to 2069. New York: Harper Perennial.

Urry, J. (2002). Mobility and Proximity. Sociology, 36(2), 255-274. Retrieved 10/3/2008 from http://www.its.leeds.ac.uk/projects/mobilenetwork/downloads/urry1stpaper.doc

Vogel, V., Evanschitzky, H., & Ramaseshan, B. (2008). Customer Equity Drivers and Future Sales. *Journal of Marketing*, *72*(6), 98–108. doi:10.1509/jmkg.72.6.98

Yamamoto, G.T.(2008). İnternette İletişim ve Secondlife (Internet Communication & Second Life). Pi Dergisi, 2008.

YouGov. (2006). The Mobile Life Report 2006: How Mobile Phones Change the Way We Live. London: YouGov.

KEY TERMS AND DEFINITIONS

Anime: It has been taken from half of the Japanese pronunciation of "animation". It is used in television series, films, video, video games, commercials, and internet-based releases, and represents most, if not all, genres of fiction (http://en.wikipedia.org/wiki/Anime 5/1/2009).

Cell Generation: It is a generation also named Generation Z born from 1991 to the present. These are the kids that are born under a computer and have no past memory of life during the Cold War. They grow up with cell phones, iPod's, MySpace, YouTube, and homeland security (http://www.urbandictionary.com/define.php?term=generation+z 5/1/2009).

Creativity: The process of developing new, uncommon, or unique ideas (http://www.nagc.org/index.aspx retrieved 10/3/2008). It also refers to the act of generating new and useful ideas, or of re-evaluating or combining old ideas so as to develop new and useful perspectives in order to satisfy a need. It is the capacity to select, re-arrange, combine, or synthesize existing facts, ideas, images, or expertise in original ways (http://www.quantum3.co.za/CI%20Glossary.htm#C retrieved 04/10/2008).

Cyberspace: Public space where individuals can meet, exchange ideas, share information, provide social support, conduct business, create artistic media, play simulation games or engage in political discussion with the use of the network of information technology infrastructures. It is the notional environment in which digitized information is communicated over computer networks (http://www.dtic.mil/doctrine/jel/doddict/data/c/01473.html retrieved 10/3/2008).

Generation D: Digital Generation for those who have grown up in a digital world.

Hybrid Space: The environment which assembles physical place and cyberspace.

I Generation: It is the Ipod generation that prefers to listen to music over the iPod. The iPod is a digital media player from Apple Computer (whatis.techtarget.com/definition/0,,sid9_gci1189222,00.html retrieved 10/4/2008).

Individualization: The act of individualizing (http://en.wiktionary.org/wiki/individualization retrieved 10/3/2008).

Mobile Culture: A new culture which is brought by mobile devices, mobile and digital communication.

Mobile Customer: Persons or institutions that use the mobile device, purchase via mobile device and/or see mobile devices as a shopping/purchasing medium.

MoU: Minutes of use

Non-Place: 'non-place' designates two complementary but distinct realities: spaces formed in relation to certain ends (transport, transit, commerce, leisure), and the relationships that individuals have with these spaces. (Augé, 1995: 94)

Secondlife: A virtual site that a person joins in a virtual world. Second Life is a massive multiplayer universe (MMU) set in a 3D virtual world created by Linden Labs and major technological corporations have used Second Life to market products or services to a niche tech-savvy audience (http://whatis.techtarget.com/definition/0,,sid9_gci1235244,00.html retrieved 5/2/2009).

Theory of Reasoned Action (TRA): A theory developed by Martin Fishbein and Icek Ajzen, (1975, 1980). The components of TRA have three general constructs: 1) behavioral intention, 2) attitude, and 3) subjective norm. TRA suggests that a person's behavioral intention depends on the person's attitude about the behavior and subjective norms (BI = A +SN). If a person intends to do a behavior then it is likely that the person will do it. Furthermore a person's intentions are themselves guided by two things: the person's attitude towards the behavior and the subjective norm. Behavioral intention measures a person's relative strength of intention to perform a behavior. Attitude consists of beliefs about the consequences of performing the behavior multiplied by his or her valuation of these consequences. Subjective norm is seen as a combination of perceived expectations from relevant individuals or groups along with intentions to comply with these expectations. In other words, "the person's perception that most people who are important to him or her think he or she should or should not perform the behavior in question" (Azjen & Fishbein, 1975 retrieved from http://en.wikipedia.org/wiki/Theory_of_reasoned_action 10/3/2008).

X Generation: Generation X can technically be defined as the generation following the Baby Boomers. The X Generation was born between 1965 and 1980, 1961 and 1981, 1964 and 1979, 1963 and 1979, 1965 and 1975 or since the mid-1960s, depending on which source you use. For practical purposes we will say that Generation X was born between 1965 and 1980, now ranging in age from 17-32 and usually judged by characteristics assigned to them by the media (Jochim, 1997). Now ranging in age from 28-43 is in this group.

Y Generation: Most commonly, Generation Y is defined as including those born between 1980 and 1995 (http://www.promarinternational.com/pdfs/IndustryStrategicStudies/ConsumerInsights/Gen%20Y%20snacks%20management%20summary.pdf retrieved 10/4/2008).

Chapter 6
Mobile Promotion

Promotions are various activities which are realized in order to create and maintain demand for products or services. Promotions are created to reach customers and to gain some profit from them. Therefore, these activities are very important for marketers. In mobile marketing, mobile promotions mean that the companies' promotional activities are done with mobile devices. Mobile promotion can also be defined as the conduct of promotion activities via mobile instruments. In its most common sense, mobile promotion considered as the promotional method for the customer in an appropriate and permitted manner by benefitting from a live media via mobile devices such as PDA. In this chapter, mobile promotions are evaluated. Mobile advertising and consumer reaching technologies such as SMS and MMS are discussed in detail. Types and the user profile are also examined and some current activities are stated. One-to-one communication and sales promotions such as coupons and other motivational material are broadly composed in this chapter.

PROMOTIONAL ACTIVITIES

Promotion is an instrument maintained through communication between the transmitter of the given message and the customer who receives it one-on-one. Deleon (2004, p.15) regards the mobile phone as a seductive advertising instrument compared to conventional advertisements. All promotion instruments which actually have a mobility feature can be evaluated as mobile, and they create a different category than other promotional instruments. The intended mobile promotion in this study is the promotions realized via the mobile instruments people carry on themselves. As mobile promotion is among the strong

DOI: 10.4018/978-1-60566-916-8.ch006

Figure 1. US mobile advertising spending

US Mobile Advertising Spending, 2007-2012 (millions)

	2007	2008	2009	2010	2011	2012
Mobile message advertising*	$810	$1,470	$2,380	$3,060	$3,830	$4,500
Mobile display advertising**	$34	$85	$186	$327	$453	$541
Mobile search advertising***	$35	$107	$242	$531	$910	$1,484
Total	**$878**	**$1,663**	**$2,808**	**$3,917**	**$5,194**	**$6,525**

*Note: numbers may not add up to total due to rounding; *spending on placement in text messages, includes direct spending on message campaigns as well as spending on promotional coverage of end-user messaging costs; **spending on display banners, links or icons placed on WAP, mobile HTML sites or embedded in mobile applications such as maps or entertainment services (eg games or video); ***spending on sponsored display ads and text links that appear alongside mobile search results, as well as spending on audio ads played to mobile phone callers making a directory inquiry*
Source: eMarketer, March 2008

092634 www.eMarketer.com

alternatives of mobile advertising together with the new technological expansions on this issue, it creates an impact with short-term changes of promotions on the basis of providing discount or additional products, which is used in conventional marketing. The messages that contain mobile coupons and mobile discount news are also among the mentioned mobile advertising promotions. According to MMA Global (2007), the mobile coupon is an electronic ticket solicited and/or delivered by the mobile phone that can be exchanged for a financial discount or rebate when purchasing a product or service. They are often distributed through SMS, MMS, Bluetooth and other mobile instruments. Mobile coupons are very useful promotional materials for stimulating a personal shopping desire as other advertising messages.

According to the E-marketer (2008), the mobile message advertising value will increase by five times from 810 million dollars in 2007 to 4.5 million dollars in 2012 (see Figure 1).

According to the 2007-Q3 research of Dutch Mobile Promotions (Overview of mobile promotions in The Netherlands during 2007-Q3), a mobile promotion aims to achieve the objective of creating encouragement for the purchase of a product or service by mobility, by changing retail understanding with a mobile phone or a similar mobile instrument and with a mobile operator as a result of the motivation created through presenting certain discounts from retail channels, or extra bonuses, SMS's or other certain valuable products as gifts to encourage the use of the mobile operator. Due to that, both advantages for the sales of the product are created, and the demand for the instruments and domains of the mobile services is reinforced.

Three different promotions can be created in the mobile environment. They are promotions based on the common attitude and consideration of both demand and supply, which is realized with the cooperation of company-led promotions to increase the demand, and customer-led promotions to increase the supply. Mobile promotions directed by companies are massively spread especially as SMS announcements. The functions of the mobile phone evolve towards marketing, and so, turn into a network marketing instrument. Furthermore, it does not only remain as network marketing, but also convert the mobile instrument into an entertainment platform through promotions such as m-advertising and m-entertainment. Technological and social systems converge and become blurred. It can also turn into m-advertising in time. However, in order to prevent it from turning into craziness, serious responsibilities are required to be undertaken by companies that realize the promotion. Otherwise, the conclusion of promotions may deviate from the purpose and their values may decrease.

Customer-led and supply-directed promotions are greatly important both in the production and marketing activity preferences of companies and in their long-term plans. As the first objective, they influence short-term sales and clarify the direction of the demand, and provide data to act accordingly. Somehow interrelated to mobile research, it may lead into the formation of results suitable for a good customer analysis in a short time.

Regardless of the promotional type, the important thing is to present the promotions to the customer at the time and place he or she wishes to watch them. This encounter means the performance of one-to-one marketing, though it is remote-controlled. In this method, results provided through a full time information exchange and delivery is very important in terms of reliability. This system enables not forcing information out aggressively, but asking people whether they want the promotion, and enabling the performance of arbitrary actions. Mobile promotions require extensive customer relations management.

MOBILE ADVERTISING

Employment of mobile advertising channels provides miscellaneous advantages during the promotion of a mobile service or product. Leading one of these advantages is that the customer is highly conscious on using technology. The consumer who uses or watches a mobile advertising channel is able to use the mobile service more extensively than individuals in the generic population. Simply, even seeing the mobile advertisement can be considered as evidence of the customer's correct tendency towards mobile products or services because these people are able to open mobile web pages, see written messages and benefit from various conveyor interfaces, and even regard and consider their mobile phones as an instrument of benefits not only for services other than speaking (4INFO, 2008). Performed researches indicate that through a central advertising management system, multimedia files with advertising content have become a provider of communication to a mobile instrument.

According to Juniper Research (2008), mobile advertising is mainly divided into 7 categories. We also added 2 more categories to this sequence:

1. SMS Advertising
2. MMS Advertising
3. In-Content Download Advertising
4. On-Portal Advertising
5. Mobile Internet Advertising
6. Idle-Screen Advertising (screen savers, etc.)
7. Mobile TV and Video Advertising
8. Mobile Search Advertising, and
9. Mobile Display Advertising.

Elements such as photographs, music, video and texts can be used with the above-mentioned multimedia files. As advertising content, attention can be drawn with discount tickets, movie trailers, telephone melodies, mp3 or mobile games. Certain researches outline the differences in SMS marketing. With mobile systems, it is possible to organize campaigns which are at such an extent so as to include the person at the furthest end of the system to the campaign and delivery to a wide mass of mobile phone users' at the most suitable place, and which present the advertisement with the most suitable conditions for buyers.

Via this system, product and brand is enabled to not exclude anybody and make promotion much easier. Especially for the young generation of consumers, as the desire of benefitting from the mobile environment is clearer and more tangible, it is possible to say that customer awareness will gradually develop further in this direction. As these campaigns increase the abilities of content developers, they are more suitable to perform more effective activities and presentations. Furthermore, access to target users and the communication environment which considers their demands is a highly accessible environment for both the buyer and the seller.

SMS MARKETING

SMS (Short Message Service) has become one of the most frequently used text messaging services at present. Short Message Service (SMS) is a wireless service available on digital mobile networks. It enables the transmission of text messages between mobile phones and other systems such as electronic mail, paging and voice mail (Yamamoto, 2009). With this service, alphanumerical or small non-text content can be transferred from a mobile instrument to the other. Similar to messaging between individuals, messaging from an individual to an application (poll applications) or from an application to an individual (information services) is also possible. According to Deleon (2004), an SMS message can transmit very short data on very low bandwidth. Therefore, a feature of a message to provide maximum benefit in minimized form should be paid attention. GSM, TDMA and CDMA based mobile phone networks provide an extensive support for the usage of this service. SMS has grown into a major communication tool for consumers. Accessing them via SMS marketing has become an increasingly important strategy for businesses.

The first SMS message was sent on December 03, 1992 from a personal computer to a mobile phone via Vodafone network (Michael & Salter, 2006, p.18), and this transmission for the purpose other than voice communication on phone enabled the performance of a secondary action on phone. SMS messages can contain a maximum of 140-160 characters and they are text messages without any graphics or images. In terms of timing, there are three types of SMS communications.

1. SMS messages that are created on mobile phone and end on mobile phone.
2. Those that are created on the mobile phone and sent to another telephone line or computer from the mobile phone.
3. Those that are terminated on the mobile phone but sent from another network which has the ability to sent text messages to a mobile phone.

The best way to deliver text messages to the highest number of people under current technology is by cell phones because at present, 99 percent of the cell phones are able to receive and send text messages. The newest technology as an alternative to mobile web downloads and videos is used on cell phones. There are certain factors which prevent advertisements and promotions to a wide mass in the communication environments other than cell phones. The duration, scope and reached mass with the campaigns

Table 1. Advantages and disadvantages of SMS marketing (Source: G. T. Yamamoto, 2009. Understanding SMS Marketing. http://www.articlesbase.com/marketing-articles/understanding-sms-marketing-806088.html)

Advantages	Disadvantages
Cost-effectiveness maximization	Character restriction
One-to-one personalized communication	Negative response to unwanted texts
Shortened time to market	Fraudulent messages
Ensuring message delivery and reaching customers at any time and place	
Effective targeting for new and existing customers	
Time sensitive information transmission	
Customization capabilities	
Marketing messages are always welcome	

via mobile phone are not unlimited. Therefore, it is recommended to use rather more permanent and relatively more familiar types of text messages. Known as the easiest way to reach the largest number of consumers on mobile phones, the hegemony of SMS is predicted to continue until a higher number of consumers begin to use more advanced devices (Tarhan, 2008).

The advantages of SMS marketing are cost effectiveness, direct communication, effective targeting, shortened time to market and permission-based applications, etc. On the other hand, character restriction, unwanted texts and fraudulent messages are the most prominent disadvantages for SMS marketing (see Table 1).

According to Gauntt (2007), other mobile messaging types are also developing in addition to SMS. Multimedia messaging service (MMS), mobile instant message (MIM) and mobile e-mail are gradually becoming more widespread.

SMS marketing also has some different applications such as purchase context, technical context and user context. These contexts are mostly given preference while companies implement SMS marketing efforts.

According to the Mobile Marketing Association, 89% of the companies are expected to use text messages or multimedia messages to reach their target masses in 2008. One third of these companies express that they would allocate approximately 10% of their advertising budgets to a SMS marketing environment. Especially for developing or small scale establishments, sending SMS messages is regarded as one of the lowest cost advertising and promotion methods at present. On the other hand, businesses are willing to create a permission-based structure and present discount coupons for their loyal customers (http://www.textsmsmarketing.com/ retrieved 10/12/2008).

After recently-added video contents and visual communication features of cell phones are used for advertising, it will be possible to make more enhanced presentations and a different media approach will evolve. Accordingly, the use of mobile phone in planning the future of direct marketing with a contribution to the real sector in terms of customers and with mutual monitoring of the advertised product or advertiser, campaigns may become more convincing and advertising value may increase further.

If SMS message-based transactions are reversible, then the user may support the program or people with financial contributions. Going one step further from the means of the Internet, marketing - the delivery of targeted messages to the customer can be provided even when the individual is not sitting in front of the computer but when he or she is mobile (http://www.orangespotmg.com/sms-marketing.php retrieved 10/12/2008). So, mobile marketing will be able to provide customers with real-time acting opportunity, turning into a communication instrument that allows for higher quality and more influential campaigns in the near future.

Marketing with SMS does not only remain with keeping the messages related with certain activities of organizations for products or services they are willing to market or with the promotions they realize, but also includes much miscellaneous information in the content of a portfolio and creates a databank. Considered as a new media communication style, SMS reaches well-known points all over the world and its data pool becomes richer with each mobile phone user.

However, SMS messages are sent without discriminating between interested and disinterested parties among the receiver mass, which means swimming against the current. Banks and mobile operators have entered into quite an aggressive attitude while trying to take the lead in SMS use with an attempt to turn individuals into their permanent customers. They continuously send messages on different subjects to contribute to their images or to remind them of their images. In this style of communication, the idea that it might probably be accepted is insufficient because banks should have obtained the mobile phone number and the messaging permission of the customer at a point. Even if permission is obtained, the very frequent transmission of messages on irrelevant issues may discomfort the customer and has the risk of creating a customer relationship problem.

Customers should be reached via SMS, and the companies which need to please the customer should produce more different solutions. Similar to the obsolescence of faces that are continuously seen in the media, the name may become obsolete. Aggressiveness may turn into a negative image or persistence may reinforce creating contrary results.

The SMS approach is based on the service concept that consists of multi-components. (Bartolomeo, Casalicchio, Salsano & Melazzi, 2007). Actually, in order to properly conduct this service, companies are required to revise their stands in the communication environment that they use with short intervals. The concept of SMS might have been very attractive for customers in the beginning and customers may be glad because of the message communication which enables their perception as a reputable customer. However, after many companies begin to call people every day from their mobile phones with disturbing frequency, then all of the messages may go to the trash can. Different strategies require being developed in order to for the customer to distinguish messages from each other. Mobile advertisers should initially think that the majority can be disturbed from overloading and they should base their strategies on that. For example, in sent messages, there is not even a single company which says "good morning" or "good evening". Although messages received by recipients in very early morning or at around 05:00 a.m. hit the target, they may cause serious problems of personal disturbance (http://www.farketing.com/fikirler/2005/12/rahatsz_ederek_.html).

It is known that announcements sent through SMS Marketing and SMS are benefitting by a new media. However, with the SMS method, it is not possible to sent MMS (that is photograph, music, and trailer). SMS marketing indicates a structure absolutely accessible to the mobile producer, marketer, manager or customer. SMS messages are sent to the entire mass whose customer mobile phone number is known, and a marketing communication which distinguishes a customer from others hits the target is not in question. SMS is a system which can send messages to long distances. However, spending more credits may be required for long-distance use. In addition to text, SMS also enables exchanging a certain number of symbols recorded in the memory of a mobile phone.

With the development of MSN system via mobile, the scope of chatting through SMS will increase. However, SMS marketing and promotions allow for movement in a static framework. Software and hardware support are not required in order to send a planned promotion via SMS. In these applications, CRM efforts sometimes turn into teasing people or urging the situation, which in a sense, lead to ethical criticisms. Simple offers should be proposed to customers while going towards mass marketing through SMS marketing. During the acquisition of information through subscription or payment, it is natural for subscribers to request reaching a valuable content in this structure. Therefore, subscribers find some of these contents worth money, while others prefer those free of charge.

In the future, life is required to be fictionalized as indexed to mobile from many perspectives. The development of the features of mobile activities other than SMS turns mobile communication into a more striking form gradually. The recently developed Bluetooth system is used as a healthier method in delivering promotions via mobile phones, and brings dynamism to mobile communication. This system also builds secondary communication between advertisers and those in the close vicinity, benefiting from their physical existence; therefore, it can be highly preferred. Every person, organization or establishment can obtain the flexibility and ease of presenting their own values. Bluetooth is a more suitable communication environment for the development of elections suitable for the social environment and cultural structure. This system may run as Internet-compatible and Internet-controlled. PR professionals benefit from the potential of mobile instruments.

SMS is suitable for the commissioning of various marketing applications; For example, digital buzz marketing, viral promotion efforts, mobile auction, etc. In these applications there are some various forms such as Premium SMS (PSMS).

PSMS can be defined as the applications related with logo or melody in contest programs or transfers on gaming applications in general. On SMS applications, GSM operators generally try to encourage its use by creating a service based on a different pricing than the standard price list.

However, mobile marketing can still not go beyond just sending SMS messages. According to the GSM Association, mobile phone users send more than 10 billion text messages every month, and SMS is considered as the most popular data service in the world (Dickinger et al. 2004).

In mobile business literature, there are many researches that try to suggest the critical success factors (CSF) in SMS-based campaigns. (Carroll, Barnes and Scornavacca, 2005; Scornavacca, Barnes and Huff, 2005). Here, the issue that appears as the strongest in the most considered conceptual frames and empirically experimented models is the perception of the customer towards SMS-based campaigns (Barnes and Scornavacca 2004; Dickinger et al. 2004).

Actually, setting off from the definition of Fox (2007) who regards marketing as the gainful identification, attraction and retention of good customers, the inclusion of elements such as identification, attraction and gain in one-to-one mobile phone and related text messaging has begun to be obligatory in order to attach the values presented by virtual dimension to marketing.

SMS MESSAGES TRANSMISSION SYSTEM

The transmission system of SMS messages is provided via an intermediary operator (See Figure 2). SMS messages, which are sent to the short message service of the operator, are transmitted to registered masses by the operator. A message can reach all registered telephone numbers excluding those which are not registered. Under the light of feedbacks obtained from the line, whether the registered person is

Figure 2. SMS Message Delivery System

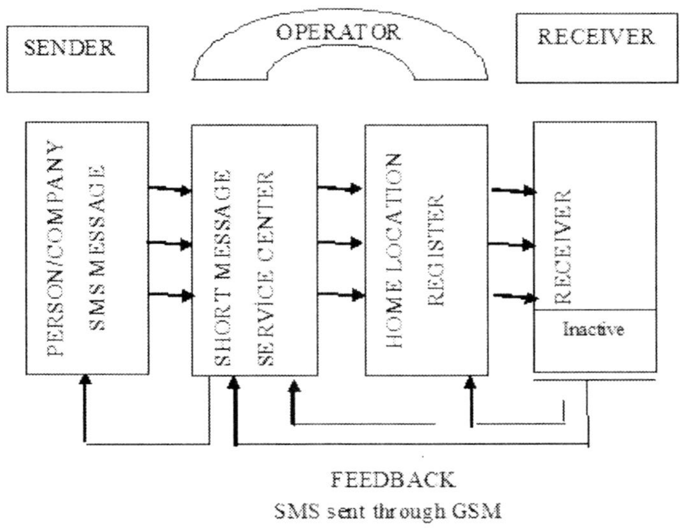

reached can be understood, and it enables measuring and evaluating decisions to be given for continuing the messages. The transfer of news related with different countries or very far, large cities seems functional most of the time in terms of attracting customers.

A major difference of SMS messages from the mass media is that they directly target the individual. Here, an important difference can be explained as the perception of communication-building status as opt in or opt out. With innovations in SMS and related technologies, fast and very significant developments have occurred in communication. For example, the first SMS message was sent in 1992 from a PC to a mobile phone (Saunders, 2002). At the end of the 1990s, businesses in Europe and Asia began to monitor the telephone traffic of customers who communicated through mobile phone numbers. A massive SMS campaign period was observed in the 2000s when the first stage of mobile communication was experienced. In that period, the number of SMS messages exceeded 10 billion per month and SMS began to turn into a new marketing and advertising channel.

With the addition of voice message sending and ringtones, SMS began to be used not only with text characters but also as binary SMS's. In 2002, MMS messages also began to be used with the development of MMS capable telephones. In 2003, a new subscription structure was created with systems that enable not only communication but also downloading the latest news in the communication and broadcasting sector. In 2006, mobile phones turned into a channel where millions of messages are sent, and the new mobile communication system which was created with other mobile devices took its place among the essential pieces of life (see Table 2).

A marketing approach structured on the main idea that "everybody reads messages received on the mobile phone" lies behind the success of SMS marketing.

Table 2. Advances in SMS and its Related Technology Source: NTT DoCoMo (2003), Sadeh (2002), Trends (2006), etc. in Okazaki, S. & Charles R. Taylor (2008). What is SMS Advertising and Why Do Multinationals Adopt It? Answers from an Empirical Study in European Markets, Journal of Business Research. 61, pp. 4–12

Year	Advances
1992	The first commercial short message was sent from a personal computer on December 3, 1992 by Neil Papworth of the Sema Group to Richard Jarvis of Vodafone on the Vodafone GSM network in the UK.
1994	Nokia began featuring the classical work of "Gran Vals" on its wireless phones which became known as its standard ring-tone. The ring-tone was converted into one of the most successful and profitable mobile tools in combination with SMS.
Late1990s	The rise of SMS (Short Message Service) advertising began in Europe and in some parts of Asia when businesses started to collect mobile phone numbers and send off commercial content.
1999	NTT DoCoMo developed the i-mode which included a SMS-equivalent service or short mail.
2000	As firms began to send out massive SMS message campaigns, users were informed that there would be a fee for each SMS sent. This led to an immediate and protracted decline in SMS usage.
2001	Nokia began a new service for ring-tones that allow users to change the way their mobile phone rang, using its smart messaging protocol that was built on the binary SMS rather than the standard text SMS.
2002	The first MMS-capable phones started to appear. Different manufacturers introduced MMS technologies in different ways. For example, Nokia included MMS directly into its latest phones while Sony Ericsson introduced phones with EMS (Enhanced Messaging Service). EMS is halfway technology between SMS and MMS, providing some of the features of MMS (e.g. formatted text, simple pictures, simple audio such as ring-tones, and even some animation), and was designed to work with existing networks.
2003	News service providers such as BBC had a simple SMS-based, mobile news service available for some time. Vodafone UK and Fauna& Flora International began to offer a full-featured website, providing a wide variety of conservation news stories (wild life), discussion boards, field diaries, etc.
2006	As of February 2006, SMS-based mobile marketing is still the most popular method in the industry with several hundred million advertising SMS messages sent out every month in Europe alone.

BENEFITS OF SMS AS A MARKETING INSTRUMENT

The benefits of SMS can be considered based on two sets as personal and company-oriented. For individuals, SMS messages are considered reliable, fast, easy, cheap and a safe means of communication. From the point of both companies and individuals, SMS messages are those which give an idea about the targeted subject or which are sent as summary information. Companies are able to organize affective marketing campaigns that reach very extensive, but targeted masses with these messages, and they can make use of a one-to-one marketing opportunity. In a relationship with customers, the message gains a call value through showing an individual approach and making the customer feel that he or she is special. When this call draws the attention of a potential customer, it may trigger the mood for shopping and enable the person to develop a positive behavior. As it provides companies with an opportunity to reach customers anytime, anywhere and almost under any condition; it especially also creates the opportunity of accessing the customer and delivering the messages at hot points (http://www.mobilesmsmarketing.com/benefits.php).

By increasing the speed of entry into the market, enabling the immediate delivery of information and a document related with any product and service to the target masses in the market, it responds to the time sensitivity of companies racing against the time, and increases the ability of acting fast. Also by presenting time-related offers to customers during suitable days and times, it can create an opportunity to increase customer satisfaction. Mobile communication increases individualization characteristics of the offer to be presented by person to person, company to person or person to company. There is the idea of using a system that rapidly helps enable the possibilities of developing structures which help form interfaces to rapidly meet corporate requests and gain control at every point through mobile communication instruments. Some messages can be transferred directly via mobile phone to all subscribers or spontaneously to the list recorded in the mobile device. Information can be obtained regarding whether sent written messages reach the opposite side or not. This possibility turns into data that guarantees reaching the customers.

Especially in campaigns oriented towards customer portfolios developed upon permission, the benefit of the message to send it to the correct people is maintained and companies can be benefitted by obtaining the expected result according to the purpose and so as to provide it in a manner to converge to that purpose. This can be taken into account as an element which can ensure a higher success level for campaigns. Targeting can be achieved with an influential forecast of campaigns performed through this channel. In the meantime, in order to create a more loyal customer mass and enable the customers to assist manufacturers with their ideas and creativity through their feedbacks, it is also possible to reflect the contribution of the customer to every aspect of marketing.

Mobile communication can be considered as a cost effective and high-income channel. The pre-identification of customer requests and requirements enables both developing the sales of existing products, and obtaining sales information related with sales. Approximately 20% of a rate of return can be obtained with a well-organized SMS campaign. This means approximately 10 times higher return than a newspaper ad and 5 times higher than direct mail. The cost of the message sent per person is calculated as around 10-12 cents (http://www.businesslink.gov.uk/bdotg/action/detail?type=RESOURCES&itemId=1075390385). As SMS messages can be used in parallel to other advertising and promotional instruments, they can contribute as being complementary and increase efficiency in every phase of marketing.

Companies that use Internet and mobile communication instruments eliminate the consumption of paper used for brochures, etc., and contribute to sustaining an environment-friendly marketing activity. Thus, unnecessary paper consumption for large scale campaigns is eliminated. For example, as envelopes and inserts of daily papers etc. wastes may turn into an environmental problem, the use of mobile phone and Internet in sending text messages may contribute to protecting the environment by including organizations in the systems developed against environmental pollution.

SMS also creates a difference in mobile marketing strategies and brings advantages (http://www.mobilestorm.com/sms-marketing/ 10/12/2008):

- Using a short code to generate subscription without having to pay a high monthly fee to own or maintain it.
- Enabling people to sign up for the created mobile club or network of the company directly from their cell phone as well as via Web forms.
- No longer requiring subscribers to give the supplier their carriers as this is automatically determined by the new SMS services.
- Providing bi-directional interaction of registered members; or example, by asking questions to registered users or calling in return for a message.
- Increase the impact by creating a communication chain among members.

Although SMS messages do not enable sharing the first contact in a campaign, they can be used as a very good awareness-creator. With this feature, they can be considered with a very strong complementary environment. This environment can be benefitted from very easily in order to trigger any incident. For example, the first contact or inclusion in the audience can be obtained by creating the opportunities to monitor the newly introduced environment or to obtain SMS updates after training or with the messages such as "If you send as an SMS, you may get the last seat to the concert free or win a concert CD", and then the latter connections can be made via e-mail or SMS (McNabb, 2005).

OPT-IN/ OPT OUT

According to Michael & Salter (2006), opt in is registering into a mail list to receive an e-mail through clicking on an icon or generally checking a dialog box in order to meet the information requirement on certain issues. In mobile, a further step is, to deliver messages within the framework of personal permission of individuals through their inclusion among recipients of messages after they send a message to a certain address, or checking a dialog box in order to meet their demands on issues of interest other than information, such as news, etc.

Opt in is the acceptance of the content by the targeted mass for content upon the recipient's own will. So, an activity which means giving value to sent messages by the customer is realized. In the ideal situation, they are single message campaigns or easy and open-to-issue messages. The targeted mass in sending these messages are both apparent, and they are the messages where the sender is clearly defined (McNabb, 2005).

Also, according to Michael & Salter (2006, p.98), the user, who decides to stop being included in the target mass of communication should also be able to stop receiving the messages (opt out).

Types of SMS Marketing

SMS marketing differs according to the number of targeted people and features of the message. SMS messages used by work places are SMS message, TV motivated voting, or moment-based warnings for registered customers, customer messages broadcasted on interactive TV, messages related with sales or promotion campaigns, spot-sales oriented messages and even messages used for GPS-based mobile advertisements as a further step (Scannell, 2008). SMS messages can also be separated as written and non-written SMS messages. When considered from person to person, person to organization or organization to person, the elements of written messages are the device, person or persons. However, different environments are expressed as it indicates quantitative differences. Non-written messages can be panels and pictures such as screen coupons. With respect to the number of people to whom the message is sent, SMS messages are separated as individual messages and bulk messages with singular or plural characters. SMS marketing is considered in terms of giving more advertisements and they are considered two types as: text messaging advertisements and pictured message advertisements while the third generation mobile instruments has also gained the opportunity of creating picture messages.

Here, the general focus is on the issues such as the evaluation of any type of virtual content, location data and awareness at content level. The advertising servers of companies are considered as decision-making engines, and advertisement categories can be developed regarding which advertisement messages and when they would be presented to individuals, and the classification of the delivery according to user's GPS coverage area and search query becomes possible.

Harper mentions the worries on the failure to clearly distinguish which messages are spam and which messages are necessary according to retailers and others depending on the situation (Scannell, 2008). This ambiguity is a situation which would also create bad feelings in terms of customer satisfaction and the failure to create a sufficient benefit with the messages. Furthermore, there are risks of acting according to the rules of a game based on an unsettled law due to the ineptitude of the manufacturer, consumer and intermediary functions. Prevention of the problems that develop in a borderless environment may not be very easy. Protection-oriented layers have been tried to be created in order to prevent problems which reach to the level of giving up, and features to protect personal confidentiality and privacy of the person have been tried to be introduced.

Although the imposition of confidentiality policies and information use rules are against the spirit of mobile communication, they still require being taken into consideration in order to create warning systems which would come into stream in order to prevent the information of third parties about the transactions and the delivery of messages from persons who are not granted with the recipient's permission. Also psychological information should be considered during the development of the programs which identify the time between calling and decision-making, and when differences could arise in thinking. However, in addition to developing mobile devices which might reduce the technological problems regarding the elimination of the confidentiality complaints, people using these devices should also be educated and such efforts are required to be integrated in education.

USER PROFILE

The SMS user profile is very wide due to the massive or personal use of a communication style which has various features such as being low-cost, full-time, answerable and recordable. It is used by both public and non-profit organizations, and profit-organizations in manufacturing (producer, factory, stock market), science, information, communication (media, mail, informatics), service (health, education, finance, public relations, army), marketing (services, distribution, advertising, human resources etc) with a borderless profile which can be considered according to countries, regions, financial means and coverage area of technologies.

Mass SMS is especially considered and used as an essential requirement of public and nonprofit institutions. Public organizations and institutions have taken their places in the system as good users. Performing even the slightest announcements via SMS allowed all non-profit organizations such as associations, chambers, unions, funds etc. to complete their work fast and with low cost; therefore, these organizations have also been included in the customer profile. In the private sector, people and organizations at each point between the manufacturer and the consumer use mobility for personal or massive communication. The instant announcement of campaigns by retailers and wholesalers of manufacturing companies to related customers with immediate replies from the customers creating an information-dense environment as well as creating customer loyalty are also the important aspects of the user profile. Mass SMS began to be observed as an essential requirement in education, which is demanded by parents; thus, people and organizations that are willing to benefit from the domain of information have also been included in the customer profile. The absence of the student and results of tests have begun to take part among the reasons of preferring private schools.

In addition to retailers and wholesalers, executives and marketers also create a segment with high customer density. The mobile service usage rates of certain segments may vary from one season to another. For example, towards summer or holiday times, tourism companies begin to use SMS at a high rate while department stores also use them more frequently during national days and festivities. Customers create profiles according to the number of people included in the system and the usage frequency. A mobile system has customers from every section of society from churches to kiosks and restaurants with the inclusion of associations, federations, schools, banks, health organizations and their employees in the customer profile. Everybody can take part in the customer profile by being involved in this system on his or her own account. Bulk SMS has users from almost every section of society.

Bulk SMS

Bulk SMS can be expressed as sending SMS messages directly to a group through a massive communication technique. Its feature is presenting a single SMS message to a wide mass on a single occasion. The sender initially identifies the number and identity of each recipient and creates a group accordingly. The recipients are with a number which might begin from one and reach to a very big number. A sender identity tag is created for each user, and the same defining ID always expresses the same brand when using mass SMS. The most important feature of bulk SMS is that it can be grouped on demand. A message can be sent on demand to created SMS groups with a single command. Many major companies present their products and services through low-cost bulk SMS services.

According to Hensley (2008), approximately a 30% savings have been obtained in the printed media campaign expenditure through bulk SMS's. Wide network coverage can also be obtained in the wide network zone where bulk SMS is used. This system, which can be comfortably used for convergence and information purposes, provides a deniable contribution for saving time and reducing costs. At present, the senders of bulk SMS are also able to follow up if their SMS's are received or not. An important function of bulk SMS services is the ability to learn whether the messages are received by the customers on time. Bulk SMS guarantees dispatch to recipients without any geographical borders. Furthermore, they can be sent manually or automatically on special dates or times set by the sender.

Campaigns, discounts, promotions and advertisements can be announced through bulk SMS. A telephone chain can be used for issues that require alerts and warnings. Notices, reminders, greeting and felicitation messages can be sent. Order delivery confirmation data can be sent.

Also, on the modular system developed with B2Mobile instruments, previously used lists can be seen and price modification information can be announced by companies to dealers via SMS messages, and the details of campaigns can be communicated very fast. With this function, B2Mobile accelerates marketing activities and campaign management, ensures the transfer of the entire workflow to mobile environment with automatic SMS messages of order status along with shipment data, and payment warnings, and it is used in producing complete solutions. B2Mobile, which enables dealers to enter their orders with SMS messages, both ensures obtaining information through interrogation of the requested price, stock, campaign, shipment and order data or new developments, and it also creates a highly reliable source to obtain momentary data.

Premium Rated SMS

Premium services are the procurement of services involved with content and recommendation through a phone call, sending a text message or connecting to the Internet via mobile phone or via data connections in return for a service fee. The rate of the fee includes advantages which are announced to users via campaigns or obtained through agreements.

Premium SMS services create a service system that distributes useful and convenient information for buyer via the same network through standard SMS conveyors. At present, it enables people or organizations to obtain the required information for them at any desired moment and through the shortest way. They can provide the latest news on stock share prices, flight information and delays, World Cup and the results of other sports events or games of luck, or provide a search system on any issue which is considered as important. Premium SMS messages are actually similar to standard SMS messages. However, they provide a different service concerning three main issues (Svalasen, 2004).

1. Immediate delivery of urgent notices to telephone screens.
2. Interactive chat or immediate delivery of vote counts in polls to the other side.
3. Summing up and presenting users messages on identified issues on demand.

The configured subscription system for SMS messages and the cost paid for each message can be automatically collected via line. During the calculation of prices for such applications, the distances which identify connection time, connection type and utilization periods are taken as the basis. The calculation units in a connection are based on a price list and they are always demanded according to the price list which is valid at the beginning of the calculation unit. The price differences between services are correlated with the content and size of the service and the distance of the communicated point. Emergency services (police, rescue, fire brigade, etc.) do not require payment. Special circumstances such as the creation of a personal profile page, participation in blogs and forums, uploading media contents such as photographs, texts and games, newspaper feeds, gaining virtual items and/or other services are among the issues of pricing. Buyers accept the service fee in order to obtain a special service in a rather special circumstance with information and activities.

There are certain lines among the funny, handy and informative premium services created by third parties:

- Fortune telling and horoscope lines
- Weather forecast
- TV shows or contest selection lines
- Dating and chatting services
- Ringtones and fax back services
- Test result lines, etc.

Premium SMS services are on demand as they require the payment of a certain rate by customers for the mobile content. Applications such as subscriptions, polls, contests and transfer rates indicate a specific cost for mobile users.

There are two types of premium SMS services in general: sent messages and received messages.

- Customers are priced according to a certain price list for each SMS message sent to the premium line on mobile system.
- The customers are priced according to a certain price list for each SMS message received from the premium line on mobile phone.

Premium SMS text messaging campaigns have certain shortfalls. Therefore, sometimes they can fail. Nevertheless, they have recently turned into the most preferred indicating platforms. For example, perhaps American Idol can be considered as one of the most successful companies in this type of messaging. Here, sending votes through both test messages and calling is requested through the contest and among the votes from one to ten, 500 million votes were received by phone call and 50 million as text messages. The difference is probably because phone calls are simpler, more familiar and easier compared to text messaging. The results of the American Idol contest are considered as proof that texting preferences have underdeveloped (Becker, 2007).

In SMS text messaging, there are two important factors which motive acting (Becker, 2007). These are micropayments and costs.

Micropayments are certain conveyors who perform SMS interaction and can also undertake invoicing. This enables fast and easy payment to the audience and this figure is added to the mobile phone invoice or it is deducted from credit.

As the publisher always does for calls that stimulate the masses to act, the participant is invited to take part in the idea that he or she would take a share according to the highest rate. SMS services, which are presented as the opportunity to call certain numbers through unpaid telephone numbers, can actually be considered as cost-increasing elements.

The main problems concerning premium SMS messaging are as follows (Becker, 2007):

- Limited Income Means
- Income Margins
- Reaching the Market
- Invoicing
- Interaction Area
- Reliability
- Transferability

Telephone or SMS voting system has been introduced for most of the big and multinational contests. The most adored contestants are scored by the individuals from 12, 10, or 8 to one, and the individual contributions are converted into the common value of a country. The participations obtained according to the vote of all countries become determinant in leading the singer of a county to the top. The important issue for these polls is that real-time delivered votes are evaluated and concluded in a period as short as 10 minutes. This listing suggests a fair result as it is not interested in the motivation behind the acts of the audiences like in Eurovision Song Contest.

E-Mail to SMS

These are the systems which enable sending e-mails to mobile phones in seconds as SMS text messages via gateways. Similarly, SMS replies can be converted into e-mails. They are also defined as E-Mail to SMS and SMS to e-mail. (http://www.sms2email.com/ retrieved 11/2/2008). Such a service is provided through the conversion of electronic information in the e-mail accounts of automatic and intelligent communication systems as compatible with SMS mobile instruments (http://www.ipipi.com/email_to_sms.htm retrieved 11/2/2008).

These systems provide the following contributions:

- Analysis of email and the latter information which are indicated as important
- Strong protection against SPAM
- Sending messages via simple and integrated web functions (web 2 SMS)
- Using a strong and free Webmail customer
- Creating and delivering messages on mobile devices
- Scheduling and programming of mobile notices
- Benefitting from POP and SMTP servers

MMS MARKETING

MMS allows users to add to their mobile phone messages various multimedia elements (text, picture, sound, video, animation, etc.) and send the message synchronously with all of its elements in a global messaging standard. MMS was designed in order to benefit from the high transmission potential of third generation mobile networks. Nevertheless, it can also be used via existing GPRS networks. The EMS (Enhanced Messaging Service) standard created a transition between SMS and MMS. EMS can be considered as a great leap in sending sound, graphics and pictures with a content combination that consists of text, melody and simple pictures.

With the addition of picture, music and video files into text through MMS Multimedia messaging or MMS technology, it is now possible to say that mobile phones have become impressive in terms of cinematographic values, and mobile technology has reached a high level. The most evident and interesting difference between MMS and SMS is that it is not limited with text. With MMS, pictures that are taken with digital cameras and transferred to telephones via any way can be sent to recipients on the other side with the same quality. In addition, sound and video can be added to MMS messages. As messaging is possible between MMS compatible telephones, accessing multimedia messages via Internet may also be possible.

Users can see the MMS messages sent to them also via Internet. For example, if the pictures shot on that day are sent to a person via mobile phone, but the receiving telephone is not MMS compatible then it cannot receive the picture. At this point MMS – C comes into stream, warns the receiving party with an SMS and gives the Internet address from which the sent content can be obtained (http://www.bilisimterimleri.com/bilgisayar_bilgisi/bilgi/84.html).

Having appeared as an enhanced or enriched version of SMS, the infrastructure of MMS was prepared in order to provide more functions together. Therefore, MMS can be considered as an enriched rather than a differentiated version of SMS. Adding picture, music and video content to text with this technology has enabled watching short videos on mobile video cell phone.

MMS messages differ from SMS messages in that they can communicate the information with a logic based on a different interaction. One of the most interesting aspects of MMS is the language it uses which is named Synchronization Multimedia Integration Language or SMIL (Hansson, 2008).

Comparison of SMS and MMS

- SMS is a very simple service. It is very easy to use. But attention and effort is required when the sender writes the messages. It has a high impact degree as it is based on monitoring through the use of non-audible and non-text elements. MMS has been introduced with a more complex approach. These two systems have size and content differences in terms of marketing.

Table 3. Average Number of Monthly Calls vs. Text Messages Among U.S. Wireless Subscribers

Average Number of Monthly Calls vs. Text Messages Among U.S. Wireless Subscribers		
	Calls	Texts
Qtr 1, 2006	198	65
Qtr 2, 2006	216	79
Qtr 3, 2006	221	85
Qtr 4, 2006	213	108
Qtr 1, 2007	208	129
Qtr 2, 2007	228	172
Qtr 3, 2007	226	193
Qtr 4, 2007	213	218
Qtr 1, 2008	207	288
Qtr 2, 2008	204	357
Source: Nielsen Mobile		

- SMS has a 160 character text limit. Character limitation has been removed with MMS.
- SMS can be delivered to any device in the market which is able to receive and send messages. MMS can only be sent to the latest technological devices.
- As not all mobile users have MMS featured phones, they cannot benefit from the given service at a large scale. Due to the problems arising from different aspects, their use by media campaign creators is slowly progressing. (http://www.adverblog.com/archives/000789.htm). This situation can be recovered only after all phones in the market turn into higher technological phones. On the other hand, the customer who is fed up with intense campaigns based on mobile change may also be reluctant to obtain a new phone set.

The telephones, which are supported with a MMS program content, can be used to message with each other through MMS while they can also continue to reach other telephones through SMS. MMS represents the next step in the evolution of messaging. At present, people do not only take photos, but they are also willing to send them, share them with others and they support the operators' services in this regard with their demands. MMS adds a new and richer aspect to the short message concept with color pictures, animations, audio-video clips, and face-to-face video messages.

Tendencies towards new contents mean the opening of new income gates not only for mobile operators but also for content and service providers. With its features which make communication more colorful and meaningful, which enable users to express themselves and their emotions in a better way, MMS gradually reinforces its place in communication.

Text Messaging and Speaking

Text messaging is the method which began to be preferred after overcoming the ineptitude in telephone use. According to the data of Nielsen Mobile, a typical US mobile phone user sends and receives more SMS messages than making calls on mobile phone. It is stated that registered mobile phone users in the US sent and received 357 text messages, but made only 204 phone calls on average in the second quarter of 2008. This figure increased by 450% compared to the figures of two years before. It can be considered that similar tendencies are valid for all mobile phone users in the world. SMS is preferred especially due to its low cost, ease of reach to the recipient, speed, ability to store in memory and ability to repetitively read the message. It can be expected to reach the designed dimensions in time and even exceed those dimensions (see Table 3).

According to Nielsen Mobile, age appears as the important factor in the study. According to the results from the invoice data of prepaid mobile phone subscribers of four major service suppliers, subscribers between 13 and 17 years old receive 1742 text messages a month on average while they make only 231 phone calls. However, these figures for those at and over 65 years old are 99 calls and 14 text messages per month on average. According to the results of the same research, it is noted that users at and over 45 years mostly prefer telephone calls instead of messaging, and they prefer the habits from the past. These data also indicate inter-generational differences in mobile approach.

REFERENCES

Barnes, S. J., & Scornavacca, E. (2004). Mobile marketing: the role of permission and acceptance. *International Journal of Mobile Communications*, *2*(2), 128–139. doi:10.1504/IJMC.2004.004663

Bartolomeo, G., Casalicchio, E., Salsano, S., & Melazzi, N. B. (2007). Design and Development Tools for Next Generation Mobile Services. International Conference on Software Engineering Advances (ICSEA 2007).

Becker, D. (2007). Participation TV: Premium SMS vs. Toll-Free IVR. *International Journal of Mobile Marketing*, *2*(1).

Carroll, A., Barnes, S. J., & Scornavacca, E. (2005). Consumers Perceptions and Attitudes towards SMS Mobile Marketing in New Zealand. In Proceedings of the 4th International Conference on Mobile Business. Sydney, Australia, (pp. 434-440).

Deleon, B. (2004). Marketing on the Go. Ft. Lauderdale, FL: DGXMedia.

Dickinger, A., Haghirian, P., Murphy, J., & Scharl, A. (2004). An Investigation and Conceptual Model of SMS Marketing. In Proceedings of the 37th Hawaii International Conference on System Sciences, (p.1-10), Hawaii in Scornavacca, E. & McKenzie J. (2006) Perceieved Critical Success Factors of SMS Based Campaigns: A Managerial Perspective. Paper presented at the European Conference on Information Systems, Gothenburg, Sweden.

Eurovision: the voting results during the show, (n.d.). Retrieved from http://www.upstreamsystems.com/spotlight/eurovision.htm

Fox, J. J. (2007). Pazarlama SüperStarı Nasıl Olunur? (How to Become a Rainmaker?), (Trans. E. Kafalı). Istanbul: Nokta Yay.

Gauntt, J. (2007, November). Mobile Message Marketing. eMarketer.

Hansson, A. (2008). How to Go Mobile. Retrieved 11/9/2008 www.infonu.com/downloads/HowToGo-Mobile.pdf

Hensley, D. (2008). Bulk SMS as a Marketing Tool. Retrieved 10/24/2008 from http://www.articlesbase.com/technology-articles/bulk-sms-as-a-marketing-tool-410220.html

http://en.wikipedia.org/wiki/Short_message_service#Premium-rated_short_messages (n.d.). Retrieved 11/8/2008.

http://www.2020systems.com/internet-ad-glossary-e-q.html (n.d.).

http://www.adverblog.com/archives/000789.htm (n.d.).

http://www.bilisimterimleri.com/bilgisayar_bilgisi/bilgi/84.html (n.d.).

http://www.businesslink.gov.uk/bdotg/action/detail?type=RESOURCES&itemId=1075390385 (n.d.).

http://www.farketing.com/fikirler/2005/12/rahatsz_ederek_.html (n.d.).

http://www.girisim.com.tr/bankatek/sayi15/b2mobile.htm (n.d.). Retrieved 11/1/2008.

http://www.ipipi.com/email_to_sms.htm (n.d.).

http://www.mobilesmsmarketing.com/benefits.php (n.d.).

http://www.mobilestorm.com/sms-marketing/ (n.d.).

http://www.orangespotmg.com/sms-marketing.php (n.d.).

http://www.sms2email.com/ (n.d.).

http://www.textsmsmarketing.com/ (n.d.).

http://www.vodera.com.tr/blog/?p=1#more-1 (n.d.). Retrieved 10/12/2008.

4INFO. (2008). Marketing Mobile with Mobile: Lessons in Strategy. Retrieved 5/2/2009 from http://vmware.eyeline.mobi/wp-content/uploads/2008/12/4info_marketing_mobile_20080327.pdf

Juniper Research. (2008). Mobile Advertising -Because I'm Worth It. WhitePaper Extract from Mobile Advertising Strategies & Forecasts 2008-2013.Hampshire, UK: Author. Retreived from http://www.juniperresearch.com

McNabb, A. (2005). Making Sense of SMS Marketing. Retrieved 11/23/2008 from http://business.maktoob.com/reportdetails-20050226082324-Making_Sense_of_SMS_Marketing.htm

Michael, A., & Salter, B. (2006). Mobile Marketing: Achieving Competitive Advantage Through Wireless Technology. Oxford, UK: Butterworth Heinemann.

MMA. (2007). Retrieved 5/23/2009 from http://brandedcontent.adage.com/pdf/2007MobileMarketingGuide.pdf

Okazaki, S., & Taylor, C. R. (2008). What is SMS Advertising and Why Do Multinationals Adopt It? Answers from an Empirical Study in European Markets. *Journal of Business Research*, *61*, 4–12. doi:10.1016/j.jbusres.2006.05.003

Saunders, C. (2002). SMS, Ten Years Later. Retrieved 11/01/2008 from http://www.instantmessaging-planet.com/wireless/article.php/10766_1553321

Scannell, T. (2008). GPS-Based Mobile Ads: Where Does Privacy Fit? Retrieved 10/24/2008 from http://news.earthweb.com/mobility/article.php/3780591

Scornavacca, E. Barnes, S. J. & Huff, S. L. (2005). Mobile Business Research, 2000-2004: Emergence, Current Status, and Future Opportunities. In Proceedings of the 13th European Conference on Information Systems, Regensburg, Germany.

Svalasen, S. (2004). Premium SMS: Interaction Made Easy. Retrieved 11/1/2008 from http://www.computingunplugged.com/issues/issue200401/00001208001.html

Tarhan, B. (2008). Retrieved 6/13/2008 form http://www.marketingturkiye.com/BilgiBankasi/Detay/?no=479.

Yamamoto, G.T (2009). Understanding SMS Marketing. Retrieved 5/2/2009 from http://www.articlesbase.com/marketing-articles/understanding-sms-marketing-806088.html

KEY TERMS AND DEFINITIONS

Bulk SMS: It is mass SMS which is a fast and effective mobile text messaging application to send messages to an intended number of people at the same time.

Premium SMS: Subscribers are charged extra for receiving this premium content, and the amount is typically divided between the mobile network operator and the VASP either through revenue share or a fixed transport fee. (http://en.wikipedia.org/wiki/Short_message_service#Premium-rated_short_messages retrieved 11/8/2008)

MMS: Multi-media Message Service.

SMS: Short Message Service. It is commonly known as text messages with a limited number of characters that can be sent to a mobile phone.

Opt In: A process whereby a user voluntarily agrees to start receiving email, usually commercial, about a topic of interest (http://www.2020systems.com/internet-ad-glossary-e-q.html retrieved 11/8/2008).

Opt Out: A process whereby a user voluntarily agrees to stop receiving email, usually commercial, about a topic of interest (http://www.2020systems.com/internet-ad-glossary-e-q.html retrieved 11/8/2008).

SMS Advertising: Employment of mobile advertising channels via text message to promote a service /product/person/idea/group etc.

SMS Marketing: Marketing with text message that enables information to be sent to groups of people quickly and conveniently in much the same way as e-mail.

Chapter 7
Mobile Distribution, Procurement & Sales

Distribution, procurement and sales are powerful tools for businesses to obtain a sustainable competitive advantage. Distribution can help companies find some solutions for many challenges in producing better or cheaper products and reaching customers in highly competitive environments (Heidelberger, 2009). Almost all buying decisions include factors such as delivery and handling, marginal benefit, and price fluctuations. Combining existing product, price and brand advantages with investment in channel programs and technology is a proven strategy for establishing success with distribution. Mobile procurement has gradually become more and more important in our era. This also brings entrepreneurs new opportunities to reach customers especially in banking and press businesses but these sectors should not be forgotten that these are the easiest sectors to apply this kind of distribution. These sectors' implementations mostly depend on communications; on the other hand, this doesn't mean that other sectors could not employ these systems. Notably, the retail sector should be considered with all its distinctions. On the other hand, mobile sales systems are also very promising.

However, currently mobile distribution is explained in two distinctive aspects. One of them is based on the distribution of products and services side and the other one is distribution of the contents side which is very frequently used in Bluetooth and LBS systems in mobile marketing.

In this chapter, mobile distribution and sales are generally examined. LBS applications and Bluetooth marketing are taken in to account as their content delivery convenience and disparity.

MOBILE SALES

Mobile sales are frequently not in usage in our current situation. These will be the new systems for customers and purchasers. When these systems are realized, facilitating business systems give several

DOI: 10.4018/978-1-60566-916-8.ch007

opportunities to customers they will consider important for businesses. However, such sales can be conducted in parallel to a call/data center system in an organization facilitating changes of mobile instruments incoming to the market and easily accepted within these systems. Within call/data center based sales and order follow up applications, the product is ordered through a call center representative, IVR and SMS channels, and then the order is routed to mobile distribution vehicles for fulfillment. Applications can be made from distribution vehicles and they support hot sales. So, the customer communicates with the order line via mobile devices of the delivery staff and accelerates order and delivery. Also, order automation and a CRM database can be created for such a structure. Mobile devices communicate with the center and receive delivery orders via GPRS and update the invoice data with the center.

Similar call-center based mobile sales applications are developing in the entire world. At present, many companies benefit from call centers for customer communication and complaint management as an out source. The movement of order automation to this source is also possible. As the ordering of products by people via various media and their home deliveries have shifted outside of the core businesses of companies, the system has appeared as an ideal solution for increasing both the efficiency and managing of costs. Mobile-sales related issues can be considered from three basic points:

- Developing multi-product, campaign-supported and distribution-optimized applications
- Developing a hot sales and order follow up application on mobile industrial devices
- Enabling more effective work of each unit in their own areas of expertise through modular activities

The advantages of mobile sales can be listed as follows:

- Mobilization of all key sales and company resources
  - Obtaining a fast and mobile performance on the vehicle
  - Palm top optimization
- Wide Platform support
- Fast spread and low cost
- Continuous accessibility

While creating these kinds of systems, businesses prefer some modular systems to face the new situation and understand the systems better. The system cost, quality, access and support will also be another concern to businesses. The main decisions would go around with these circumstances. Therefore, modular systems should be evaluated by the ordering, distributing and managing of the transaction and financial dimensions.

The most correct time to deliver a marketing message is the time when the consumer is ready to buy. Therefore, determination of location presents very important opportunities to marketers. Via mobile phone, it is possible to know where the consumer is and to reach the label information via service providers. With this method, consumers can be reached at the shopping center that is when they are more ready to buy, and messages can also be delivered (Gökmenoğlu, 2008).

MOBILE PROCUREMENT

The issue of mobile procurement comes forth further with the advancement of technology and development of location-based service systems. According to Yamamoto and Karaman (2007), the next step after e-procurement will be m-procurement. Mobile technologies are currently in the stage of development and the use of these technologies is not up to capacity; however, many products mix in with each other and turn into devices that enable m-procurement. The most important ones among them are, of course, mobile phones and PDAs. In order to maintain m-procurement, financial, technological, legal and communicative infrastructures are required to be established.

Mobile purchasing creates environments where vehicles and individuals can serve at close points when they are mobile. After the customer places the order, the vehicle nearest the mobile device with the ordered product is able to place the delivery in a very short time as a result of the developed system. Here, the pivot is the time and labor savings. The required documents are procured as mobile and they can be invoiced with a mobile signature. So, each mobile device could turn into a sales point.

Additionally, solid data bases, customer tendencies, and profiles gain importance and services can be conducted in parallel to LBS systems. Speed and correct delivery for the first time are also among the elements that increase customer satisfaction.

MOBILE DISTRIBUTION

These purchasing efforts should be supported with some distribution activities. Mobile distribution is the systems subsume mobile devices. These are the distribution systems that begin and/or end with a mobile connection. These systems are also in their preliminary stage at the moment. According to Raymund (2006) the monthly communication between distributors and their channel customers involves approximately 1.5 million inbound and outbound phone calls. Transactions take place immensely over the Web, including order placement and round-the-clock information access. Distributors also handle approximately 200,000 technical support calls per month for channel sales. And the credit services we offer keep product flowing. Some of these transactions will be skipping over to mobile systems soon. The readiness for the handling multiplying orders and to meet increasingly specialized customer needs, including the development of robust e-business capabilities are getting more importance. Oursourcing is one of the facilities that enable to cope with these issues.

With outsourcing concepts which have come forth in the business concepts of the new economy, marketing any product via various media channels, taking the order for the product and outsourcing its distribution are among the rising inter-business trends. On the other hand LBS practices and Bluetooth marketing are taken in to account as their content delivery convenience and disparity.

Location Based Services

Location Based Services (LBS) were initially used by the US Department of Defense in 1978 with a GPS satellite system consisting of 24 satellites for military purposes. In 1980, the US opened this system to the world (OECD, 2006, p.253). This opening marked the beginning of commercial applications. Although LBS navigation systems are generally considered popular instruments, they can also provide many innovative service options in countries such as India where these services do not address only to urban customers, but also to the low-income sections of cities and the rural population (IAMAI, 2008). Examples are the weather forecast information given to fishermen in deep-sea or the audio reply technology for illiterate individuals. Such attempts are also very important for the growth of countries.

With corporate solutions based on location data provided by location-based services, questions such as place of product, delivery time, place of sales team, and place of sales support team can be replied while they can also be used as personal solutions to locate "the nearest dealer, service point, or place of a friend or child". Such location based services include geographical location assistance, emergency spot identification or the identification of the location of victims, creation of social networks to unite friends and local map assistance. In order to carry out preparations with LBS, a correct positioning technology and a geographical information system that consist of the region and location maps where mobile service providers can easily work on complex eco-system is required.

In its early days, LBS could not be very successful due to insufficient location technology, lack of standards and barriers between operators and networks, or certain other infrastructural limitations. However, in current investments, it is considered among the fundamental business models of operators. The "location" now maintains its importance with the creativity of content providers and inclusion of application developers in the game (Francica, 2007).

LBS create new styles of interactions with other users during movement in any place, region or city. With this new spatial perception, mobile phones do not remain only as instruments of building communication, but they have also turned into micro computers which are portable in social places (de Souza e Silva, 2006).

According to Juniper Research (2008), LBS are the provision of personalized services according to the membership principle in its own geographical region. LBS are not fully 'Mobile Web 2.0' (and as massive application has not started yet that is while pilot studies are ongoing). If it is used in combination with other applications and databases, then it enables the user to share their own location details with third parties or other applications, and to share movement of information in its own local surroundings.

In LBS, the devices are in a wirelessly connected structure that uses signals to communicate. As distances may change according to place, they are considered in three basic categories (Quinn, 2008). These are Personal Environment Networks (1-30 feet), Local Regional Networks (up to 200 feet) and Wide Area Networks (almost accessible to anywhere).

According to Fields (2005), LBS are the services that benefit from the location or place in presenting meaningful information to users. In mobile marketing, LBS applications are realized by using certain regional location information in order to communicate the right message related with a certain advertising or marketing services to the right person at the right time and place.

According to the same author, LBS index applications are generally those such as mobile search, effective search and local search. LBS index applications present meaningful local data similar to those in the online search results of Google. The basic difference of LBS index applications is that it can provide real information search and present this information in valid business locations. For example, when a customer who uses this service goes to a different city and stays at a hotel, he or she can learn the place and directions of the nearest flower shop or hospital via his or her mobile phone.

LBS index applications should not be considered just as a city or postal code because such applications are stagnant and general. However, information obtained with LBS index applications is information that describes a certain location, the address of a street or the place of a crossroad. One of the most important issues related with LBS is the ability to add special location information which could yield better results. So, it enables adding some specialized attributes into the information. Frequently, LBS can also

Figure 1. Types of location-based advertising

Adapted from Kölmel and Alexakis (2002) and Carat Interactive (2002)

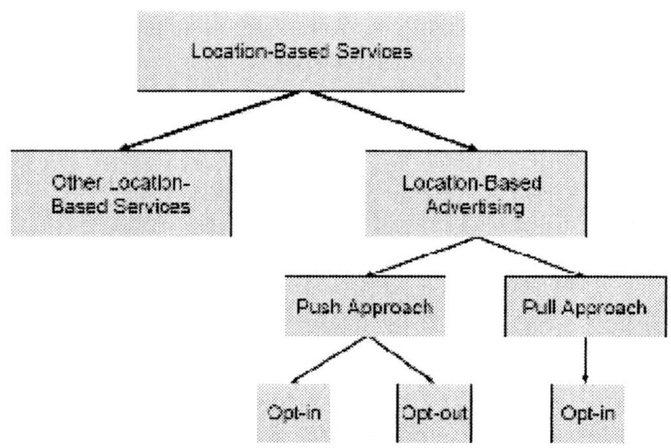

Source: Bruner II, G.C & Kumar, A. (2007). Attitude Toward Location-Based Advertising.

http://www.jiad.org/vol7/no2/bruner/Attitude%20toward%20location-based%20advertising.pdf,

p.3-4. retrieved 6/27/2008.

be associated with applications that are described as automatic location information (ALI). ALI is very useful in terms of facilitating index application; however, it is not considered as a critical element in the success of LBS. LBS index researches can actually be developed with the evaluation of existing or future locations created by its users in various ways. The strongest LBS applications shall be those designed to facilitate the works of users from the source location of the user to the destination location.

There are several important drivers that will bring about a mass market breakthrough for LBS at long last. These include increasingly user firendly handsets, accelerated implementation of high accuracy, high performance positioning technology and an increasing emphasis among operators on data services to compensate for lower voice revenues (http://www.mapforums.com/next-five-years-european-lbs-market-4798.html).

According to Figure 1, location-based services are divided into two: as location-based advertising and other location-based services. Here, location-based advertising has two further applications as push and pull.

For services where place and location is important, such as emergency notices, assistance services, description of interesting places, traffic data, social services and payment services, these services can be provided more effectively with LBS (Driscoll, 2001).

According to Okazaki & Taylor (2008), location based services occur if the new media is suitable for the customer as to their features explained below; for example, topics such as the weather forecast, hotel maps, address retrieval, and the situation of traffic (Barwise and Farley, 2005; Barwise and Strong, 2002; Okazaki, 2004; Pagani, 2004).

According to Okazaki (2004), enhanced technologies such as FeliCa and Bluetooth increase the productivity of GPS technology applications. This enables advertisers to propose real-time offers to subscribers of such services who are in the required proximity. This is also appealing for customers in terms of becoming LBS users (Pagani, 2004). Therefore, companies are provided with extensive means to implement their pull strategies.

The following results have been obtained in the researches performed by In-Stat (Potter, 2007):

- Present mobile LBS users are corporate subscribers who have a higher income level compared to the average income.
- The most popular LBS ideas are related with local information and navigational applications.
- Respondents of the research are technical engineers and students with a high and middle income in the 35-39 and 18-29 age groups, and they are highly interested in the topic.

Such applications cause the development of social networks also lead to the appearance of brand new ideas and persons such as micro-celebrities (well-known and famous people in a certain group or network). This introduces a new expansion to the issue of advertisement presentations (Gade, 2008).

Another finding is that it allows the user to procure the product or service requirement suitable for the self-specific situation of the user and valuable information tailored for the user from the service provider. The service provider uses this information for local presentations and to present advertisements. Besides, the service provider can provide user information to third parties in order to help the presentation of special offers that merge user information and information from other certain applications. So, offers can be customized. It points to e-business models, which are created by social networks that emphasize the advantage of using network information (this information can be demographic, behavioral, or situational or position depended).

According to Batista (2006), created applications draw attention and lead to interesting questions. The interest data of users monitor "micro- celebrities" from the individual user's perspective and draw attention as to how an individual brings that person to a "micro-celebrity" position. The same type of attention data are also based on which products or services draw the users' interest. Furthermore, should the "micro-celebrity" be associated with a certain brand and other similar subjects, this is included among these data.

However, the entire approach here is based on only features such as attention-drawing data and accessing, managing, control and owning our data. According to Young (2006), marketers or social network operators (they are the platforms with all the user data) do not fully address to individuals and they are not close to the customer.

Most people are included in a social network and the participation distance when the data is not the network varies according to time and situation. Instead of regarding the social network operators as the only source of user data, marketers should monitor the actual users. Setting up from this view, we can say that an infrastructure to store the individuals' attention data related with sites, services and platforms is just being developed. However, it is thought that "micro-campaigns" will be more influential at this stage. It is thought that these campaigns would be more effective in the beginning if related data are obtained from wide resources.

LBS Value Chain

Location-based systems encompass service provision in their structures. This type of services creates a brand new value chain where a new one is added to the existing value systems. According to Crocker (2008), those who contribute to the LBS value chain include content developers, application platform sales machinery or software transporters and operators. Content providers provide wide scale data which would reinforce LBS applications. The related content consists of digital maps, weather forecast, real-time traffic, real-time application points, and location brackets. LBS application platforms can be considered as the platforms that provide technologies and infrastructure which enables developing the applications of location based services. In Mobile LBS applications, ready-pack applications that enable special requirements-oriented service delivery can also be mentioned.

Content Providers

Contents such as readily available digital maps or weather forecast services actually express a value on their own. However, the actual value they provide in LBS as mobile content is their integration with other data sources and applications; for example, the integration of mapping and traffic information or the integration of the interested point with search information and navigation applications. Therefore, it is possible to carry out a very strong navigational application. Data, which is created by merging data taken from many different resources, creates a higher value than the user would obtain from the pieces one by one, and creates a kind of synergy.

Many service suppliers, who provide LBS applications, present the content in a similar way to the Internet and other institutional applications. As mobile location based services support the location based services, they are gradually turning into a growing market. Actually, although it is not possible to describe the mobile-content related market as highly-developed at present, the content market has developed with Internet and gained a highly institutional, private and complex form. Such content providers are obliged to act with many partners. These partners are the organizations that collect the data, add value to them and return them to the partner that initiated the action as processed and in the new form.

Some of the special content providers which reinforce LBS are:

- Digital map providers - these sellers have digital map databases and they license map information for LBS. The leaders of them are Navteq, Tele Atlas and Pharos.
- Traffic data providers - the providers of these services cluster or filter the traffic data they obtain from various resources and provide them to users via application providers. Inrix and Navteq are the leading providers.
- Weather forecast data providers – the providers of this service present the weather forecast obtained from a wide geographical area in real time. Weather Bug Accuweather and Intellicast are the leading providers in this respect.

Application Platforms:

Application platforms are considered as the platforms where providers introduce intermediary software that interrelates, merges and integrates the two different applications. Application platforms also produce the software on the customer's side, provide the associated kits (equipment, etc.) and convert them into familiar user interfaces.

Apart from the developed environment, LBS platforms also provide the rendering, fixing location and management of database mapping. Autodesk (www.mapguide.com), deCarta, ESRI, MapInfo and Wavemarket are some of these platforms.

Application Vendors

Application vendors contribute to the management of work flow and processes in this value chain and provide services, and they create a superior value as a result. They provide a serious service although they are mostly considered as small niche structures. The ability to convert such work processes and requirements into mobile software applications is considered as the main competency of these application vendors. Such vendors provide applications such as follow up applications, staff or employee tracking applications, gap filling applications, and relay navigation applications. Networks in Motion (NIM), Gearworks, and TeleNav are among these vendors.

LBS APPLICATIONS

LBS applications can be considered in three main segments (McGeough, 2002). The leading one of them is the applications provided by public organizations and municipalities especially towards emergency. In the public field, LBS should be considered as emergency evaluation assistance services, location/direction information services traffic data, social services and payment services (Driscoll, 2001).

From the consumer's point, applications oriented towards subscribers include child care service, vehicle telematic service, road site assistance, navigation service, tourist visitor service, ground-based invoicing and friend and family finding, and last minute ticket finding. The company and attempts to cover issues such as tracking assets, tracking employees, tracking work order and tracking distribution and supply chain, and fleet management (see Table 1).

LOCATION BASED ADVERTISING (LBA)

According to Nysveen et al. (2005), mobile advertising includes wider applications than the ads sent to mobile phones of individuals. It should even be considered wider than the definition of mobile channel. Mobile advertising should be considered as a versatile and interactive network through which various companies, people and organizations could communicate with individuals via various mobile delivery technologies.

LBA should be regarded as rather a subsystem of LBS. According to Bruner II & Kumar (2007), LBA is the presentation of marketer-controlled information at a special and certain point where users can receive the advertising medium. Here, issue location has the most important attribute because the given advertising and the location are connected in a way; for example, being very near to the place of discount or being in the store.

Such advertisements can turn into portable elements if the customer does not like the advertisement or adopt or read it. Because, when the customer is moving, the device is also moving. Therefore, as defended by Oliver, Rust, and Varki (1998), LBA has reached an aspect which would maintain real time marketing. These systems which, for example, have the ability to offer last minute tickets to those

Table 1. LBS applications

Government	Business/Enterprises
E-911	Asset Tracking
Emergency Response	Enterprise Worker Tracking
Police and Fire Response	Work Order Dispatch
Virtual Toll	Location-Based CRM
House Arrest	Supply Chain Visibility
m-Municipality & Assistance	Fleet Management
m-Fingerprint	Cargo Tracking
Other	Electronic Fencing
	Fraud Detection Services
Consumer	Resource Planning
Child Minder Service	Real Estate Service
Vehicle Telematics Service	Smart Shipping Services
Roadside Assistance	Other
Navigation Services	
Tourist Visitor Services	
Location Based Billing	
Friend and Family Finder	
GeoMedicine Services	
Traffic & Dimensions	
Service Finder	
Job Finder	
Parking Place Finder	
Last Minute Ticketing	
Other	

who are willing to visit a festival and/or a show, to work in parallel to the Internet, to expand the given service area during which the organization can present the advertisement at the right place, at the right time and to the right person. The cost of such services can be paid by advertisers, or they can even be presented free of charge depending on the adopted technology.

Here, the person may achieve lists that enable taking action, or prefer to answer warnings received from the surroundings.

In this context, the access point becomes a very special point. Kiosks are one of the available access points. According to Rowley and Slack (2005), kiosks tend to be more service-oriented than general purpose work stations (see Table 2). On the other hand, some kiosks provide Internet access and e-mail type of services while others present a limited number of services or realize a limited transaction. Here, the given duty or service acts as an intermediary in delivering the service to the mentioned place; for example, ticket booths at railway stations, or information maps at touristic places.

The mobile advertisements at these points are of a more influential type than the other general advertisements that address to everybody.

Table 2. The kiosk taxonomy (Source: Developed from Rowley, J. & Slack, F, 2005. Towards a Taxonomy of Kiosk Based Self-service Technologies, E- Business Review, Vol.5)

Dimensions		
Place	*User*	*Purpose*
Travel Public Services Retailers-outlets-shopping malls Banks and other financial institutions Entertainment Catering General Public field	Passengers Urban residents Shoppers Investors and buyers Those who want to relax Those who want to eat Traffic flow, emergency, etc.	Information • High Customization, High Complexity • High Customization, Low complexity • Low customization, High complexity • Low customization, Low complexity Relation • Personal communication • Marketing communication • Loyalty scheme kiosks. Operation • Goods delivery • Digital content delivery • Service delivery • Remote delivery • Ticketing • Money transactions

BLUETOOTH MARKETING

Bluetooth is one of the alternative communication channels that has been presented in the world in the domain of mobile marketing dated from 2005 (Patel, 2008) at the time Bluetooth enabled mobile phones release. Bluetooth is a technology which can work everywhere without outdoor or indoor differences, and which can provide great facilities. The technological base of the data received or sent to these mobile instruments may provide distribution in the Bluetooth area with additional technological instruments, and they can be made effective on a secondary level.

Bluetooth enables a regular connection which can be used in the short range environment in order to build a regular connection between the users and their surroundings through wireless technology. According to Buttery and Sago (2004, p.67), Bluetooth is considered more as an independent communication technology than global license by working at a 2.4 GHz ISM (industrial, scientific and medical) band with rapid frequency hopping. Therefore, Bluetooth undertakes a highly evident role in various applications (Buttery & Sago, p.79):

- Ad hoc networking
- M-commerce
- Information services
- Public wireless information

Bluetooth wireless technology is a short range wireless standard which has been developed for personal connection with an electronic device. The current use of mobile technology is in the structure defined as the fourth version, and it is continuing to develop on issues such as small form factor radio, low energy, low cost, built-in security, robustness, ease of use, and ad hoc network utilization. Almost every business day, five new Bluetooth featured products is launched and approximately 13 million Bluetooth units are loaded to transportation vehicles every week. It is indicated that there are approximately 2 billion

Bluetooth enabled and Bluetooth installed devices in the market (http://bluetooth.com/Bluetooth/Press/SIG/BLUETOOTH_SIG_OPENS_NEW_DOORS_TO_SMALL_BUSINESSES.htm 10/19/2008). Actually, mobile technology has completed the features expected from third generation instruments, and third generation devices began to take their places in the global market.

It was not expected to have such a wide scope during the first time Bluetooth's awareness and acceptance appeared; nevertheless, the result obtained five years after its introduction does not surprise anyone. However, the commercial uptrend of Bluetooth can only be explained with the proliferation of new mobile phones. With these telephones, companies have obtained new ready-made advertising and promotional methods based on technological means with new opportunities of accessing the right customer at the right place and at the right time.

The same activity is realized with certain apparatuses added to the mobile computer. However, the most practical way to perform marketing activities is by using Bluetooth on mobile phones as they have the highest ease of carrying.

As of May 2008, Bluetooth has completed its 10th year of entry into the market. In the researches performed by Millward Brown in 2003 and in the autumn of 2004 on 1300 people between the ages 18 and 70 in the US, UK and Japan, a certain growth is observed in awareness and recognition of the utilization of Bluetooth wireless technology. Awareness has increased from 47% to 77% in the UK, from 43% to 61% in Japan and from 22% to 41% in the USA. (http://bluetooth.com/Bluetooth/Press/SIG/iBluetoothi_Awareness_Nearly_Doubled_Among_Consumers.htm retrieved 11/3/2008).

According to the research carried out every year by Millward Brown regarding global awareness on Bluetooth technology, awareness reached the highest of all times with 85% in 2008. Researches for five subsequent years indicate that the ratio is increasing.

Customers in China, Germany, Japan, Taiwan, USA and UK indicated that they heard about Bluetooth technology and they are aware of its logo (68%) and they use devices which have Bluetooth technology. The devices used by these people are (http://www.bluetooth.com/Bluetooth/Press/SIG/CONSUMER_AWARENESS_OF_iBLUETOOTHi_WIRELESS_TECHNOLOGY_AT_ALL_TIME_HIGH.htm retrieved 11/9/2008):

- Mobile telephones (79%)
- Earphones (61%)
- Computers (56%)

In the research carried out on 2500 people of ages between 18 and 70, at minimum 60% of the respondents in four of the mentioned six countries accept that they have Bluetooth enabled devices. The awareness of Bluetooth technology has been identified as 85% while that for Wireless USB is 57%, for WI-FI (53%), IrDA (53%) and Wimedia UWB (16%). Also, 78% of the customers who expressed that they heard a lot about Bluetooth technology indicated that they would pay more for these products while 57% says that Bluetooth technology facilitates life. Foley (2008) indicates that the technology spread through mobile phones and earphones is currently used in almost every industry (http://www.bluetooth.com/Bluetooth/Press/SIG/CONSUMER_AWARENESS_OF_iBLUETOOTHi_WIRELESS_TECHNOLOGY_AT_ALL_TIME_HIGH.htm retrieved 11/9/2008).

This technology first came into the agenda to produce wireless earphones in order to use mobile phones as hands free. At present, Bluetooth technology is used on laptops, PDAs, cameras, printers, mobile phones, mouse, keyboards, MP3 players, loudspeakers, automobiles, and medical devices.

Bluetooth which is made ready for systematic communication in marketing, enables contacting a user near a billboard, kiosk, banner, or shop window. When a person who has a mobile phone turns on his or her Bluetooth, a Bluetooth message approval is sent to the mobile phone via servers. A piece of the content such as "From the .... Brand" is sent to the personal mobile phone, and it is requested to be confirmed if the content is accepted. If the user confirms, the Bluetooth Marketing campaign will be directly loaded on the mobile phone and that content will present information (price discounts, special campaigns, product introduction information etc.) about the store near the customer at that moment. Bluetooth does not cause an extra cost as it is a free of charge data transfer method.

The most commonly used mobile content among those sent via Bluetooth is SMS text messages. These messages aim to be influential concerning shopping requests with the inclusion of customer directing information. These messages can distribute short videos, wallpapers, screensavers, text files, java games or mobile phone melodies to customers through a nearby Bluetooth.

For example, in a wide scope study held simultaneously in many countries by Opel, billboards, advertising boards, bus stops and mega boards were used. Customers who desired to download C'MON dolls to their mobile phones and use them as background picture, logo, screen saver or theme saw the text "Turn on your Bluetooth, download C'MON dolls" on the billboard and turned on their Bluetooth devices. In the meantime, when they received the text "YOU HAVE MESSAGE FROM OPEL, ARE YOU ACCEPTING?" they confirmed with a Yes button and downloaded funny contents with various images of C'MONs, the latest commercials and melodies of C'MONs to their mobile phones in front of a billboard.

Customers are gradually becoming aware of such technologies and they develop the ability to use this type of technologies when they want which has pushed forward the idea of producing new approaches in order to develop the ability to address the conscious customer both for the telephone set manufacturers and for the marketers. They are trying to use this technology on their products in order to create differentiation and increase profit margins. The experts indicated that actually the cost of installing a Bluetooth chip to a mobile phone is not very low, and they have begun to act with the strategy of expanding their areas of use by adding them into many products. Even added on motorcycle helmets, sunglasses, gaming controls and some fitness equipment, Bluetooth signals its future use in new application areas like home entertainment and fitness devices. (http://bluetooth.com/Bluetooth/Press/SIG/2008_MARKS_TEN_YEARS_OF_emBLUETOOTHem_WIRELESS_TECHNOLOGY.htm retrieved 11/9/2008).

All previously-mentioned applications such as SMS and MMS, which are included in mobile systems, are realized by connecting to a customer's mobile phone via an operator network. Wireless technologies such as Bluetooth and wireless create an alternative to the mentioned general structure and allow for benefitting from the media with free of charge applications. For example, in the operations performed with a users' telephone number (such as GPRS and SMS), the usual applications consider point targets while they are away from the right place and right time concepts. With alternative connections through Bluetooth, campaigns can be made more mass-specific.

These campaigns are the permission-based transfers although they are transmitted to the customer's phone from transmitters installed at privately-owned broadcasting organizations and areas at fixed locations, and they are based on on-demand communication. For example, a supermarket may inform its campaigns or informative messages about service sections to all mobile phone users that enter the supermarket or that are in the close vicinity via a Bluetooth campaign instead of sending an SMS which would be costly or the people around an entertainment center can be informed about the program of the day or night through creatively-thought invitation files. As these messages are opt-in, they will be monitored by only the users who are willing to see them, and they will not create "message pollution" or "spam" on the mobiles of those who do not want to receive them.

Bluetooth advertising is strong because:

- It reaches on the spot and on time.
- It provides high-quality relationship building and long-time saved advertising. The customer accepts the message when the advertisement message is delivered to the customer, but the customer can watch it anywhere and any time he or she prefers.
- Having introduced a new advertising approach, Bluetooth enables the customer to reach the advertisement and explore the product or service at the closest point.
- It has high inducement value as it promotes a place or as it is acquired at a close point to the source. For example, receiving hotel advertisements at airports or beer advertisements at bars can be influential in making a selection.
- It has the ability to increase sales point activity and to provide interactive information for all possibilities the customer may ask for.
- It allows making promotion campaigns with promotional content distribution.
- It allows for benefitting from a free of charge communication service and reduces the cost of marketing campaigns.
- It is a more effective approach than conventional marketing in terms of reaching the customer faster.
- In foreign countries, the issues such as finding a destination, road, description are rapidly clarified and the person can act easily according to the provided information although he or she does not know the local language.

As Bluetooth is a free of charge way of data transmission, it is suitable for developing a very low cost marketing activity. It is a very simple and easy to use technology which can work and which can be applied everywhere regardless if it is indoors or outdoors; furthermore, it can be enabled with a small mechanism and work in parallel to a media that helps convey the message. Bluetooth can add advertising value to billboards, kiosks and posters. A person can enable Bluetooth function anytime and follow the received messages, and become interested in the content of the message received. At the moment the message is approved by the user and when Bluetooth begins to direct shopping behavior, the Bluetooth marketing campaign reaches its first goal. The interest paid to Bluetooth can later be maintained with other mobile content such as SMS, downloadable short videos, wall papers, screen savers, text files, java games or mobile phone melodies, and other downloadable elements of sports, music, lifestyle and dining.

PLACES TO USE BLUETOOTH MARKETING

Bluetooth marketing can be included in activities with a rather social content such as concerts, fairs, festivals, sports contests, museums and archeological excavation areas for individuals who create society. They can also be used at rapid circulation points such as bus stops, railway stations, and airports. During activities at such places, Bluetooth can be used as an effective instrument to present the video of the activity, the video or commercial of the sponsor, the music or melodies related with the activity, the wallpaper of the organization, the screen saver which shows the activity logo, the promotion and marketing of future activities, the map and place information of the area and information in a fair.

Bluetooth marketing allows for communicating detailed information and prices of the products sold on site, the introductory files of sponsors, and virtual business cards. During sight-seeing, the option to select the language for the explanation of historical places can be provided, thus the selected audio-visual explanation videos, the means of transportation, the times of tours and information of other nearby historical places as well as other special information such as museums or art galleries, and a download of a map of the museums can be offered. Also, at places such as exhibitions and expositions,

Table 3. Some Bluetooth utilization points and presented promotions

Place	Promotions
Airport	1.Flight information 2. Information about the city 4. Free Shop discount coupons 5. City-wide and/or inter-city transportation routes and information 6. Restaurants and sightseeing information 7.Other
Cafe, Bar and Restaurant	1. Visual presentations of special menus 2. Description of special activities 3. Reservation for next visit via Bluetooth 4. Invitation of the potential customers in front of the door with continuous discounts and gift coupons 5. Increase of appeal of consumption with discounts and gift coupons presented to the customers inside 6. Food/beverage gift coupons 7. Other
Entertainment centers / Cinemas	1. Trailers of current movies 2. Food/beverage gift coupons 3. Show time 4. Next opening shows 5. Wall papers and screen savers of movies 6. Last minute special discounts or discount coupons which allow sales of tickets in a very short time 7. Other
Hotels	1. Hotel introduction and information 2. Local access information 3. Password transfer for wireless internet access 4. Order via Bluetooth from room 5. Delivery of information on activities of hotel 6. Certain touristic information 7. Food beverage coupons 8. Other
Kiosks, Stores, Shopping Malls	1. New products, bar codes, customer loyalty programs 2. Brand introductions 3. Gift checks and multimedia contents 4. Where is it? Information 5. Advergaming 6. Special last minute sales discounts and other discount information 7. Lotteries 8. Other
Stadiums	1. Anthem and songs of the supported team. 2. Statistics and match schedules 3. Small movies with pictures of players or goals of the supported team 4. Next activity schedule 5. Concert or match details 6. Map showing place 7. Other

the introductory videos of artists whose works are exhibited, audio-visual videos of the works, and the introduction of the next program can also be presented. Similar introductions of musicals or concerts, maps and drawings, melodies, video clips and songs of the artists, questionnaires or contests about the artists, desktop background and screen savers may also be effectively used. In sports games, the goals of the supported team, the anthem and songs of the club, and many other meaningful elements for the user can turn into subjects of Bluetooth marketing.

Provision of planned and detailed access to not-easily-accessible information such as statistics and match schedules also help individuals to focus their interests on these issues. This creates an impact that goes beyond an existing mass, and allows every interested and curious individual to be included in the subject.

With Bluetooth technology, which is now available on almost all mobile handheld devices in the market, communication can be established between two instruments in order to transfer data and files. The distance between the device and the Bluetooth broadcaster can vary between 10 and 100 meters, depending on equipment features of the communication devices and the environment.

Advertising content or aimed multimedia files presented via Bluetooth are instruments such as pictures, music, video, and text, and it refers to the eye and ear, but additionally it also requires using the sense of touch as it is downloaded. This actually means benefitting from multi-perception having a high possibility of influence. Also, a customer can be directly addressed in the multi-tasking structure. It allows for the delivery of discount coupons, ringtones, movie trailers, mobile games with advertisement content, mp3 and an advertising network centrally directed to mobile device users (mobile phone, PDA, SmartPhone, etc.). Different system projects will gradually be developed. Mainly, three elements are dominant in system projects: Bluetooth Server, Advertising Management Server and Web Panel.

1. **Bluetooth Server:** It is the sub-system that allows for the delivery of related multimedia advertising files to the end user's -Bluetooth supported- mobile device. These servers constitute the nodes of the advertisement network system distributed to places where the end user spends much time such as shopping malls, mass transportation stations (subway, airport, etc.), and cinemas. They are the structures with high memories and generally run on Linux base. For example, the system used by Mopromo (http://www.mopromo.ro/) is a Bluetooth equipment integrated 200 MHz processor with a 128MB cache memory with 180x180x100 mm dimensions known as the Linux embedded system. Bluetooth servers work as connected to the Internet in a way (LAN, WAN, Wi-Fi, GPRS etc.) in the located place, so the content of the Bluetooth server can be changed from the center.
2. **Advertising Management Server:** It is used as a sub-system developed in order to manage the Bluetooth servers of the system. Client-server architecture is established between the advertising management server and Bluetooth servers. For example, the advertising server processing the advertising campaign uploads via a Web Panel and sends the advertising files to related Bluetooth servers, and processes the information received from Bluetooth servers (the download statistics, etc. or information which show the performance of the advertising campaign) then transfers them to the Web panel.
3. **Web Panel:** This is a sub-system that allows advertisers to subscribe in the system and directly conduct their own advertising campaigns. Advertisers can log into the system via a promotion web site and benefit from a publisher who can upload the related multimedia file to the system, and publish them via Bluetooth servers configured at the places where targeted customers are densely located. In the latter versions of these systems, the end-user can be provided with an analysis

performed with data-mining algorithms. They can provide information of "the locations where related advertisement could be more successful if it is broadcasted" by basing it on the obtained data. So, the requirement for questionnaires, face to face interviews and customer satisfaction or customer information acquisition-related efforts would decrease.

This system does not bring any costs to the end user because the files are sent via Bluetooth free of charge. Accordingly, the user would not need any software/hardware support. The integration of this system in Near Field Communication (NFC) in order to develop a network structure where the user would make a direct payment can be considered. Hence, advertisers who use the system would not only promote their own products or services, but also have the opportunity to sell their products and services and receive payment via the system.

Through the subscriptions of individuals who are included in the conventional customer portfolio mobile device users can indicate and include their own preferences in the system via portal, stating the advertisements and discount opportunities on issues they would be interested in. It allows advertisers to reach customers that they are directly targeting, and also the inclusion of new or fresh users of Bluetooth to be included in the system. So, realizing a brand new application that allows reaching the directly-targeted customer is possible.

The leading questions that come into mind about the system are the security and reliability issues. It is thought-provoking that a virus can be sent to a mobile phone when Bluetooth is on by taking advantage of certain gaps in the Bluetooth protocol. If the end user attempts to open a virus file "only by looking at its name" without knowing where it came from, then he or she may activate the virus. New mobile applications are being developed in order to eliminate this problem. Also, the user may be

Figure 2. MOPROMO system schema

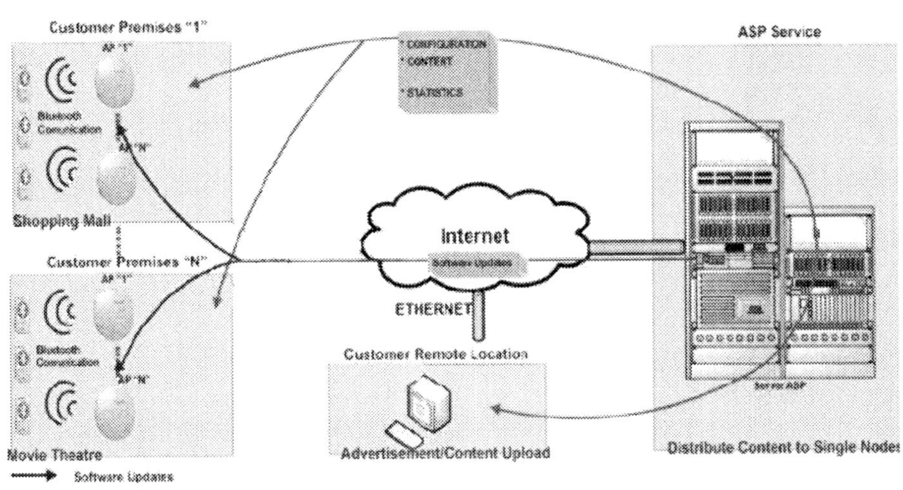

Source: Asil Ersoydan (2007). "MoPromo - Bluetooth tabanlı reklam agi projesi hakkında" 12/10/2007 tarihli e-mail attachement Mopromo ppt.

informed that the file is sent from the network of a promoted company and therefore, it is a trusted file. Also with mobile application, the user of the mobile device can enter the system, browse the content of the Bluetooth server and download the file to the phone rather than wait for the file to be sent to the user. Furthermore, a bi-directional communication opportunity is also created when they download the mobile application. Besides, mobile device users can be warned on the website about activating the Bluetooth function of their mobile devices near the locations of promoting Bluetooth service and turn off the function after receiving the files. Also, continuous communication can be established with the mobile phone which downloads just once.

One of the weak points of Bluetooth is the narrowness of the covered area and customer. Operating in a 15-30 m. diameter area may both be considered as a weakness and as an advantage. In a selected coverage area, the potential customer is very near the promoted product and has a higher chance to evaluate the promotions and given information compared to normal conditions. There is no roaming. If the receiver is active, the advertisement presented to the receiver who exits the coverage area is cut off. If the files are big, then the download period increases and the download may be interrupted while moving. This may help to keep the customer in the vicinity or frequently attract the customer to that point or area.

Also, as 3G telephones provide video-calls, solving many problems arising from the remote manager and employee and from the proximity of the activation of providing devices will be possible via video-phones.

The systems which first began with infrared transfer technique which could be converted into a different system later adopted the Bluetooth technique, and they have enabled the utilization of an interactive advertising and promotion network quite different from the conventional mass advertising and promotional approach. The receipt of advertisements and promotions on the mobile phone as permitted turns the push concepts of the force and aggressiveness of SMS marketing into pull with Bluetooth Marketing. The influence of promotion is high as the customer is very near the purchasing or service place. Push is a method used by a company to send informative messages to people in the permitted database. That is, the company provides people included in the database with promotion, place, ticket features, type of information upon their requests, and creates a pressure for buying those products or services. The customer can act anytime when he or she thinks that the pressure is a suitable option for him or her. In the foregoing structure, push can be considered as force to open Bluetooth to activate the customer. Pull is considered as the communication of a campaign to the consumer through a classical media (TV, magazine, outdoor, internet etc.) or Bluetooth, and the consumer's response upon its receipt. For example, the activities which include sending a keyword to a short number in order to participate in a lottery or to win a gift are considered in pull marketing. Bluetooth can be used as both push and pull.

Bluetooth is an effective marketing tool within its coverage area. Therefore, proximity appears as the most important factor. However, the statement that proximity influences shopping will be better understood with measurements and evaluations after the system is densely used. In addition to being a solution that includes and carries forward concepts such as e-commerce, it is also possible to use the social environments of users of the system during advertisements and promotions. This approach which has emerged as a new approach based on creating influence in the close vicinity may also change the buying behavior of the customer in time. In the mean time, marketing activities in continuous contact with an individual should be considered as new opportunities of benefit in the social environment.

Here, the feature of creating a social network or bringing together individuals who are in the same region and who have the potential to communicate with each other creates an important difference.

In the activities to be realized via Bluetooth, a campaign can also be supported with visual elements. These are billboards at the entrances of shopping centers, or promotions such as local posters or leaflets. They describe the product or service and promise various "digital gifts", discounts, promotion, coupons, sweepstake or descriptions which can mean material benefits for the customer. Every activity to be held in order to lead those who are not included in the system to activate their Bluetooth devices should be realized with the perspective that one customer means one thousand customers.

REFERENCES

Barwise, P., & Farley, J. U. (2005). The State of Interactive Marketing in Seven Countries: Interactive Marketing Comes of Age. *Journal of Interactive Marketing, 19*(4), 67–80. doi:10.1002/dir.20044

Barwise, P., & Strong, C. (2002). Permission-Based Mobile Advertising. *Journal of Interactive Marketing, 16*(1), 14–24. doi:10.1002/dir.10000

Batista, E. (2006). Retrieved 11/22/2008 from http://gigaom.com/2006/06/19/of-social-networks-and-business-models/

Bluetooth Marketing Nedir? (What is Bluetooth Marketing?) (n.d.). Retrieved from http://www.vodera.com.tr/blog/?p=1

Bruner, G. C., II, & Kumar, A. (2007). Attitude toward Location-Based Advertising. Retrieved 6/27/2008 from http://www.jiad.org/vol7/no2/bruner/Attitude%20toward%20location-based%20advertising.pdf

Buttery, S., & Sago, A. (2004). Future Applications of Bluetooth. In P. Smyth (Ed.), *Mobile and Wireless Communications: Key Technologies and Future Applications*, (pp.67-79). London: IEE.

Crocker, P. (2008). *Value in the Location Value Chain*. Retrieved 5/3/2009 form http://www.directionsmag.com/article.php?article_id=2901

de Souza e Silva. A. (2006). Re-conceptualizing the Mobile Phone: From Telephone to Collective Interfaces. *Australian Journal of Emerging Technologies and Society, 4*(2), 108-127. Retrieved 10/3/2008 from http://www.swinburne.edu.au/hosting/ijets/journal/V4N2/pdf/V4N2-4-Souza-Silva.pdf

Driscoll, C. (2001). Are Consumers Interested in Wireless Internet Location-Based Services? *RCR Wireless News, 20*(12), 56. Retrieved 4/26/2006 from <rcrnews.com/lockland.cms?articleId=37351>

Ersoydan, A. (2007). MoPromo - Bluetooth tabanli reklam agi projesi hakkinda. 12/10/2007 dated email attachment Mopromo ppt.

Fields, L. (2005). *Location Based Services – a Marketer's Dream*. Mobile Marketing Association. Retrieved 6/27/2008 from http://www.mmaglobal.com/modules/article/view.article.php/21

Foley, M. (2008). *Bluetooth Sig Executive Director Named Wireless Influencer of 2008*. Retrieved from http://www.bluetooth.com/Bluetooth/Press/SIG/BLUETOOTH_SIG_EXECUTIVE_DIRECTOR_NAMED__WIRELESS_INFLUENCER_OF_2008.htm.

Francica, J. (2007, October 27th). Mobile LBS Applications are Growing and Networks in Motion is Benefiting. *Directions Magazine*. Retrieved from http://www.directionsmag.com/article.php?article_id=2587&trv=1

Gade, S. (2008). *Business Models Thoughts*. Retrieved 11/22/2008 from http://lbsblog.net/

Gauntt, J. (2007, November). Mobile Message Marketing. *eMarketer*.

Gökmenoğlu, E. (2008). *Mobil Pazarlama*.http://pazarlama2.blogspot.com/search/label/Mobil%20 Pazarlama.

Heidelberger, C. (2009). *Creating a Distribution Advantage*. Retrieved 5/3/2009 from http://www.crm2day.com/highlights/EpZpykukukyWExPixf.php http://bluetooth.com/Bluetooth/Press/SIG/2008_MARKS_TEN_YEARS_OF_emBLUETOOTHem_WIRELESS_TECHNOLOGY.htm (n.d.). Retrieved 11/9/2008.

http://bluetooth.com/Bluetooth/Press/SIG/BLUETOOTH_SIG_OPENS_NEW_DOORS_TO_SMALL_BUSINESSES.htm (n.d.). Retrieved from 10/19/2008.

http://bluetooth.com/Bluetooth/Press/SIG/iBluetoothi_Awareness_Nearly_Doubled_Among_Consumers.htm (n.d.).

http://en.wikipedia.org/wiki/Location_based_advertising (n.d.).

http://en.wikipedia.org/wiki/Proximity_marketing (n.d.).

http://ezinearticles.com/?Virtual-Business-Cards-that-Make-a-Difference&id=269400 (n.d.). Retrieved November 16, 2008.

http://usa.autodesk.com/adsk/servlet/index?id=6546938&siteID=123112 (n.d.).

http://www.accuweather.com/premium-benefits.asp (n.d.).

http://www.acs.com.hk/knowledgebase.php (n.d.).

http://www.bluetooth.com/Bluetooth/Press/SIG/CONSUMER_AWARENESS_OF_iBLUETOOTHi_WIRELESS_TECHNOLOGY_AT_ALL_TIME_HIGH.htm (n.d.).

http://www.decarta.com/solutions/index.htm (n.d.).

http://www.gearworks.com/ (n.d.).

http://www.mapguide.com (n.d.).

http://www.navteq.com/ (n.d.).

http://www.teleatlas.com/index.htm (n.d.).

http://www.vodera.com.tr/blog/?p=1#more-1 (n.d.). Retrieved from 10/12/2008.

IAMAI. (2008). White Paper on Location Based Services on Mobile in India. Retrieved 5/3/2009 from http://www.iamai.in/Upload/policy/LBS_Draft_Indicus.pdf

Juniper Research. (2008). *Share, Collaborate, Exploit Defining Mobile Web 2.0.* White Paper extracted from Mobile Web 2.0: Leveraging 'Location, IM, Social Web & Search' 2008-2013. Retrieved from http://www.c2mweb.com/bouncer?t=files/Whitepaper_Mobile_Web2.0.pdf

McGeough, J. (2002). *The LBS Value Chain Partnerships for Success.* Retrieved from http://www.wirelessdevnet.com/library/geomode_me2002.pdf

Mobile Web 2.0: Leveraging 'Location, IM, Social Web & Search' 2008-2013.White Paper Retrieved from http://www.c2mweb.com/bouncer?t=files/Whitepaper_Mobile_Web2.0.pdf

Nysveen, H., Pedersen, P., & Thorbjornsen, H. (2005). Intentions to Use Mobile Services: Antecedents and Cross-Service Comparisons. *Journal of the Academy of Marketing Science*, *33*(3), 330–346. doi:10.1177/0092070305276149

OECD. (2006) *Information Technology Outlook.* Retrieved 5/24/2009 from http://www.oecd.org/document/10/0,3343,en_2649_33757_37486858_1_1_1_1,00.html

Okazaki, S. (2004). How Do Japanese Consumers Perceive Wireless Ads? A Multivariate Analysis. *International Journal of Advertising*, *23*, 429–454.

Okazaki, S., & Taylor, C. R. (2007). What is SMS Advertising and Why Do Multinationals Adopt It? Answers from an Empirical Study in European Markets. *Journal of Business Research.* doi:10.1016/j.jbusres.2006.05.003. also in http://ctl.scu.edu.tw/scutwebpub/website/DocUpload/CourseTeaching/cyc20078391819_1.pdf

Oliver, R. W., Rust, R. T., & Varki, S. (1998). Real-Time Marketing. *Marketing Management*, *7*(4), 29–37.

Pagani, M. (2004). Determinants of Adoption of Third-Generation Mobile Multimedia Services. *Journal of Interactive Marketing*, *18*(3), 46–59. doi:10.1002/dir.20011

Pagani, M. (2004). Determinants of Adoption of Third-Generation Mobile Multimedia Services. *Journal of Interactive Marketing*, *18*(3), 46–59. doi:10.1002/dir.20011

Patel, H. J. (2008). Bluetooth Marketing for Digital Networks. Retrieved 5/3/2009 from http://www.digitalsignagedirectory.com/component/content/article/1-latest-news/66-bluetooth-marketing-for-digital-networks

Potter, E. (2007). *High Demand Foreseen for Mobile LBS in China.* Retrieved from http://www.instat.com/press.asp?Sku=IN0703906CWW&ID=2046

Quinn, C. (2008). *M-Learning Devices: Performance to Go.* Retrieved 9/5/2008 from http://www.learningcircuits.org/0708_quinn.htm

Raymund, S. (2006). *The Advantages of Distribution.* Retrieved 5/24/2009 from http://www.channelinsider.com/c/a/Commentary/The-Advantages-of-Distribution/

Rowley, J., & Slack, F. (2005). Towards a Taxonomy of Kiosk Based Self-Service Technologies. *E-. Business Review (Federal Reserve Bank of Philadelphia)*, 5.

www.maps-gps-info.com/maps-gps-glossary-kl.html (n.d.).

www.wimob.com

Young, R. (2006). *Of Social Nets & Business Models*. Retrieved from http://gigaom.com/2006/06/19/of-social-networks-and-business-models/

KEY TERMS AND DEFINITIONS

Automatic Location Information (ALI) Applications: They are applications whereby various technologies automatically locate the cell phone (and thus its user) and then use that information to expand.

Bluetooth Marketing: It is Mobile Content Delivery via Bluetooth to any mobile phone. Transmissions can be received by individuals in that location who wish to receive them and have the necessary equipment to do so (www.wimob.com and http://en.wikipedia.org/wiki/Proximity_marketing retrieved 11/16/2008).

FeliCa: It is a multi-functional electronic wallet with contactless electronic IC chips developed by Sony. In combination with NTT DoCoMo's Java-based applications, users can use FeliCa for diverse transactions, such as commuter pass, e-money, membership card, and movie tickets, among others, simply by waving their phone in front of enabled sensors (http://ctl.scu.edu.tw/scutwebpub/website/DocUpload/CourseTeaching/cyc20078391819_1.pdf).

LBA: Location-based advertising is a new form of marketing communication that uses location-tracking technology in mobile networks to target consumers with location-specific advertising on their mobile devices (http://en.wikipedia.org/wiki/Location_based_advertising retrieved 12/4/2008).

LBS: Location-Based Services is a service which uses knowledge provided about where a user is located in order to deliver information relevant to the user's position.

LBS Directory Applications: These applications typically provide information related to a precise location – not just a city or even a zip code – but a street address or intersection.

Micro-Campaign: These are small-scale campaigns which include a certain social group by knowing the features of that group and directly addressing to that group.

Micro-Celebrity: They are the people who are famous and admired in a society as those who affect others in that society.

Mobile Business Card: These are introduction tools which include the name, surname, business place address, phone number, e-mail address and mobile phone number information of people and other important information which can be transferred from a mobile phone to other mobile phones. Due to their structures, business cards have already been produced for mobile use. Here, the difference is the method of transfer from one mobile phone to another device.

Mobile Distribution: It is the distribution systems that begins, and/or ends with a mobile connection.

Mobile Purchasing: Creates the environments where vehicles and individuals can serve to close points when they are mobile.

NFC: Near Field Communication is a short-range high frequency wireless connectivity technology which enables the exchange of data between devices over about a 10 cm (around 4 inches) distance (http://www.acs.com.hk/knowledgebase.php retrieved 5/24/2009).

Permission-Based Marketing: If marketers ask permission before advancing to the next step in the purchasing process, this is defined as permission marketing. Companies ask permission to send advertisements to prospective customers. It is also used by some Internet marketers, email marketers, and telephone marketers. It requires that people first "opt-in" rather than allowing people to "opt-out".

Prepackaged Applications: These are application bundles featuring innovative software products allowing users to take advantage of advanced services.

Pull Marketing: It occurs with "Send the keyword, join the lottery", or "Enter the store, join the lottery" type of messages. In pull marketing, users participate in the campaign via their mobile phones; campaigns are previously announced via conventional channels such as sending a keyword from the package of a product via SMS. As this type of campaigns can be directly linked to sales, they are frequently preferred by brands.

Pull Services: It delivers information directly requested by the user.

Push Marketing: The push method can be defined as reaching individuals' mobile phones via databases of companies. The most common form is delivering the campaign and discount data to users via SMS.

Push Services: It delivers information which is either not or indirectly requested from the user.

Virtual Business Card: Along with a signature, it tends to have the sender's name, their job position, and what company they work for in e-mails. Sometimes, people go an extra step and also add phone numbers, websites, and an extra splash of color. Although color business cards are the trend in the real world, you can actually customize it so that it has its own unique style. You will no longer have to worry about copying someone else's layout. Nevertheless, e-mail signatures are becoming a virtual success (http://ezinearticles.com/?Virtual-Business-Cards-that-Make-a-Difference&id=269400 retrieved 11/16/2008). Other ways to promote companies services through a virtual business card could include message boards. If a person is a member of any professional service, perhaps he or she could also add a signature there. By having a custom business card online, his or her information may attract more people than they think.

Chapter 8
Mobilizing World

The human being benefits from his or her ability to communicate and turn knowledge into action in order to sustain its ability to survive on this planet, the earth. As a result of the fast life conditions imposed on humanity, point-to-point relationships have begun to be established in a faster way and the idea to use technology to acquire and share knowledge has become widespread. Doing the right thing leads to improving and advancing the standard of life. The products of the mind can now be produced easier than ever by scope of technology. Intercommunication between people begins with talking; humans first talk and then express his or her emotions and opinions.

Mobile telephone is the name of the latest technology which creates a worldwide area to talk in. One can easily notice how much the sector and its applications have developed only by looking at the first mobile phone which was launched in the world. Motorola Dyna TAC 8000x is one of these telephones. Its dimensions are 13x1.75x3.5. This is a brick-size device and you had to pay US $3,995 to own it in 1983. In return, what you would get was just a telephone which provides just voice communication and which could be used while moving. This affected concurrently users, families, types of entertainment even health issues and payments. The positive and negative effects have appeared in the evolutionary stage. Like the virtual environment (Han, Kim & Lee, 2005), the mobile environment which is used in order to cover customers' needs for communication, information and entertainment is related to marketing with its different spirals and own sanctions.

In this chapter we would like to give short notices for future researchers about the present conditions of major important topics and some new trends of these subjects.

From its giant size and price since 1983, mobile devices have been circulating rapidly and permanently. At present, the costs paid for these telephones with enhanced technology and reduced dimensions have sharply fallen and almost disposable mobile phones have begun to take their places in the market. These developments showed the people of the world the ease of mobile life, and enabled the appearance of

DOI: 10.4018/978-1-60566-916-8.ch008

new and different demands. The diversification and increase of demands arises a different market and understanding of market based on mobility.

Mobile telephone operating systems, which have turned into one of the investment areas of giant capital, are gradually gaining value. The value of rights won through tender for operation license has reached billions of dollars in Europe. The income provided by those who use the system flows through a different channel and it is collected in certain accounts. These revenues, where many a little makes a nickel, have reached appetizing sizes. The more the number of users (clients) the bigger the income equivalence is which indicates that the sizes of business will reach enormous levels. Although the shares to be obtained from this growth are related with the level of development, they are shared among those who are included in the technological production or hardware/software stages. In this regard, supporting local developments and keeping open the ways to rightfully benefit from sharing may greatly contribute to world peace.

THE MOBILE FAMILY

Telephone communication which was initially realized from city to city and then from country to country has caused new advancements in the daily lives of people. However, at present, in addition to communication with far distances, the requirement of the family and the individual to communicate with the mobile phone has gradually increased interaction via mobile communication and made it more influential. The formation of the mobile family seems as one of the elements that keep the family together. With the establishment of mobile networks, inter-city and international access has become easier which has in turn affected both business life and family life. Mobility, which affects new generation behaviors, enables the creation of certain common family forms other than the nuclear family. Big families come together at certain times of the week or month and create a different togetherness, or groups which come together with certain purposes continuously communicate with each other similar to a family and they are used in developing structures against individualization. Mobile devices provide significant benefits to the mobilization of the individual in the society because mobile communication can be enabled under every condition: home, travel, etc. Possession of mobile phones by everyone regardless of age and gender creates certain flexibilities in life. Leading flexibilities include the monitoring of children and young people such as understanding and identifying their needs just by pushing a button. On the other hand, a continuously carried mobile phone is considered as a reliable tool to follow up children especially by parents, and it helps the security of the family. Knowing who does what, where and when is important in terms of people who are under danger. Nevertheless, their frantically frequent use by curious people due to the monitoring function may result in the consideration of mobile phones as trouble-making instruments.

Family and business issues have started to be more connected and blended to each other. The mobile family could be created due to the new protecting systems both in the house and on the members of the house. But family concerns and intercommunications have become different. The value of the mobile phone has increased further for the family as it also provides a kind of interaction especially for family members who have long travel programs for business or for introverted people. Therefore, phone users and their tendencies affect general and special applications.

MOBILE PHONE USERS

The mobile phone is among the fastest communication instruments, but it has gained meaning beyond communication in our lives and turned into a platform that fulfills the functions of all devices which deliver activities such as informing, entertaining, measuring and evaluating. At present, the most important purpose of the mobile phone is to reach others either through texting or calling. Although the mobile phone is used to cover needs in general (to talk and communicate), it also addresses the different purposes of use; for example, the chronometer, clock and agenda, photo camera, gaming, entertainment, Internet, map, telephone index, voice recorder, remote control, music, radio, notebook, communication lists, etc.

The habit of using mobile phones has become widespread for almost every age. Sometimes it is even given to babies like a toy. Children who have not even learned how to read and write turn on the mobile phone and play various games while they can also change the melody of the phone or use many functions of the phone.

From a customers' point of view, services provided to customers in parallel to both the extra features of mobile phones and to its ability to build communication are important. People are now paying attention whether the mobile phone has the features to facilitate their daily lives, to add entertainment to their times, and to provide information to them. Individuals may, for the time being, prefer to use certain functions available on mobile phones with other tools. Most of the respondents benefit from the speaking and messaging features of mobile phones rather than their large-scale features. Looking at utilization areas, most people can actually be satisfied with phones which can call and message to a desired number, or mostly have music broadcast, clock and some other small functions.

Users may also build an emotional connection and a deep relationship with these instruments which are also considered as close friends by users (Wehmeyer, 2007; Jarvenpaa & Lang 2005).

In personal communication; sight, hearing, touching, tasting and smelling senses are used first. Speaking allows expressing the receptions received from all senses. Therefore, speaking has a comprehensive feature which is used in transferring any kind of perceptions. Speaking, written, image or video messaging can easily be made by using mobile phones. The feature of increasing the speed of communication of campaigns that are realized by service providers should not be ignored. The young segment, which covers an important place among mobile phone users, also pays attention to the cost; therefore, they prefer messaging in general. Options and features such as listening to music, watching cinema, and playing games are among the activities which can be carried out during a journey even while walking.

A study oriented towards mobile phone use was made in 2007 in Istanbul, Turkey as a face to face questionnaire with 1370 mobile phone users (Yamamoto, 2007). 32.4% of the mobile phone users also use mobile phone for other purposes than mobile communication. This indicates that the non-communication features of new telephone production influence the consumer. In this research on the mobile phone, although speaking and SMS took the largest share, those who use it only for talking and texting account for 49%. Mobile phone is emphasized not as a toy but as a necessity. On the other hand, it is normal to see that the usage ratios are separated among different features because everyone has different requirements and hobbies with regard to other features of mobile phones than talking and texting. Features such as sending SMS, listening to music, games, Internet, and chat are used by people to fill in their free times and to meet their entertainment needs that have remained in their personal preferences.

Also, as the high portion of talking is followed by SMS and Bluetooth, it can be considered as a proof that this service and technology has been settled in our country.

In Turkey, young people benefit from the speaking features of the mobile phone by 27%. This supports the opinion that the function related with communication changes and differentiates towards more economic channels. The reason of the fact that the extra features of mobile phone cannot yet hold an effective portion is probably because these activities remain outside of the actual purpose of the mobile phone.

The ratio of donations through mobile phone appears low. It can be due to not finding this method reliable, or that people are unaware of such a use.

When a customer selects a mobile phone service provider, the intra-network and inter-network costs are considered very important. The importance attached to cost is 41.5%. Service providers that offer lower cost lines are mostly preferred because frequently speaking customers make use of different price lists and advantages. Prevalence is also considered as a very important aspect in selecting a mobile phone service provider (by 36.7%) because the more the customers use the selected GSM operator, the higher the benefit will be (especially through discount campaigns). The ratio of being influenced by the price when buying a mobile phone appears very high. High prices can lead the buyer to think that the telephone is nice, because of associating quality with price. People are influenced by the price of the mobile phone, and sometimes, the higher price makes the mobile phone more attractive.

Regardless of the income level, people are willing to make mobile phone calls at a lower price in a wider coverage area with uninterrupted access. Other features of mobile phone are considered as secondary. The phone that has the price that is the most affordable and the features sufficient to cover a user's needs is considered as the ideal phone. When replacing their telephones, people generally tend to wait for campaign periods thinking that the price would decrease in time rather than immediately buy at very high prices. The costs of mobile phone calls are calculated according to the unit rates identified by operators, and they are collected from the user under various names such as credit cost or calling invoice cost. Most people are concerned about high cost request and change their operators if an opportunity to adopt a low-cost network is provided.

It is not possible to say that mobile phone users attach importance to the feature of upgrading to a higher model apart from those innovators who are willing to create prestige among mobile phone users by renewing their telephone models. In fact, most people do not have concerns about owning new mobile phone models. Those who think that they cannot keep pace with the speed of the technology by upgrading to a higher model tend to replace their mobile phones when they run out of order. People that directly support the aspect of upgrading to a higher model can achieve their goals only when they reach a stimulating scale.

Some mobile phone users may replace their mobile phones because they are worn out due to very frequent use or due to insufficient power-up time or the lack of functions they require, or due to the quick breaking down of mobile phones. These are also opportunities for them to upgrade their mobile phones when they consider buying a model with more advanced features.

GSM operators with very wide coverage areas are taken into consideration first. Actually, if really serious disconnections occur during communication, then there would be no point of using these devices. However, 3G technology and further developments will be directed towards rapidly eliminating these problems on communication networks.

From the customer's point of view, lines are important in terms of delivering service. From the mobile phone users' point of view, the function of a line is more important than the brand name. The brand name gains importance when it is identified with functionality. Brand, which is considered as one of the most important value increasing elements in certain societies, is not given a prominent status here.

People have more pragmatic opinions about using second hand or lower model phones. They can also change their GSM operators depending on the discount packages. The use of a present GSM operator by everyone in the family is mostly influenced by advantages which may be obtained through the same line. The same line owners would not think of changing their operators so long as they talk at a lower cost and pay less money. The free text messaging or bonus credits on the same operators are applied as solutions to reduce the cost.

Many brands introduce more beautiful and stylish models in order to make use of aesthetical elements. The thing the mobile phone user who is located on a different scale wishes is the personal identification with the mobile phone after buying it rather than the image of the mobile phone during its entry into the market. Here, personal identification means finding the device which represents the features and elements the consumer likes, and anything that reflects on the consumer.

From an individuals' point of view, the existence of differences arising from style and aesthetics is natural. After the elements based on financial opportunities are excluded, the ratio of being influenced by style increases. The reason can be based on people's devotion to design, elegance, and etc. During relationships with outsiders, physical appearance is always paid attention by everyone. The stylish and affordably priced telephones are considered as ideal telephones. Appearance-related elements can be shaped according to the personal tastes of people such as color, voice, voice-related toning, bearing apparatuses etc. Individuals emphasize the importance of having an appearance in accordance with their personalities.

As a high level of being influenced from aesthetics is considered as the expression of different tastes, it is considered to be reflected by the personalities of people. The rapid development and decreasing cost of technology gradually moves mobile phones away from the symbols of certain statuses, and requires other additives to raise status. Mobile phone producers are probably aware of that as they are spending efforts to develop more styles, artistically-designed smaller mobile phones. In time, with the inclusion of jewels into mobile phones, raising the prestige with really high-valued telephones has become possible. The presentation of highly priced telephones to groups which do not abstain from paying high prices depending on admiration values will contribute to marketing.

Groups that still remain novitiate are not influenced by the data download speed. This is probably because they are not familiar with the technology or because these technologies have not reached them. Groups that have just begun to use mobile phone perceive the interaction mostly by talking and texting. Mobile phones featuring new functions and other services of operators are not widely used yet, and consumers refrain from any additional costs that might incur by given services.

Those who use mobile phones for games are mostly very young and young users. In addition to its innovations, mobile phones have also reflections in our lives with their social influences.

Mainly the following factors come forth when using mobile phones:

- Ability of use
- Expansion date and size of mobile phone
- Usage frequency and continuity

Mobile phone advertisements and promotions also emphasize these features. Besides, instead of the features which cannot be caught up with technologically, mobile phone producers are now trying to emphasize the allure of the mobile phone.

Mobile phones have become a piece of our lives and have some meaningful things of life such as entertainment, games and education which also change with these mobile devices.

MOBILE ENTERTAINMENT (MOBILE MULTIMEDIA SERVICES)

As mentioned in our previous sections on changes and customers, mobile phones have obviously changed our lifestyles in the last ten years. In this sense, the structure of entertainment, which has an important place in our life, is also changing and evolving. Mobile entertainment is a new concept which mixes entertainment, education and telecommunication industries together and they can be considered as entertainment-oriented activities with the development of personal technology while moving or at any geographical location (http://en.wikipedia.org/wiki/Mobile_Entertainment#cite_note-6 retrieved 12/12/2008). From this perspective, targeting the correct people among the users of mobile technologies is essential from the mobile technology producers' point of view. Especially creating a rich environment in terms of targeting methods in the entertainment sector, mobile instruments have the abilities to employ many entertainment-oriented functions and features. From the point of view of entrepreneurs in the entertainment sector, this will imply making better use of the mobile environment and the creation of projects oriented towards entertaining customers (Yamamoto, 2008). Really, when entertainment is moved to the mobile environment, it really is activated with certain new methods such as the separation of a whole into pieces or the reformation of the whole (Shiffman, 2008, p.7). Many things from the buzz spread through mobile phone to mobile games which can be included in mobile environment. Actually, coming together to play games in the social network environment is also included in this concept. Mobile entertainment can be separated into various categories in terms of individuality, age, number of users, and simulation:

- Individual entertainments which are directly realized via mobile phone (mobile phone games, trivia, etc.).
- Entertainments where the mobile phone is used as an instrument and which bring together members of the social network (social network games, dating etc.).
- Entertainments where the media is employed by many users and entertainments which provide feedback as mobile (contests and games, gambling, trivia, etc.).
- Information and entertainment provided by replacing other media instruments with mobile phones (TV, radio, music, video, web site, sports news, puzzle, sudoku etc.)
- Games and entertainments created by the mobile users themselves.

The services given via mobile phone can actually be classified as personal entertainment services or bulk entertainment services within themselves.

Mobile phones can be defined as a mass media instrument which is created with general individual participation. Most mobile phones feature TV. The number of mobile phone subscribers is much higher than that of Internet subscribers. 31% of the global expenditure consists of delivering music on a mobile phone basis, and the number of mobile phones in the world is nearly five times higher than that of automobiles. Video games account for 20% of the software income obtained from mobile phone. Advertising, news, TV, Internet, social networks, even Hollywood movies and new publications also take their share from this expansion (http://www.amazon.com/Mobile-Advertising-Supercharge-Exploding-Wireless/dp/0470185988).

According to Juniper Research, mobile entertainment products and services are expected to reach 64.8 billion dollars in 2012. The entertainment services of Juniper Research are divided into seven categories:

- Mobile gambling
- Mobile adult services
- Mobile games
- Mobile music
- Mobile TV
- Mobile video
- Mobile user-creative content
- Mobile infotainment

Here, the biggest growth is expected in mobile gambling which could climb to 3.2 billion dollars. Also, social networking services are expected to reach 5.7 billion dollars. Ringtones and wall papers, which have turned into conventional services, are considered as bulk mobile services. However, the methods of use change rapidly and create a more flexible and complex mobile entertainment environment where various elements such as multi-player games, social network services or videos can be used together. (http://www.mobileeurope.co.uk/news_wire/113709/Mobile_music%2C_games_and_TV_to_generate_%2434_billion_by_2010%2C_claims_Juniper_Research_.html 12/12/2008).

Mobile websites (WAP sites) can also be shown as another way of reaching mobile entertainment. Websites, on the one hand, present entities of brands, and on the other hand, they can be considered as a separate entertainment channel for users (4INFO, 2008). Mobile entertainment websites serve for delivering news or information rather than text messaging. These channels can diversify and suggest their own entertainment values through the creation of a geographical branding or a clear benefit for the customer.

At present, mobile video still constitutes a small market. However, if such types of structures are conducted with the social sharing logic, then they will result in a brand new and different type of creative consumer group similar to the previously-mentioned micro-celebrities. People like the efforts to show many things of entertainment, and humor, and they also aim to influence each other through physical appearances. This type of approaches may also cause the development of a buzz-type of new-age advertising. On the other hand, rapid sharing of movies which are not shown only by professionals but also by amateurs as adult entertainment is also possible. The viral impact in the entertainment media can rapidly spread around either in a positive or negative manner. Negative links may grow and lead to social problems. Therefore, the issues of revising the moral and legal aspects of mobile entertainment, and increasing them to the norms which do not exceed legal dimensions will keep people busy.

According to Merisavo et al, in the study of Bauer et al. on 1028 people, the entertainment and informative values are considered the strongest impulses in the adoption of a mobile advertisement. Mobile advertisements with such high entertainment and informative elements are approached positively and it is thought that they can be used in order to increase behavioral tendency to use mobile services. In the research of Pura (2005), it was identified that the organization of location-based SMS services accord-

Table 1. Mobile phone services and delivery methods (Source: Developed from Mobile Sport & Leisure Essentials (3rd ed.) White Paper Juniper Research http://juniperresearch.com/shop/viewwhitepaper.php?id=79&whitepaper=38 *.)*

Content/Service	Delivery Methods
General news	Text messages and warnings
Financial news	Picture messages and warnings
Telephone guides and other guides	Voice messages and warnings, sound clips
Weather forecast and travel information	Video clip download
Traffic data	Streaming video, watch while ringing videos
Astrology, signs	Wall paper graphics
Comics, cartoons	Voting
Sports services	Chat
Monitoring Services	Location
Navigation services	Betting and games websites
Celebrities related group services	
TV, movie related group services	
Social services	
Games	
Gambling	
Other services	

ing to environmental conditions was convincing and with financial value highly influential in mobile service use.

Television, which has been added among the functions of mobile phone, gradually turns into a very important issue. Actually, these devices have been tried in Japan only for approximately two years and certain standards are required to be created in the world. On the other hand, users of mobile devices are very keen on downloading music, image and video. These aspects are especially drawing the attention of the young generation (www.connect.de in http://mobmarketing.wordpress.com/ 9/5/2008).

According to Patrick Parodi, the global president of Mobile Entertainment Forum (MEF), in mobile entertainment, there is a tendency from the "producer-push" oriented approach to the "user-pull" oriented approach in order to create a mobile entertainment value chain with all players. Also it indicates that the mobile entertainment venture, which is funded with advertising, would revive not only mobile entrepreneurship but also many other contributing fields.

Even pranks on mobile phone are now offered by mobile service providers. For example, if someone wants to make a prank to his or her friend by changing his or her voice, then he or she can call the related number of the service provider who gives this service, and select the voice of a child, woman or man and speaking with the desired voice tone is enabled (Sabah, 2008). Such ventures show how much the entertainment sector has developed and differentiated. Also voices of secret witnesses are modified in order to prevent their identification in courts and on TV.

The content created by the user gradually gains importance in the mobile world. Owners of mobile contents may be a cinema, or music studio, a publisher or a newspaper. These organizations or institutions can own the title of the content. In order to present private content on mobile devices, the consent of their titleholders is required (StrategyAnalytics, 2008).

Mobile phone first brings into mind the following general groups and content as well as delivery methods. Delivery methods and contents are different according to the universality or specialty of the content and service (See Table 1).

Mobile Music

As a result of technological developments, the use of music exceeds beyond accustomed channels, and a certain new music broadcasting benefitting from the Internet, GSM networks and digital media players is observed (http://www.mu-yap.org/getdata.asp?PID=291 12/13/2008). Mobile music has a feature based on any type of music consumed through the use of the mobile environment. This music may be melodies or tunes either directly stored in the software of the mobile device, supplied from the Internet via mobile operators or sent from personal mobile device to personal mobile device.

With personal or local value, these pieces of music may also create a value while they are distributed via mobile line. The use of the music environment on the mobile system turns into an important aspect of marketing. Increasing the frequency of use, publication rights, legal sanctions, etc. are the issues that are required to be solved within music marketing.

According to Takeishi and Lee, owing to technological innovations in music creation, production, recording, playing, distribution, promotion and advertising, the structure of the music industry is also changing and new population music styles as well as ways of performance and methods of use are forming (2006).

At present, a piece which is compiled, arranged or composed by an artist can be marketed as tens and even hundreds of different products. At present, a multi-optional feature arises for individuals for building communication between them and the artists or works of artists they like. If they desire, customers may buy a piece of music and download it from the mobile environment as they are presented, they can obtain music work from the mobile system as a CD, put pictures of their favorite artist or album as a wall paper to his or her mobile phone, use favorite melodies as ringtones, watch them as music videos, be a friend with other fans in a social network, or make a subscription or send them as a gift. In short, many customers can obtain these different products and services from different platforms and use them in different ways (http://www.ifpi.org/content/library/dmr2008.pdf 12/13/2008). Therefore, the mobile

Table 2.Global digital revenues of industry (Source: IFPI, PWC Global Entertainment and Media Outlook.)

Global Digital Revenues of Industry (2007)	
Industry	Digital %
Recorded Music	15
Newspapers	7*
Films	3
Books	2

*US only

communication environment is an environment that enables acquiring music and being involved in social environments which are formed around music.

There are certain important issues which put a brake on the digital music industry. For example, the understanding and ability to serve other systems with systems in the music sector or receiving services from them in order to provide common active operations of exchanged services has not sufficiently developed yet. Digital rights management (DRM) standards of different service and equipment providers are different. According to DRM standards, registered software can be used only with leading equipment brands. Also, marketing investment is required for mobile music services as a new style. The related offerings to the customer in the mobile sector have not fully appeared yet. Opportunism or piracy, which aims to exploit work and obtain unjust earnings by benefitting from the broadcasting area is in question (http://www.ifpi.org/content/library/dmr2008.pdf 12/13/2008). Researches indicate that excluding games the most developed area among other creative performances in the entertainment industry in terms of digital income earning is mobile music. In the global digital income spread, registered music accounts for 15%, which is almost twice of newspapers (7%). Movies and books respectively account for 3% and 2% (Table 2).

Mobile music has significant marketing potential. At present, the most popular music format is master tones, which present a fragment from the original music. In 2007, the sales of complete song downloads have increased. Compared to the same period of 2006, its sales increased from 6% to 12%. The share is expected to increase further according to the created images or melodies of favorite artists as a mobile personalization component demanded by the customer.

Meanwhile, mobile music faces certain challenges. Especially a limited number of marketing activities, preventive data responsibility and penalties, gaps and weaknesses of customer-oriented presentations, low penetration in developed instruments, the factors related with the transformation between mobile phone generations are considered as some of the barriers that affect the mobile music market.

Various issues such as the quality of service provided by the network of mobile music applications, ability to roam around various networks and the management of location are gaining importance. Also pricing, protection of legal rights, matching the customer's requests with the content provided in the mobile environment, and overcoming synchronization problems between the user's equipment and the system are among the other important issues (see Table 3).

Having taken its potential place among new mobile business modes which have appeared after 2007, mobile music addresses to potential expansions from a marketing point of view. Among them, various models can be mentioned such as a la carte download music services, subscription services, and advertising supported music services (see Table 4).

Table 3. Mobile music applications

Mobile Applications	Importance of Networking Requirements	Certain Points
Mobile Music/Music on-demand	**Quality of Service** (bandwidth requirements, delay and delay variation, tolerable loss ratio) **Roaming Across Multiple Networks** (continued connectivity, network availability, wider wireless coverage) **Location Management** (location tracking for determining the location of an object, location database for storing and location information)	• Pricing issues (on demand, subscription, free with mobile advertising, etc) • Legal issues (copyrights & undefined copyright owners) • Matching contents to user preferences • User synchronization

Table 4. Emerging business models for mobile music distribution

A-la-carte download services
Subscription services
Consumers offered more non-DRM download services
Advertising-supported services
Record labels go direct to fans
Brand partnerships

A-la-carte download services appear as the most prominent among these models. As an example iTunes has become the third largest music retailer of the US, catching up with the sales of Amazon and Target. Subscription services of companies such as Napster are also growing because iTunes provides this service only for the devices it has defined while it fails to provide services for other devices. This limitation causes the rise of certain new services offered by mobile operators and ISP's. Certain problems occur because inter-operators operation is not possible in the music industry and certain major record companies and major technology developers have not joined or are reluctant to join it, and the possibility of offering download services that remain outside of Digital Rights Management can be in question. Advertising-supported models suggest various methods to promote new pieces of music, and they work in parallel to social networks. Record companies sign license agreements with social networks such as MySpace, YouTube and Imeem and obtain the right to broadcast their music videos via those channels and get stake from advertising revenues.

A fan can send listened music or watched video to other people or groups online at the same time. The number of sharing increases by multiples and sometimes, an artist can reach thousands of fans in only a couple of seconds (Scott G, 2005). This can be used equally in the same manner with mobile devices after 3G, and the ability to share music via different chains will be benefitted.

Records with new music videos can directly be sent by through the web sites of artists to their fans. Record companies among music producers make use of the artist and brand cooperation opportunities through the pieces of music sent via mobile system.

Here, the purpose is to highlight those music interactions in mobile environment and mobile music, which uses portable technology. This should be perceived as a new domain. Also, mobile devices, which now mean more than just being portable devices, create a chance to evaluate personal performances by covering issues such as music production, sharing and mixing, and those who share temporary music, create new user groups in P2P and distributed network structures. People and groups that benefit from this system are able to distribute oral, musical or written elements, which they have obtained from strangers in the same location, to their friends at different locations, and perform distant actions without requiring central supervision.

The conversion of electronic devices into miniature sizes and high capacity warehousing capability develop the ability to keep music collection at a single place, and mobile phones with gradually enhanced MP3 features turn into musical platforms that fulfill the purposes of other devices with these features. Mobile telephones facilitate music transfer as they are ready for network communication anytime.

There are two different opinions about mobile music content share.

Figure 1. Basic structure of mobile music

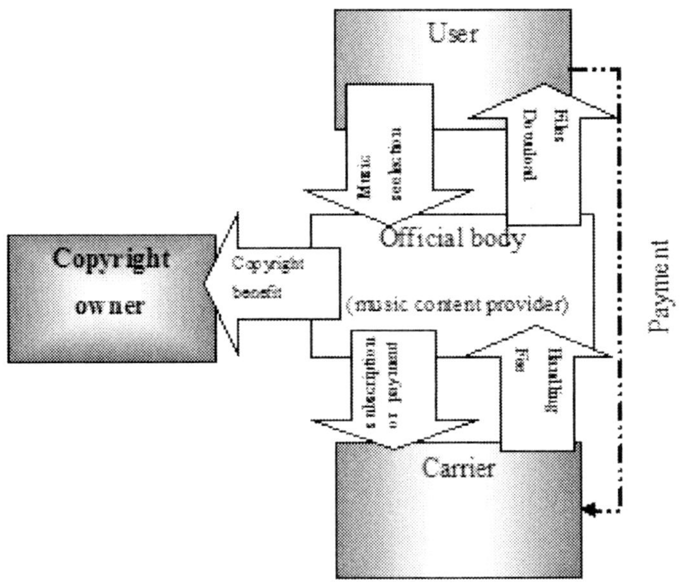

- The first one is free of charge through using the same features of free downloading and finding everything, same as the primary applications of Internet.
- The other opinion is valid use in return of the fee paid for copyrights.

However, the system is currently unclear regarding the payment of value for music to be used. Figure 1 simply explains the copyright payment system with a scheme. Here, the users download mobile music in return for free, and the copyright benefit is paid to the copyright owner via an official body.

According to Schmidt (2008), information products are intangible products distributed via digital forms and electronic networks. Free rider phenomena and "privacy" will lead music or art work creators and distributors to develop copy protection or digital rights' management systems, and the existing service will be converted into a special and physical product.

Sports and Mobile Applications

Sports have been a very important event of humans since ancient times. It is good for body and mental health. Even when not doing it, watching or being a fan are important circumstances for many people's lives. Therefore, while mobile phones are penetrating our lives, the impact on sports should not be forgotten. Sports draw the attention of many people in the world, and news related with sports or issues such as subscribing to sports clubs or supporting a team or a sportsman constitute an important part of our lives. Although people do not deal with the sports themselves, they may watch them, follow up and provide material or moral support in order to fulfill their motives as a supporter.

According to Juniper Research (2006), the total share of mobile sports in the entertainment and information market (content and messaging) activities such as SMS, MMS, chat, wallpaper and image, video clip and streaming video services is approximately 4.17 billion dollars. The figure is forecasted

Table 5. Sports related contents and services

Content/Service	Delivery Methods
• Match results, • Goals, • Score tables, • Injuries, • Transfers • All interesting news, • Striking comments from professional columnists, • Live Narration, • Goal or latest minute videos, • Games such as Olympic race, FIFA, Pro-evolution Soccer, • Quizzes and games related with football or any other game that receives high interest at the location, • Quizzes related with sports, • Other.	Text messages and reminders
	Picture messages and reminders
	Voice messages and reminders, voice clip, live narration
	Video download
	Streaming videos
	Wall papers, graphics
	Voting
	Chat
	Location

to reach 9.5 billion dollars in 2001. However, the mentioned figure should be considered as an average of the total expenditure. Mobile sports are one of the fastest growing sectors in the mobile industry. 1 billion dollars of assets of sports content and services in 2006 are expected to reach 3.9 billion dollars in 2011 after the expansion of 3G technologies to all regions and enrichment of mobile sports content with an increasing number of licenses. This estimation suggests that sports content and services will increase by 41% by 2011.

Table 5 indicates sports related content, entertainment products and certain methods of delivery in mobile content. These types of services, in addition to the information and results related with match scores, goals, score tables, players' conditions, and comments, information and performances requested by a more different mass who is interested in contests and games can also be included within communication services. The ability to play sports-related games on mobile phones individually or in groups is also in accordance with considering mobile phone as a different platform for betting.

Leading barriers related with mobile sports services are the users' lack of experience, localization and optimization, and limitations related with digital rights. If a visual presentation is realized, then other challenges such as delays, network or device limitations can also be in question (http://www.juniper-research.com/shop/viewreport.php?id=79 retrieved 11/9/2008).

Sports mobile presentations are mostly those which are prepared for the supporters of football teams. Clubs who wish to attract supporters' interest and sponsors come together and benefit from mobile communication. For example, in the USA National Football League, the New Orleans Saints have seriously benefitted from mobile efforts, which brought great efforts for the team with the support of its fans. A sponsor, a mobile service provider and a TV channel were also involved in this practice in order to create a mobile database. The created database facilitated identifying the potential to watch matches on the sports channel. The supporters were asked to send the name of their most favorite player as text messages, and so, the material support of the supporters was obtained. (http://www.sportsline.com/wireless/alerts 5/27/2008 and http://www.mobilemarketer.com/cms/news/messaging/1573.print)

Mobile sports is not only limited to football, but it has the potential to grow and expand with software to serve and satisfy people on issues related with baseball, basketball, golf, tennis, car races and other sports branches.

Mobile Game

Digital games, which have entered our lives with Internet and computer technologies, can now be offered via mobile instruments. Games were considered as the leader of services provided to people, who want to spend time by using the mobile phone, and they are becoming more diversified and the mobile game industry is gradually developing. A very significant piece of online time is spent on gaming. This trend will continue to increase in 2009 with new games entering our lives. With the inclusion and joint use of more different brands and offline channels, users are expected to become more involved. Game fictions that include social network features (comparison of the results by the users themselves, competing with friends, etc.) may attract a higher interest (http://www.antifit.com/?p=1455).

Video games which can be played on mobile phones and which are produced by some famous brands are developed in a manner to address those who demand different styles of games to vary in different ways with single players, multi-players or teamwork. Mobile game modules have various payment options such as pay for the download, free game, and pay for single game. In terms of subscription, games are separated as free subscription and advertising-supported.

Those who develop mobile games make plans mostly according to the following features and design of different games:

- Technology games, (games embedded to telephone or device, SMS/MMS games, demo games, J2ME games, BREW games, independent Operating System (OS) games etc.)
- Number of players (Multiplayer/Single player)
- Style (excitement, logic/puzzle/skill games, sports and race games, arcade, role playing games, card and casino games, FIL games, adult games, lifestyle games, etc.)

Mobile games are among the most popular and most consumed mobile services.

They are developed by considering the systems of the games that are established in mobile instruments. Therefore, the content may be different in terms of size and complexity. For example the Nintendo's GameBoy, Sony's PSP, Nintendo's DS are game instruments based on a denser wireless and media utilization capacity and Internet. Although the programming of Internet and media communication devices is difficult, it is possible to make use of a stronger media interaction in activities such as games, entertainment, sports and music, etc. (Quinn, 2008).

According to Finn (2005), if a generalization is to be made, then mobile games can be separated as conventional computer and console games transferred to mobile platforms and the games that employ SMS. SMS games which turn into an interactive channel to respond to TV shows and contests can be used in many situations and areas. The main evidential point of these games is the turning of the mobile phone into a game and entertainment instrument. Therefore, mobile games can be defined as those which people can benefit from or participate in willing to play the game at any location. However, obviously, there are certain differences between the behaviors of mobile players and classical players (http://www.mobile-weblog.com/50226711/changing_mobile_gaming_habits.php).

In a research conducted by Nielsen Entertainment for Nokia in six countries and on 1800 people, 80% of the mobile phone game players play a minimum of once a week, while 34% play every day. The average daily gaming time is identified as 28 minutes; the average time is 39 minutes in India, 31 minutes in the US, and 29 minutes in Thailand. 61% of mobile gamers stated that they play it when they are mobile and 62% when they are at home. The ratio of mobile phone players who stated that they play mobile games instead of waiting in their free times is 56%.

According to the study, people pay more attention for new mobile games to cover their game requirements. Nokia multimedia game director Jaakko Kaidesoja confirms and expresses that users expect better graphics, better content and better games.

According to the research of M:Metrics about gaming and game downloading by consumers in the US and in Europe, due to the reasons such as pricing preference limits and being uninterested, mobile phone users do not buy any other games after buying one. The study -made in 2006 - indicates that among the registered mobile phone users, those who downloaded games is 4.2% in the UK, 2.5% in Germany and 2.7% in the US. Buying games is directly related with mobile phone shopping, and there is no regularity associated with the selling stages. It has been identified that 30% of game downloaders are first time downloaders, and the purchasing ratios of new registrars has been identified as low, and the ratio of repeated downloads do not exceed 20-30%.

M: Metrics senior analyst Seamus McAteer indicate that there is no restrictions regarding the addition of new games to operator portals, and these data indicate that operators should spend more effort in order to download a mobile game. In the same research, it is stated that the ratio of those who played mobile games in March 2006 among the total registered mobile phone users is 23% in German, 32% in the UK and 24% in the US (http://www.mobile-weblog.com/50226711/research_reveals_mobile_game_use_trends.php).

ADULT CONTENT

When divided according to ages, it is quite natural to have differences between the contents of children and young people, and that of adults. There are certain differences between information and culture games that refer to every age and adult content. According to Juniper Research, the global market for adult content is very big. This figure includes books, magazines, strip clubs, toys, movies, DVD, telephone and online content that totaled approximately 100 billion dollars in 2006. Game-type materials with audio-visual features are considered as the most popular and profitable products. The study highlights that the media where these games are presented is used in making the mobile device or communication channel more popular with mobile content or is considered as the main income generating element.

Mobile Adult Content Product Types

They can be separated as purchasable contents and natural contents as a factor in preferring certain mobile devices and features. However, the content obtained from both types can be classified as follows in general.

- Text based adult content (printed media and publication products, word guess and mathematical operation games, encyclopedic information, etc.)
- Visuals (Photographs, drawings, holograms, clocks, calendars, map schemes, diaries, telephone books, daybooks, etc.)
- Sounds (Moantones)
- Erotic games for adults

- Video-based services (transfer of short erotic movies, etc. videos and images)
- Intermediation services in purchasing (provides connection with organizations or websites which provide the above-mentioned services).

Juniper Research estimates that the above-stated figures would increase to 2.2 billion dollars in 2008 and to 4.9 billion dollars in 2013.

Certain new legal regulations about adult content are being prepared in many countries. Certain private conditions are required to provide adult content in the UK. Such conditions were evaluated within the 2006 Independent Mobile Classification Structure and the major standards were identified as Independent Mobile Classification Body [IMCB] rules. Applications are allowed only if they are in accordance with these standards. The main element of these rules can be summarized as the classification of 18 plus commercial content without damaging the independent classification structure (http://www.dma.org.uk/_Attachments/Resources/1897_S4.pdf).

The following services are those included within the framework of IMCB classification.

- Pictures,
- Video and audio materials,
- Mobile games which also include Java games,

Certain services have also been included in this group in 2006:

- Premium-paid text, audio and only audio services formerly organized by ICSTIS (regulated phone-paid services in the UK) phonepatplus.org.
- Gambling services (these services are subject to UK laws).
- Moderated and non-moderated chat rooms (non-commercially moderated chat rooms if they are for 18+).
- Location Based Services
- Contents developed by members such as web logs.
- Contents accessed via Internet or WAP where the connection is provided by mobile operator only

The following content is allowed for contents related with the 18 plus age group. Bad speeches which repeatedly and frequently include the issues of drinking or gambling type of activities that are illegal for non-adults, real or real-like sexual activities, nudity and drawings containing violence, illegal drug taking, horror content or content which encourages activities containing violence such as head butt or pulling ear.

The expected thing from mobile operators in this regard is to identify the age range of users and prevent the delivery of the foregoing content to those who are below 18, and to prevent the delivery of subscription-requiring content to everyone.

The market of adult services with a 1.4 billion dollars income in 2006 is expected to grow slightly over 2.7 billion dollars in 2009 and reach 3.3 billion dollars in 2011. Also, it is possible to say that the most profitable market in terms of mobile adult services is Europe which can afford high prices followed by the Asia-Pacific market. In the 2006-2011 period the income of mobile adult entertainment services is estimated at 14.5 billion dollars, Europe accounting for 39%, and the Asia Pacific for 33% (Gibson, 2006).

It is expressed that erotic games and entertainment constitute a major part of the adult entertainment sector. It is stated that the primary impulses that affect demand in this market are very stable. At present, the mobile channel is regarded as the last considered distribution channel in this respect; however, after the development of monitoring abilities with transfer to 3G, development opportunities in this regard will become more concrete. 3G has increased the abilities to transfer images and graphics through mobile video. The most intriguing and interesting subjects will be distributed in this way. These services, which initially entered mobile communication as test services and evolved in time, have now left their places to high-quality drawings and video content. Mobile adult video services are expected to account for 50% of the global market, reaching 1.7 billion dollars in 2009 and 2.4 billion dollars in 2011. The ratio of income from video based services is predicted to be 70% of the entire content income in 2011 (Gibson, 2006). In the same research, it is indicated that previous adult content modules have been successful in fields such as media, press, cinema, DVD, PPV (Payper review) and TV, and this channel would also be at least as profitable as the above-mentioned fields of the industry.

Furthermore, as mobile channel transfers adult content in a more different and personalized way than other classical channels, there is an expectation that a wider audience can be accessed via this channel. After gambling is allowed via mobile basing on an unlimited time and money spending, gambling will provide a much more different adult content than that is offered from current channels. This will require the presentation of the adult content in a more different way. Mobile is entertaining and it can satisfy immediately. The identification of up- to-date soft games that are openly presented to the public, the identification of the number of participants and bonus amounts, and with its immediate announcement, it is thought that mobile games will gradually become more effective.

It is possible to say that the development of the market in terms of mobile adult content would result in many legal issues and social sensitivities. Illegal attitude and practices are the most important factors which would restrict the growth of the mobile adult market.

With the inclusion of mobile-content related similar discussions in the academic area and popular media, and discussion of its impacts on society, mobile forms will remain on the agenda all the time, and consumers will always tend to receive games and services in mobile form. The first 3G operator of Australia became the first Australian intermediary company to present Playboy content and target the access market in 2004. On the other hand, with Three, in addition to the services to see playboy girls that have existed since 1960, the cybergirl of the month and playboy videos have also been broadcasted since 2001 (Three 2004).

With a similar approach, many software developers are working and trying to take place in the market with different products that present adult content for mobile computers (Manjoo, 2001). As they are adult-focused, they include explicit XXX rated pornography games, puzzles and videos. Nevertheless such products will be turned into more influential products only after 3G, when full motion videos can be played on high resolution displays of mobile devices.

MOBILE GAMBLING

Mobile gambling is another entertainment alternative for mobile users. The varieties of Mobile Gambling are generalized under three headings.

1. Casino games
2. Lotteries
3. Bets
4. Betting exchanges

Gambling in conventional format is gambling played in special format casinos and they are the leading of the brightest forms among gambling games. The most important point in terms of mobile communication is the adaptability of all mentioned forms of gambling online and to the mobile environment. As these games rely on fine graphical designs and challenging and high speed connections, they can be expected to become more popular with 3G networks and advanced devices. The awarded talent games market is also growing with the support of new technologies, and games with prizes are also increasing. Such games with awards create a new massive market which is adopted by wide masses.

Lotteries are among the most commonly used commercial practices of gambling. Such games have been traditionalized and they mean some sort of harmless entertainment rather than gambling for most people.

This is also related with how easy the users play the game. Mobile gambling presentations can be based on the idea of regarding these people as the ideal candidates. Sports bets are a type of gambling and they are one of the earliest forms transferred to mobile platform, where the customers could easily adapt into. Here, playing the game while watching any sports contests or at any desired time, place or situation is the thing that attracts customers.

Therefore, the leader of present-time gambling activities is sports bets. They are performed by well-known betting brands on mobile sports pages, which expand their customer portfolio through new channels on the mobile system. Due to its structure, mobile channel both has certain advantages and disadvantages regarding the delivery of mobile gambling services. They will also shape the future services and distribution speed of those services. The constraints and limitations on mobile gambling are indicated in Table 6. The constraints and limitations are almost identical to those of mobile adult content.

The increase and speed of the expansion of mobile phone use by individuals is encouraged by various factors such technological developments and personal trends and the increase in spendable income while a balance is tried to be maintained with its restriction through various factors such as legal regulations, laws, individual and corporate conservatism.

It is foreseen that the income created in the European and Asian Pacific markets in 2007 and 2008 and totaling around 5 billion dollars are forecasted to grow at a greater pace and almost double in 2009. Meanwhile, the total market in the world has increased more than twenty times and the income which was 1.91 billion dollars in 2006 is expected to reach 23.19 billion dollars in 2011 (see Table 7).

MOBILE HEALTH

Mobile health is a very broad subject and several books on mobile health can be written from different perspectives in general. Mobile health is another important subject for marketers. Mobile health should be considered from two aspects. The first one is the innovations brought by mobile on health practices (practitioner/user), and the second one is the effect of mobile devices on human health. Both are subjects that should be considered and discussed in detail. M-health (mobile health, mhealth) can be described as the performance of the facilitation of public health services through mobile phones, hospital tracking de-

Table 6. Mobile gambling related pushes and limitations (Source: Juniper Research)

Pushes	Limitations
Gambling desire	Rules
Spendable income increase	Age verification
Technological development	Payment process
Availability/Conformity	Conservatism of suppliers
Mobile penetration increase	Increasing speed of network cycle
Mobile user profile	Abilities and conformity of telephone
Personal confidentiality	Practicability

Table 7. Total revenues from mobile gambling regional forecast (casino, lotteries & betting) 2006 and 2011 (Source: Juniper Research)

Region	2006 (USDmillion)	2011(USDmillion)
North America	0	3340
South America	34	1328
Europe	950	7940
Asia Pasific	966	8818
Other Regions	40	1765
Total	1991	23190

vices, PDA, etc. wireless devices, and other health efforts convenient for remote-activated tendencies.

M-health also covers applications related with the utilization of mobile instruments for gathering social and/or clinical health data. In this context, the transfer of information related with health and care to practitioners, researchers, patients and to the entire society when required, and the full time monitoring of critical indicators, and even the performance of care (via mobile telemedicine) can be considered (Germanakos, Mourlas & Samaras, 2005 in http://en.wikipedia.org/wiki/MHealth#cite_note-mobileagents-0 1/2/2009).

On the other hand, there are several people like huge immigrant populations some of whom have a very scant command of English (or spoken language in countries). These are living in foreign language speaking countries, and they often need to see doctors, currently governments of these countries provide real-live translators to help these people out during hospital visits; however, with some mobile solutions it is possible for them to get a connection and explain themselves much better.

Besides the researches and efforts about explaining the impacts of radio frequency (RF) areas are continuing. These studies have certain gaps based on lack of information and they should be scientifically filled in. Actually, the exposure of people to RF areas is not a new issue. The developments in information and communication technologies cause everybody to be exposed to such artificial frequency fields in their daily lives, at home, at work and at other places. The human body absorbs a small percentage of RF areas arising from some electrical devices such as mobile phones and base stations, and this absorbed energy is turned into heat. Because of certain biological processes in the body, this heat increase is recovered without being moved to perceivable sizes. Certain scientists claim that the energy emitted from mobile phones and base stations could affect human health. It has been discussed in the media several times and many articles have been published on the safety of mobile phones (http://www.vodafone.com.tr/VodafoneHakkinda/kurumsal_sorumluluk_CepTelSaglikZararlimidir.php 12/13/2008).

Certain international guidelines have been prepared regarding the exposure levels of RF areas caused by mobile phones and base stations. Manufactured mobile phones and base stations are in accordance with these guidelines. With the International EMF Project, which was started by the World Health Organization in 1996, research related with mobile phones, base stations and their effects on health have been registered and requirements related with these researches have been prioritized. In 2006, new areas were included in the research scope. These areas are as follows:

- Long-term (over 10 years) exposure to low-level RF areas
- Potential effects of mobile phones on child health
- Dosimetry (measuring the dose of radioactive substances) or calculation of RF wave absorption amounts

In a joint study with the International Agency for Research on Cancer (INTERPHONE), the mobile phone utilization of people with tumors at the head and neck sections and those without such tumors is inspected. Although the findings of the research have not been completely published yet, the results that have been reported so far do not indicate a certain relationship between mobile phone use and tumor formation (www.mmfai.org/public/docs/jp/MMF_GSMA_ResearchBrief_German_INTERPHONE.pdfwww.gsmworld.com/health/programs/interphone.shtml & INTERPHONE, 2009).

The World Health Organization (http://www.who.int/mediacentre/factsheets/fs193/en/ retrieved 1/2/2009), issued certain advisory decisions about the use of the mobile phone. Accordingly, any scientific data which indicate low level radio frequency (RF) fields arising from mobile phones harmful for human health could not be obtained. Nevertheless, there are still certain gaps that should be filled in. More research is required in order to fill in thes scientific information gap.

The International Commission on Non-Ionizing Radiation Protection (ICNIRP www.icnirp.de retrieved 1/2/2009) has published guidelines regarding the exposure levels to RF fields caused by mobile phones and base stations. A certain exposure safety limit is stated in these guidelines. The exposed level is measured with Special Absorption Ratio (SAR). Special Absorption Ratio is the amount of RF energy absorbed by the human body.

Most of the mobile phones sold are in accordance with these guidelines. Also, there are many factors that affect the exposure level to RF waves which stem from mobile phones. Some of them are as follows:

- The distance between the user and mobile phone
- Distance to the base station
- Characteristics of the field between the mobile phone and base station
- The service employed
- Apparatuses which convey sound from the phone to the ear

Mobile phones require a minimum radio frequency field to connect to the base station. Therefore, signals from the base station are automatically set to a low frequency. The higher quality the signals from the base station are, the lower the RF field strength from the mobile phone will be.

According to Istepanian & Lacal (2003), the new generation "wireless e-health technologies" should be considered as a headline to upraise telemedical and telecare systems. These technologies accelerate the activity of using mobile telecommunication and multimedia technologies for health. They are also referred to as telemedical and telecare systems.

The health sector is an important sector which continuously maintains its importance about the production of information both in terms of use and in terms of producing scientific information related with devices and the utilization of devices.

PAYMENT SYSTEMS

Mobile payment (also known as mobile web payment or WAP invoicing) is another important subject that recently started to affect our lives. Mobile payment means the collection of the credit amount from the debtor by the creditor via any mobile instrument such as mobile phone, smartphone, and PDA (http://en.wikipedia.org/wiki/Mobile_Payment retrieved 1/3/2009). This method of payment is shortly known as m-payment. Mobile payment can be used for any tangible product such as books, tickets, computers or mobile phones while it can also be used for digital products such as music, games, and video films. Also voluntary payments of people for their selections of social participation or donations, and the costs required by mobile phone system are also among such payments.

Mobile payment consists of two basic models:

- Premium SMS based transactional payments
- Mobile Internet payments (WAP)

Both of these payment systems are used effectively in developed countries while they also have great potential in developing countries. Especially the micropayment system in developing countries is the leading as one of the most effective payment methods.

M-payment (mobile payment) is regarded as a sales point payment form. A mobile instrument such as mobile phone, smartphone, PDA can be used as the payment instrument at a sales point. Because M-payment is a type of payment conducted through wireless devices, it is a method of payment which can be freely performed at any shop or point of service; for example, at a restaurant without requiring any staff. For example, when a customer of a restaurant desires to make his or her payment quickly in order to catch an appointment, he or she can pay the check at the table without waiting for any waiter to come. By this method, the mobile device of the customer is exposed to the cash register through the mobile payment provider. He or she types the number of the table, enters his or her bank or identity number and approves the payment, and then, he or she can immediately leave the restaurant. According to the Danish company Orange Mobile Payment, the entire time spent for the mentioned transaction is not more than 10 seconds (http://www.bitpipe.com/tlist/Mobile-Payments.html 12/12/2008).

The initial m-payment trials through mobile phones were seen on wide area networks (WAN). It was for the payment of bills from their mobile phone invoices, which required entering long sequential numbers at every payment time. Other technologies under trial include more effortless procedures. Palm and Verifone IR try to conduct data exchange through infrared (IR). Apart from them, short range transfer systems such as Bluetooth, WiFi and RFID can also be used for payment. Although public key infrastructure (PKI) encryption is considered overall m-commerce, the isolation of m-payment is essential for customer's security (http://searchmobilecomputing.techtarget.com/sDefinition/0,,sid40_gci772807,00.html 12/12/2008).

M-payment is currently used in various parts of the world, mainly in Europe and Asia. The first trials in North America were made towards the end of 2001, and they were realized at two restaurant chains in the US. One of the important points related with m-payment is the similarity of features of this payment to credit card payment features. In general, it is physically used at the moment and at places where monetary exchange is to be made at the credit card sales point. However, credit card is not used everywhere or at every part of the world. At places where these payments cannot be made, mobile instruments and systems can be used instead of credit card. However, the infrastructure and systems to use mobile phones as credit cards are currently in the developmental stage. Here, the highest consideration for both parties should be security because both the payer and creditor wish to obtain security.

According to the data of GSM Association, 800 million mobile phones were sold in developing countries between 2003 and 2006 (GSM Association, 2006). According to Ivatury & Pickens (2006), it actually represents a significant potential for mobile banking (m-banking). It should also be considered as a financial service through which millions of poor people in the world can be provided with basic financial services. M-banking is defined as the provision of banking services through mobile instruments. Meanwhile, mobile-payment is used rather as an instrument for retail shopping or person to person transaction. Its role in the payment system will be completed when it becomes suitable for banking transactions.

According to Karnouskos & Fokus (2004), the entire payment activated, performed and/or approved via any mobile device should be considered as mobile payment.

There are many different units and organizations in the digital payment scenario (Table 8). For example, financial organizations can be individual or corporate banking-based, and they should adapt their mobile payment infrastructure and card operations to the mobile. On the other hand, independent payment providers do not currently provide a complete infrastructure, but they are able to react fast. Here, mobile operators also hold the status of a kind of intermediary organization or a guarantor. Apart from them, companies that develop the software and devices used for mobile payment operation, and other service provider companies can also come into question. Here, one of the critical points is that users do not fully accept this system. However, it is not possible to say that the customer is completely ready for this subject, which requires creating a new habit.

According to Karnouskos & Fokus (2004), two types of mobile payment models are in question. They are gaining and acquisition centered, or counter server centered. Bank-centered payments are the payment types where the mobile instrument is used, but they are types of payments which can allow the payment of personal account statements, and if they are Mobile Network operator centered, then the systems in this network can be developed and the payments can be received.

The methods to register the payments into accounts in this context are as follows:

- Those which are invoiced in order to be paid later (the payments added to the telephone invoice, and which are made via credit card),
- Prepayments: these are realized through prepayments held in a certain account or invoice,

Figure 2. Mobile payment

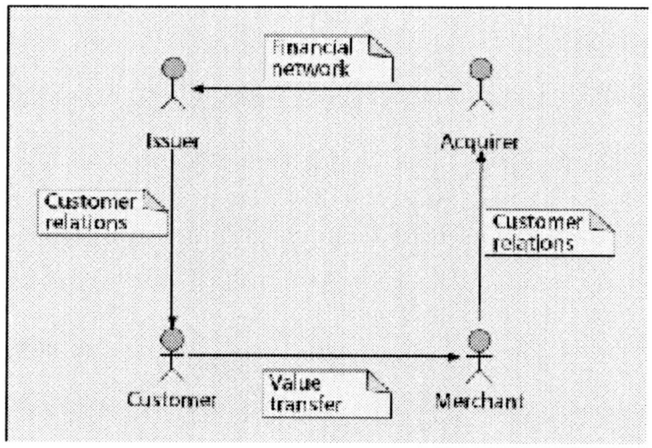

Source: S., Karnouskos & F. Fokus (2004). 'Mobile Payment: A Journey through Existing Procedures and Standardization Initiatives". *IEEE Communications Surveys & Tutorials* Fourth Quarter 2004, p.44.

- Immediate payment –real time or pseudo real time: payments only at the moment of shopping or for those which were realized a very short time ago
- Other payments: Online micropayment, Offline micropayment, Single chip, Double chip, Double Slot, e-coin based, account-based, wireless wallet, IrFM (Infrared Financial Messaging (IrFM) is a wireless payment standard developed by Infrared Data Association based, RFID based, fillable, mobile cash card, barcode paperless invoice and PhotoPay are other forms of payments.

For example, Photopay is regarded as a shopping administrator based on shooting the barcodes of products with cameras on mobile phone and sending them. Here, there are certain requirements such as: the product must have a barcode and mobile phone must have a camera. This payment system is ready for use after a short and simple installation procedure. The product is bought in two simple phases. The customer takes the photo of the barcode, the installation program rapidly deciphers the barcode and turns it into product and price information; then, the customer states the payment options and concludes the transaction. Here, any time requiring invoicing or personal details is not included (http://www.toptalent.europrix.org/ttp/2004/3004_8.html retrieved1/2/2009). Even, if the service amount is suitable, then the delivery can be made even to the customer's house.

According to Karnouskos & Fokus (2004), mobile payment is not a sort of payment method where a single procedure is suitable for everyone. Evaluations should be made according to the risk levels of different accounts. For example, micropayments require a different approach than other payments while guarantees can be requested against risks for high payment amounts. Therefore, a general mobile payment structure may not be suitable for all transactions. Besides, most people consider mobile payment as an additional access channel to existing payment systems.

Mobile finance and commercial services have taken their places among issues that should be deeply considered in recent years. For example, in the UK, o2 and Visa have demands in this regard while the

Table 8. Digital payment scenario (Source: Developed from S., Karnouskos & F. Fokus (2004). "Mobile Payment: A Journey through Existing Procedures and Standardization Initiatives". IEEE Communications Surveys & Tutorials Fourth Quarter 2004, p.46.)

Players	Basic Characteristics
Financial Organizations	• Individual based • Corporate based • Mini/macro payment infrastructure • Card operators
Mobile network operators	• Customer based • Commercial based • Mini/macro payment infrastructure • Control of end user device • Invoicing infrastructure
Autonomous mobile payment providers	• Ability to react fast • No infrastructure requirement • No customer base requirement
Other service providers	• Basic mobile payment instrument providers • Developers • Additional services providers
Users	• Not ready about security • Not ready for mobile payment logic • Possess mobile instruments
Other	• Legal infrastructure developers • Technological infrastructure developers • Threateners • etc.

security and mobile phone ranges still constitute a barrier. Central Eastern Europe based ventures' credit services have seen a keen interest with regard to P2P banking services and global cross-border remittances continue to grow via mobile money transfer scheme. This transfer is organized through the connection of millions of registered users to the money transfer network with the wallet agnostic approach (http://www.mpaymentsconference.com/newt/l/mpayments/mobilepayments?gclid=CKqr24DOupcCFUQI3wodqAQuoQ retrieved 12/12/2008).

For retail sectors, mobile payment has introduced brand new approaches and applications. Therefore, it is an influential method on the retailers' side. Payment devices, which are the products of complex engineering where various digital means are integrated, conclude the transaction for the customer in a very simple manner. These practical methods offered to the customer facilitate customer acquisition and operations at the sales point, and greatly contribute to maximizing profits and create an expectation for brisk sales. Retailers, who are aiming to grow, should spare funds from main investment for the equipment and training of employees on how to use the device. Companies, which are more eager for activities such as issuing coupons, tickets or location-based services, are more prudent about investing in payment systems.

The following points are required for mobile payments:

- Development of business strategies suitable for each sector
- Organization of technical platforms
- Security and Consent

- Identification of reasonable prices for the service and level of freedom,
- Identification of restrictions
- Regional analyses
- Cross-border money transfer applications
- Credit or prepaid systems developed for retailers
- Applications related with mobile marketing, promotion and coupons
- Training of citizens on public collections through electronic environment such as m-collection, m-sequestration, m-invoice applications.

Mobile Wallet

Mobile wallet is one of the applications about mobile payment. Its first practice in Turkey is one of the joint-projects carried out by bank(s) and service providers, who have realized many pioneering applications in their sectors. Credit cards transactions can be performed via mobile phones with the Mobile Wallet. The performance of all transactions performed through a certain credit card via mobile phones with different features and technologies is an approach that facilitates mobile payments. Such practices decrease the payment time to less than half a second. During shopping, the low limit accounts per transaction can be carried out fast through mobile phone as the cost is added to the mobile phone invoice. After the limit is exceeded, shopping is performed through entering a personal password. Mobile payment and contactless banking are the beginning of a new period in terms of the use and evaluation of money. In order to perform such projects which are expected to provide non-staffed bank services, support was provided by the Global GSM Association (GSMA) and one operator was selected from each country including the US, Australia, the Philippines, France, Ireland, Korea, Malaysia, Norway, Singapore, Taiwan and Turkey. Turkcell represents Turkey as it has the highest number of customers. It is expressed that the results of such pilot applications also contributes to the mobile payment effort to be carried out by an association on a global scale, and create a database to identify the standards in this regard. The project which has been started by banks and service providers in Turkey will be included in the widest member workplace network of the world, and it can be among the first users of the programs which allow uploading credit card data to the SIM card via GSM network (http://www.tele.com.tr/blog_comment.asp?bi=2487&m=6&y=2008&d=1&s=search 1/4/2009).

REFERENCES

Association, G. S. M. (2006). GSM Hits Two Billion Milestone. London: GSM Association. 16 June. In G. Ivatury, & M. Pickens, *Mobile Phone Banking and Low-Income Customers Evidence from South Africa.* White Paper, Consultative Group to Assist the Poor/The World Bank and United Nations Foundation.

Bauer, H. H., Barnes, S. J., Reichardt, T., & Neumann, M. M. (2005). Driving Consumer Acceptance of Mobile Marketing: A Theoretical Framework and Empirical Study. *Journal of Electronic Commerce Research*, *6*(3), 181–191.

DMA. (2005). Special Requirements Regarding the Use of Adult Content. Retrived from http://www.dma.org.uk/_Attachments/Resources/1897_S4.pdf

Finn, M. (2005). Gaming Goes Mobile: Issues and Implications. *Australian Journal of Emerging Technologies and Society*, *3*(1), 31–42.

Germanakos, P., Mourlas, C., & Samaras, G. (2005). A Mobile Agent Approach for Ubiquitous and Personalized eHealth Information Systems. In *Proceedings of the Workshop on 'Personalization for e-Health' of the 10th International Conference on User Modeling (UM'05)*. Edinburgh, July 29, (pp. 67-70).

Gibson, (2006). http://www.juniperresearch.com/shop/viewpressrelease.php?id=28&pr=37 and http://www.juniperresearch.com/shop/viewpressrelease.php?id=28&pr=5

Han, I., Kim, M. & Lee, H. (2005). Analysis of Identity and Trust on the Virtual Community. *E-Business Review, 5*.

http://brew.qualcomm.com/brew/en/about/brew_gaming.html (n.d.). Retrieved 12/14/2008.

http://en.wikipedia.org/wiki/MHealth#cite_note-mobileagents-0 (n.d.).

http://en.wikipedia.org/wiki/Mobile_Entertainment#cite_note-6 (n.d.).

http://en.wikipedia.org/wiki/Mobile_Payment (n.d.).

http://mobmarketing.wordpress.com/ (n.d.).

http://searchmobilecomputing.techtarget.com/sDefinition/0,sid40_gci772807,00.html (n.d.).

http://www.amazon.com/Mobile-Advertising-Supercharge-Exploding-Wireless/dp/0470185988 (n.d.).

http://www.antifit.com/?p=1455 (n.d.).

http://www.bitpipe.com/tlist/Mobile-Payments.html (n.d.).

http://www.gsmworld.com/health/programs/interphone.shtml (n.d.).

http://www.ifpi.org/content/library/dmr2008.pdf (n.d.).

http://www.juniperresearch.com/shop/viewreport.php?id=79 (n.d.).

http://www.mmfai.org/public/docs/jp/MMF_GSMA_ResearchBrief_German_INTERPHONE.pdf (n.d.).

http://www.mobile-weblog.com/50226711/changing_mobile_gaming_habits.php (n.d.).

http://www.mobile-weblog.com/50226711/research_reveals_mobile_game_use_trends.php (n.d.).

http://www.mobileeurope.co.uk/news_wire/113709/Mobile_music%2C_games_and_TV_to_generate_%2434_billion_by_2010%2C_claims_Juniper_Research_.html (n.d.).

http://www.mobilemarketer.com/cms/news/messaging/1573.print (n.d.).

http://www.mpaymentsconference.com/newt/l/mpayments/mobilepayments?gclid=CKqr24DOupcCFUQI3wodqAQuoQ (n.d.).

http://www.mu-yap.org/getdata.asp?PID=291 (n.d.).

http://www.safecard.ie/joomla/index.php?option=com_content&task=view&id=30&Itemid=46 (n.d.).

http://www.sportsline.com/wireless/alerts (n.d.).

http://www.tele.com.tr/blog_comment.asp?bi=2487&m=6&y=2008&d=1&s=search (n.d.).

http://www.toptalent.europrix.org/ttp/2004/3004_8.html (n.d.).

http://www.vodafone.com.tr/VodafoneHakkinda/kurumsal_sorumluluk_CepTelSaglikZararlimidir.php (n.d.).

http://www.westlake.co.uk/Mobile_Phone_Glossary.htm (n.d.).

http://www.who.int/mediacentre/factsheets/fs193/en/ (n.d.).

4INFO. (2008). *Marketing Mobile with Mobile: Lessons in Strategy*. Retrieved from 5/2/2009 http://vmware.eyeline.mobi/wp-content/uploads/2008/12/4info_marketing_mobile_20080327.pdf

Istepanian, R. S. H., & Lacal, J. S. (2003). Emerging Mobile Communication Technologies for Health: Some Imperative Notes. *m-Health. Proceedings of the 25th Annual International Conference of the IEEE EMBS*. Cancun, Mexico. September 17-21, 2003.

Ivatury, G., & Pickens, M. (2006). *Mobile Phone Banking and Low-Income Customers Evidence from South Africa*. White Paper. Consultative Group to Assist the Poor/The World Bank and United Nations Foundation.

Jarvenpaa, S. L., & Lang, K. R. (2005). Managing the Paradoxes of Mobile Technology. *IS Management*, *22*(4), 7–23.

Juniper Research. (2006). *Mobile Sport, Leisure & Information Content Opportunities & Markets, 2006-2011*. Retrieved from http://www.juniperresearch.com/shop/viewreport.php?id=79

Karnouskos, S. & Fokus, F. (2004). Mobile Payment: A Journey through Existing Procedures and Standardization Initiatives. *IEEE Communications Surveys & Tutorials Fourth Quarter* 2004.

Manjoo, F. (2001). Porn in the Palm of Your Hand. *Wired News*. Retrieved 5/5/2009 from http://www.wired.com/techbiz/media/news/2001/01/41336

Merisavo, M., Kajalo, S., Karjaluoto, H., Virtanen, V., Salmenkivi, S., Raulas, M., & Leppäniemi, M. (2007). *An Empirical Study of the Drivers of Consumer Acceptance of Mobile Advertising*. Retrieved 7/7/2008 from http://www.jiad.org/vol7/no2/merisavo/An%20empirical%20study%20of%20the%20drivers%20of%20consumer%20acceptance.pdf

Mobile Adult Content Market to Grow to $3.3bn by 2011, Driven by a Strong European Market and New Products for Casual Users. http://www.juniperresearch.com/shop/viewpressrelease.php?id=28&pr=37

Pura, M. (2005). Linking Perceived Value and Loyalty in Location-based Mobile Services. *Managing Service Quality*, *15*(6), 509–538. doi:10.1108/09604520510634005

Quinn, C. (2008). *M-Learning Devices: Performance to Go*. Retrieved 9/5/2008 http://www.learningcircuits.org/0708_quinn.htm

Şakafon ile Cepte Sesini Değiştir (Change Voice with Prankophone), (2008, August). *Sabah,* Daily Paper, 10, (p.10).

Schmidt, A. U. (2008). Free-Riding and Competition in Network Markets for Digital Goods. In *Proceedings of the 41st Hawaii International Conference on System Sciences – 2008*. Retrieved 12/13/2008 from http://arxiv.org/PS_cache/cs/pdf/0409/0409028v2.pdf

Shiffman, D. (2008). *The Age of Engage*. Ladera Ranch, CA: Hunt Street Press.

StrategyAnalytics. (2008). *Understanding the Mobile Ecosystem*. White Paper Prepared for Adobe Systems Incorporated. Retrieved 4/26/2009 from http://www.adobe.com/devnet/devices/articles/mobile_ecosystem.pdf

Takeishi, A., & Lee, K. (2006). Mobile Innovation and the Music Business in Japan: The Case of Ringing Tone Melody ("Chaku-Mero"). In S. Barnes, & E. Scornavacca, (eds.), *Unwired Business: Cases in Mobile Business* (pp.1-13). Hershey, PA: IRM Press.

Three (2004). *Playboy*. Available from http://www.three.com.au/index.cfm?pid=2217&pageid=2103&sid=2237. Date accessed 10th October 2004 from Finn 2005.

Wehmeyer, K. (2007). Assessing Users' Attachment to Their Mobile Device. In *Proceedings of the 6th International Conference on Mobile Business (ICMB 2007)*, Toronto, Canada, July 9-11.

www.icnirp.de

Yamamoto, G.T. (2007). *Mobile Telephone Usage in Turkey*. Okan University, Unpublished research.

Yamamoto, G. T. (2008). Comment on "Voice and Mobile Promo for Incredible Hulk Gets Strong Response." Retrieved from http://www.mobilemarketer.com/cms/news/messaging/1038.html

KEY TERMS AND DEFINITIONS

BREW Gaming: enables an optimal mobile gaming experience because of native performance and its standardized set of API's (http://brew.qualcomm.com/brew/en/about/brew_gaming.html retrieved 12/14/2008).

J2ME: Java 2 Platform, Micro Edition. It is the edition of the Java platform that is targeted at a small, stand alone or connectable consumer and embedded devices. The J2ME technology consists of a virtual machine and a set of APIs suitable for tailored runtime environments for these devices. The J2ME technology has two primary kinds of components – configurations and profiles (http://www.westlake.co.uk/Mobile_Phone_Glossary.htm retrieved 12/14/2008).

Micropayments: are means for transferring very small amounts of money in situations where collecting such small amounts of money with the usual payment systems is impractical, or very expensive in terms of the amount of money being collected.

Mobile Adult Content: Purchasable contents or natural contents as a factor in preferring certain mobile devices and features for 18 plus age.

Mobile Health: (m-health) can be described as the performance of the facilitation of public health services through mobile phones, hospital tracking devices, PDA, etc. wireless devices, and other health efforts convenient for remote-activated tendencies.

Mobile Payment: (MP): A payment involving the use of a mobile phone. The phone network can be used solely as the channel to authenticate the customer, or the payment can involve a debit to an account held by the customer with the mobile operator or other service provider (http://www.safecard.ie/joomla/index.php?option=com_content&task=view&id=30&Itemid=46 retrieved 1/4/2009).

OS: Operating System, historically, is the minimal set of software needed to manage a device's hardware capability and share it between application programs. Practically, "OS" is now used to mean all software including kernel, device drivers, comms, graphics, data management, GUI framework, system shell application, and utility applications. This would define Windows, Palm OS and MacOS as operating systems. Symbian provides an operating system – Symbian OS – tailored for data-enabled mobile phones (http://www.westlake.co.uk/Mobile_Phone_Glossary.htm 12/14/2008).

Pay-Now: Real-time or near "real-time" payments by mobile devices.

PhotoPay: is regarded as a shopping administrator based on shooting the barcodes of products with cameras on mobile phone and sending them.

Post-Paid: Paid in advanced [Phone-bill based, Account-based (bank/credit card)] payment by mobile devices.

Wireless Wallet / Mobile Wallet: The performance of all transactions performed through a certain credit card via mobile phones with different features and technologies is an approach that facilitates mobile payments.

Chapter 9
Mobile Affected Socialization

While mobile intrusion comes to life insidiously, it has been affecting our lives in many different ways. At first social networking and some implementations such as mobile dating comes to mind. On the other hand, mobile literacy and educational implementations are evolving and spreading rapidly with the facilities of mobile systems. Having turned into one of the most important elements that trigger and develop the social genes in the twentieth century, communication technology is also drawing the attention with its impacts that direct socialization. Internet and cyberspaces which are used in mobile communication create communication organs with their multi but non-conflicting features for person to person connections, where individuals focus on a unity through the utilization of different modes to connect (Urry, 2002). This environment is suitable to create virtual societies where many people join with various reasons and they should also be considered in terms of marketing.

The inclusion of people on the Internet as social actors evokes gathering metaphor on the background. General behaviors of any kind of gathering such as chatting, discussing, challenging and keeping secrets are also seen on this platform (Sproull & Faraj, 1997). Chat rooms, organized clubs, facebook type of websites, and virtual games are places where people spend time or perform communication-based activities.

According to Han, Kim & Lee (2005), a virtual society communicates with each other in this environment via electronic media such as Internet, and create similar areas of interest, and develop a participative intellect without requiring the interrogation of geographical location, physical interaction or ethnic

DOI: 10.4018/978-1-60566-916-8.ch009

origins. These societies can be defined as groups without boundaries. Mobile societies work with similar approaches and new formations of these societies has not completed their development.

CHANGING THINKING STYLES & SOCIAL GATHERING METAPHOR

The Internet is considered as a place that brings people together with its means. People, who use information on Internet that creates the communication network, are not only the processors of information, but also the participants and programmers of social activities. Internet encompasses not only information, but also a search for a relationship, support and approval. With new mobile technologies, relationships can be formed very quickly beforetime.

In the social gathering metaphor, Internet and mobile devices are not just used in terms of personal information production process. It also provides the participation of the person in the social process through the communication, discussions and joint evaluations and it can be evaluated as a social element where the change-causing contributions are structured and distributed. In this metaphor, communication is based on the relationship between individuals. In the environment where people are regarded as social actors, Internet can also be described as a social technology. This technology carries information to minds that are included in the application area, with a feature that can be combined with human activities and procedures. Social technologies enable people to find others according to their common areas of interest talk and listen to each other and continue their connections in time. It is a well known fact that various groups and activities such as dinners, activity or sports (such as bowling) teams, college groups are created via social technologies (Sproull & Faraj, 1997). Interaction among people on the Internet and new mobile devices creates the feeling and thinking infrastructure that allows individuals to express and reflect themselves with their individual and social identities, explain themselves and know themselves.

The connections through computer technologies create the opportunity to question personal life, goals, intentions and information, and deliver them to many secondary people in the work and living environment (Turkle, 1997). It is possible to create virtual types in a virtual environment like the characters in Secondlife (Yamamoto, 2008). Therefore, during the exposition of individuality, brand new forms of thinking do not mind a "sexual change" where men present themselves as women or vice versa arises. However, individuality options in mobile communication are narrowing down with video technologies except for Internet and games.

The subject of differences created in the thinking styles of people through computer-based communication is gradually considered with more and more serious approaches and the influences on the physical existence or stance due to the transformation of physical distance into a less important factor than before are dwelled upon (Spagnolli, Varotto & Montavani, 2003). Besides, the increasing personal interests also lead into re-questioning of relationships among people and groups (Wiertz, de Ruyter & Streukens, 2002).

Here, three main objectives have been identified regarding attempts to communicate with others. The first objective is to benefit from the database of the target society and find ways of agreement through understanding the innovative values, the second target is to suggest personal ideas and gain some sort of reputation for self-satisfaction, and the third target is to entertain and spend good time. Such similar approaches are also valid for mobile societies and mobile phone users. The communication platform, which is obtained through devices physically available in the mobile environment, began to be used in a different way compared to conventional communication styles. Mobile communication allows for

Table 1. Some developed countries mobile subscriber monthly consumption of content and applications (Source: Jaimee, Minney & Tina, Lorkin, 2007 "Mobile Social Networking Has 12.3 Million Friends in the US and Western Europe" M:Metrics, Inc http://www.mmetrics.com/press/PressRelease.aspx?article=20070815-socialnetworking retrieval date: 27/04/08.)

Activity	France	Germany	Italy	Spain	United Kingdom	United States
Sent Text Message	72.10%	80.00%	85.80%	84.00%	86.60%	41.60%
Used Photo Messaging	22.40%	21.70%	30.60%	31.50%	29.60%	17.50%
Purchased Ringtone	5.80%	5.50%	4.60%	5.60%	3.80%	9.40%
Used Personal E-Mail	4.40%	4.70%	8.90%	8.40%	7.10%	9.00%
Used Mobile Instant Messenger	3.00%	3.30%	5.70%	6.20%	5.00%	6.70%
Used Work E-Mail	2.10%	2.80%	5.20%	5.20%	3.80%	5.30%
Downloaded Mobile Game	1.40%	2.70%	3.10%	4.70%	4.80%	3.20%
Purchased Wallpaper or Screensaver	1.70%	2.10%	2.60%	2.10	1.90%	3.20%
n=	12,705	15,615	13,719	13,080	15,758	33,810

encountering with more intense but superficial transformations in human relationships. Mobile societies create a situation that facilitates convergence and interactions, and coming face to face and simplify the conversion of a virtual relationship into an actual relationship.

Therefore, mobile phones also introduce a brand new way of socialization into the lives of people. With social networks through this channel, people can more rapidly contact with groups that are suitable for themselves. Social networking is based on four basic focuses.

- Activity focus
- Interest focus
- Work focus
- Entertainment focus

According to the data of M:Metrics (August 15, 2007), the number of connections to social network websites in the US and West Europe in June 2007 was 12.3 million. The number of mobile network website followers is 7.5 million in the US which has the highest ratio. This figure accounts for 3.5% of the mobile users. The US is followed by Italy with 1.3 million users and 2.8%, the UK with 1.1 million users and 2.5%, Spain with 751 thousand users and 2.3%, Germany with 1.9% and France with 1.7%. According to the data of the same group, MySpace is the social network that attracts the higher number of mobile uses in the US and UK. Facebook has about 80 million users (Holzner, 2009).

Minney & Lorkin (2007) indicate that in most of the developed countries, the largest consumption of mobile phones is through text messages and it is followed by photo-messaging while the ratios of e-mail and instant messaging are currently at a lower ratio. The highest number of messages was sent with 86.60% while the ratio decreased to 41.60% in the US. Concerning the reception of personal e-mails via mobile phone, the US and Italy leads respectively with 9% and 8.9%. Mobile instant messenger is most commonly used in the US with 6.7%, and it is followed by Spain with 6.2% (Table 1).

It is seen that ventures related with social networks have not reached a high number as most of the devices are still old fashioned and are not compatible with 3G technology and new developments. However, this figure can be expected to increase in line with the changing level. The elements that make mobile phone practical such as easier utilization and continuous possession also encourage that.

During the development of social network use, certain new services may occur in cyberspace and more affective services can be obtained through related means. Links such as press and get married are among such interesting ventures. Websites that are configured in order to find a couple for a happy life, etc. are subscription-based. Actually, such ventures seem quite natural considering that the main function of social networks on the Internet is to find friends and spend a good time. Private interaction websites require a different type of specialization. The reason why they receive such interest can be explained with the impact of social loneliness or at least the curiosity they evoke.

Non-adolescent children and teenagers also use mobile phone in order to identify themselves and create social rings other than their families; therefore, future generations are gradually more attached to their personal devices. Children strengthen and change their ways of communication with their families through these devices (Holson, 2008).

De Souza e Silva expresses that mobile phones created a bigger influence than any of the antecedent entertainment and communication devices (2006, p. 20). Mobile phones include many forms of socialization through sportive and artistic interaction through built-in games and camera systems.

In addition, the mobile instruments' ability to connect other electronic systems is also influential on the structure of social behavior. For example, with sample applications such as the telephones to which alarm devices can transmit signals or the realization of various transactions such as taking appointments from health organizations, etc. e-government or private organizations, mobile phones have become influential on elements such as security, health, social information, emergency and even jurisdiction. Furthermore, it has also become very significant for non-governmental organizations in order to create awareness or effectuate their practices.

SOCIAL NETWORKING

Social networking can be considered as social production and organizational platform. The mobile platform has also enabled a new mode that supports creativity. According to the Juniper Research's Social Networking and User Generated Content (UGC) Report, mobile social networks are also among examples of "social networking" now. This network includes systems that allow for the sharing of content through various applications such as chatting, presence, IM and VoIP (Chard, 2008).

12.3 million people have companions on mobile social networks in the US and in Western Europe (http://www.marketingcharts.com/interactive/mobile-social-networking-has-123-million-friends-in-us-and-western-europe-1313/). Such interactions cause certain changes in the understanding of friendship, which is a determinant for social relationships. According to News Corporation (US), purchasing figures from MySpace social network reached 580 million dollars in 2006 and obtained an unexpected success, and that site also enabled obtaining the demographic data of the Web 2.0 youth and better understanding of demographical influences. Also, the situation did not remain there and the online and mobile environments that have become social network phenomena, have given clues on how social development is, and have drawn attention to the examination of a network-based culture of the future. According to Smith trusted referals are the most powerful marketing message inside social networks (Holzner, 2009, p.8).

It is known that the newly started operator portals are also used for building friendship. There can also be some portals which are established under different names on a geographical area basis, addressing differing tastes and societies.

According to Young (2006), the means provided by the mobile environment enables the creation of a situation that reinforces mutual trust through the cooperation of advertisers with users. Instead of reaching a certain mass targeted in advertising campaigns, the entire social network is used as an advertising channel as a result of the mentioned trust. In this context and distribution, campaigns which are conducted with almost no cost may appear. Such an opportunity could not be created in any existing mass media form, and such a cost-efficient advertising could not be made.

According to Leahul (2008), brands need the information to be obtained from their customers in the social network in order to win in the challenging competition environment. According to an independent research, more than three quarters of Facebook and MySpace users feel closer to the brands due to the communication in the form of a give and take in a social network environment. Facebook and MySpace have also identified that discourse with customers as one of the most important elements in creating a brand value; accordingly, the necessity for users to go beyond just being a friend or fan has arisen for the success of the brands of companies. The research of the Chartered Institute of Marketing also revealed that social networks like Facebook are very closely related with marketing and branding. A similar situation is also valid for mobile phone communication. The anytime and anywhere advantage of the mobile phone is suitable to allow for a more continuous use of social networks very soon.

In the mobile communication environment, with the addition of brand names in friend lists, the opportunity to create a feeling of admiration and a more intensified use of emotional interactions with the brand has arisen. So, brands can increase their ability to evaluate reactions and ideas of their users, and to perform more efficient marketing activities. People can sometimes open discussions about a brand or product and freely express both their positive and negative opinions related with the brand.

Actually, the leading reason of evaluations that should be made by marketers is to identify the shortcomings roaming on the communication network, and benefit from an information-based infrastructure in order to correct the inconveniences. Such attempts are the approaches that create transparency, and they are based on consumer-oriented marketing approach. Evaluated from the perspective of marketing, communication network provides extremely valuable feedback opportunities of any new venture or to identify the problems of an existing product or service because the information here is solid and correct. And it provides a deeper relationship with the customer, and lays the foundation to create a more loyal customer mass.

It is expressed that although customers that have tendencies in this style may be disturbed from the promotions sent by brands, they react differently to product and service information obtained from a fan, friend or a mass to whom promotions are sent continuously. The repeated calls and applications for marketing in a disturbing way become repulsive rather than being attractive and the promotion-aimed communication may increase negative approaches. Therefore, preliminary interrogations before shopping have become more influential than promotions. The user begins to care for balanced and calibrated programs, expedient and measured information sharing.

Social network interactions can be associated with legal and illegal groups. The situations that lead to positive and negative consequences especially regarding the Internet and mobile games appear. In social networks, the uncontrolled sharing of people's private information and labor-based artworks may also be in question. Mobile social networks turn into one of the data collection tools of various intelligence organizations, indicating features that restrict personal rights and freedom, and cause a different legal aspect. With the reflection of violence-containing games in society, violence may increase and it may have very bad consequences. For example, the popularity of Gang Wars, one of the four thousand applications of Facebook in Turkey, has almost already exceeded the popularity of the website itself (http://www.haberturk.com/haber.asp?id=100707&cat=210&dt= 10/03/2008).

SOCIAL NETWORKS AND BUZZ MARKETING

Social influence is the leading influence that encourages individuals to buy a certain product or service. This influence is required in order to be involved in the popular culture while different ways to follow it have appeared for marketers in every period. Buzz marketing is also considered as one of the strongest marketing instruments in the world (Balter, 2004). Buzz is a mass event that influences the stance and behaviors of the society and rapidly multiplies and grows in a viral structure. The fundamental feature of a buzz is inability to know the exact source. Therefore, buzz is the message which is not based on official announcements and which is circulated through unofficial communication channels (in Solmaz, 2004, Dubrin, 2000, p. 292). Word of mouth is an interim way of oral communication and it has a strong influence in creating public opinion.

Word of mouth communication is one of the oldest marketing techniques. For example, if the consumer has an experience related with the product, then he or she would transfer the acquired positive and negative information around and to his or her friend groups. Mary Kay and Amway use well-informed customers to spread the word about their products as a social channel to obtain new customers and a high level of identify-ability (Fattah, 2000). Thanks to Internet technology, digital environments have accelerated and activated the spreading and easy acquisition of information required for people, and then spreading it to other people again (Thevenot et.al, 2001).

In addition to viral marketing (Moore, 2003), marketing with the same logic in the mobile environment through the employment of an Internet-sourced person to person communication technique is required to be defined. Mobile channel and Internet are the communication channels that have viral, exciting and many other opportunities. Mobile phones can be used during viral marketing. The idea here is to distribute the message to customers and from them to the friends of customers and then to their friends that means further probable customers. Recommendation systems constitute one of the viral marketing methods. The recommended products are classified and presented in the digital environment, and the related opinions or recommendations are requested to be expressed. The people in the same environment willing to purchase a similar product can decide according to this ranking (Peni, 2007).

In this way, it is also possible to create vicious cycles created with the wrong information. In seconds, slanders and lies can be spread to millions and become discussed by those millions (Fattah, 2000). The rumor can most quickly be transferred via mobile instruments. The mentioned actions develop when voluntary people who tested the product deliver their experiences to others at anytime. As a result of technological developments, the means of delivering experiences to wide masses have appeared; therefore, the influences of buzz marketing have increased. The importance of its effective use in product and service promotion has also increased. Social and network influences are used in this marketing method. The beginning point of this method is the idea of selling not only to individuals but also to customer networks by sellers of products and services. Developments will be based on exploring and implementing methods to create influence in this regard.

The buzz type of marketing efforts creates strong influences on the customers' purchasing tendencies, and they can be more efficiently used owing to current mobile devices. The purpose of word of mouth and buzz marketing are the same with regard to their core values. Buzz marketing aims to increase sales through creating a rumor about a product or service, and to enhance the customers' credibility for the product, service or company. In such a marketing approach, certain differences arise with regard to the materials used for the product or service presentation. Actually, rumor is a phenomenon which is performed by everyone, everywhere and for everything. The fundamental point of departure for promotional activities within buzz marketing is to transfer attention-drawing activities to the customer. The environment created through customers' conversations and dialogues with other people is the suitable marketing environment. In this context, in addition to the marketer-customer relationship, customer-customer relationship can also be used to contribute to sales. With regard to buzz marketing, the already-known marketing tactics with viral influences should be improved. In this context, maybe the marketer can take action in identifying the subjects as the initial stimulator.

In buzz marketing, increasing the interest and recognition tendency for a product or service through buzzes can be defined as marketing environment creation. This generally happens when people share their purchasing experiences with others, and it develops based on personal experiences. If the customer is not satisfied with a product or service or company that offers the mentioned product or service, then he or she would spread that dissatisfaction around. If satisfied, than he or she would recommend it.

In the fields where the cost of buzz advertising is very high, this is an element to accelerate the sales of business with almost zero cost. During the spread of the buzz, sometimes almost no cost is required to be spent. The utilization of this element is also associated with the evaluation of the processes which appear in human-communication instrument relationships, which is experienced in social environments and for which a definition has just recently been sought. In this process, most of the time, individuals may act as if they are attorneys of a brand and continue to defend or attack it as much as they can afford to. Discussions are presented for opinion in a way that does not require certainty, and the present conditions are evaluated. Therefore, its success is correlated with its naturalness.

The elements of buzz marketing can be listed as follows:

- Sociability
- Person
- Group
- Mass
- Common ground
- Network
- Connection
- Distribution from correct points or hubs
- The device that provides distribution (face to face, internet, mobile)

The foregoing elements should be considered especially in the new marketing tactics that arise with the development of the technology. Viral influences or sociability develop when two or more consumers use the mobile environment in order to share their opinions, experiences, emotions and ideas related with the service, product or company. Although the distribution of certain company-related marketing news and promotions among mobile phones automatically turn the mobile phone into a marketing environment, its success essentially depends on personal actions.

People who act with a viral marketing approach enable the rapid and active distribution of the company or product or service related message among customers, and create different groups. In this approach, the transfer of the subject message of marketing by individuals to other individuals is encouraged. In this environment, the announcement power of the message increases incrementally; thus, the created impact grows faster. It can be expected that messages with higher entertaining or teaching strength would be distributed faster. Here, the Metcalfe Law shows its influence. According to this law, "The value of a network increases in correlation with the square of the quantity of its users" (Karaman, 2004). This law explains if the number of delivery increases 10 times, then the value increases 100 times. Similarly, the impact of viral marketing can be improved. The important issues in this regard are the risks that may arise with the conversion of the sent message to another disturbing message by the receivers in their minds and their prejudices about the received messages.

Creating Rumor Effect

The things that are spoken only for the sake of conversation in an environment where there is not an administrator are sent to other people and groups and rumor is developed as a result. Buzz marketing is the activity of performing public relations via customer or turning it into an excitement-creating activity through this way. Buzz is an activity which is usually defined as a nonsense thing performed by chatterboxes. However, buzz environment is usually one of the ways that help to easily convey the basic brand message. An interesting and quality telegraph can sometimes draw the interest of the media, and after people are informed about this issue, more talk can be obtained, and publication of articles by some columnists based on these issues can be provided. Pictures and texts distributed to the public opinion via media can turn into a very important public relations instrument. Customers believe in approved information from their friends, families, acquaintances or from people they care more than conventional advertising elements which are offered in accordance with the producer's and seller's interests.

In order to benefit from this method, a company should look for answers to the questions such as where do consumers obtain the information about the product the company desired to introduce, what do people tell each other while recommending the product, who are the central hubs and information connectors, what are the elements that prevent the distribution of information, what kind of other information and news are distributed on the network of buzz.

Such marketing tactics are especially very suitable for the excitement or difference creating the products or services which are removed from the market after fulfilling their lifespan. Rumors can become more effective when it is presented with a visual factor, and turns the users into visual focus points which visually present the product. Rumors complete their lives and dissolve, but leave a footprint behind. The higher the number of people who pay attention to the buzz, the more useful it becomes. It provides suitability in the social environment. It is a simple and easy to use method.

The product, service or idea should have certain qualities to create a rumor. Information related with the product should qualify as news. Topics should be based on differences people can talk about. These differences can be related with the functionality of the product, the ease of use, price or image. It can be used more frequently if it is in the developmental stage of its lifetime. They can especially benefit from this method by organizing programs directed towards children and young people. For example, some children and young people are hired by a DVD game company and they are given 10 or 20 of the DVDs, CDs, mobile game type of products, and they are requested to distribute them to their friends. The products are distributed to friends and they are asked to express the benefits of obtaining these products free of charge. So, children and young people can be influenced more positively as they gain information from presentations made by people in a reliable manner.

Direct-marketing companies such as Avon, Amway, Tupperware, and Herbalife also obtained significant advantages from such buzz marketing because these products are not sold at any store or shop. Discounts or promotions are indicated by their representatives, and their benefits are continuously emphasized. These people are also the extensive users and defenders of these products. Otherwise, performing the sales is not easy.

Buzz Hubs

The networks which enable the distribution of information are specialist-centered and social centered, and they can be structured as mobile-based and person-centered. The center of the networks which people develop by connecting to many people is flexible as a hub structure which can be continuously transferred to others. This form can be compared to a honeycomb. Each midrib of the comb keeps honey in connection with each other. But the honey is accumulated in each midrib by another bee. The networks in the shape of mega hubs are created by the press, celebrities and politicians who can influence wide masses. Specialist-centered networks are created among people who are influential in certain specific fields. Meanwhile, social-centered networks are created among people who succeed in using different channels to bring together many people in a social environment. Status is not required to create these networks. The network can begin from micro celebrities in that group, and every network around each focus point has its specific superiorities.

Individuals who can be at the center of the network represent different behavior from others. These people with the ability to organize other people and built connections with different groups are the most convenient in this regard. People who can influence others with their expertise and knowledge are those who can succeed in becoming famous and receiving interest on the Internet and mobile environment. People such as salesmen, who are considered among the convincers, are among the leading starters of buzzes. To sum up, while creating a buzz, people and groups that can be included in the sales force should always be taken into consideration.

Companies can build a system and introduce buzzes in that system. However, whether it is moral or immoral is a question of debate. It has been noted that a leading company used such a company for a new mobile phone model. During the promotion of this product, 120 actors and actresses acted as tourists at famous places of the US, and asked people passing by if they would take their pictures with the digital camera of their mobile phones. Here, the product was presented to potential customers without facing any salesmen, and people were impressed by charming, attractive and appealing foreigners, and they were not informed that it was a presentation, and the campaign was performed with the prediction that people would share their experiences in this regard with others (Nucifora, 2005). In this advertising campaign, an effort beyond the conventional instruments of marketing is in question. In this presentation, a new structure was created with a departure point from "buzz". However, after becoming aware of what was going on, the people thought that this campaign was carried out in an oppressive and deceitful way, and that they were used in creating a false world in order to cheat people. It is possible to easily organize such campaigns by distributing misleading and false data about a product or service from an unknown source. It includes its own dangers as it allows both attracting people's attention easily and misleading them.

Buzz is an instrument that leaves a long-lasting effect in marketing during the distribution of information related with the product, service or company. With this method, the spread of information and news only as rumors is not sufficient. Rumors are required to direct people to buy, and stimulate people or groups. Non-profit organizations can be effective in a very short time by using such an influence to call people for help through buzz and excitement. Buzz marketing oriented activities should come into the agenda with excitement and differentiation creating topics for both non-profit and profit organizations. The buzz, which is created with witnesses, is considered more reliable than advertisements and other promotional instruments in the close vicinity especially among young people.

In an effective buzz marketing management, first influential people are identified. These people are supported with contributions such as the product, service, or discount coupons. Then, a situation that continues without any efforts of the marketer arises and a buzz is transferred to others by these people without putting any cost on the organization. In general, provocative stimulants are used over the public benefitting from the entertainment or discussion concerning situations in general. During these provocations, a cultural capital is acquired by both the buyer and the distributor. Psychological satisfaction arising from "knowing", "affecting the trend" or "being the first person to use the product" is featured. So, influences with a scale from the smallest to the largest can be created. Businesses use this common motive and behavior for their own benefit paying a small cost or no cost at all. So, the ability to benefit from a self-created communication network is obtained. Under the circumstances that draw the attention of other sources, it can also allow for benefitting from others' advantages.

These practices generally create certain expectations on people. The expectation can first appear with influential advertising. Promotions can also be converted into rumors. For this method to yield a good result correct information should be presented to people and correct individuals to affect people should be reached. Some of these people are customers who directly contact the company in order to obtain information or provide information to the company. Companies are required to identify a unit about promising customers and benefit from the information obtained through the research studies of such units.

The buzz marketing method ends in an outcome beyond creating an expectation and leverages expectations. Therefore, products or services to be launched should always be beyond expectations. Unless expectations are met, the situation may turn into a negative buzz and risk the product or service. Positive buzz brings success to the product, while negative buzz may cause death to the product. Buzz marketing can also be used when de-marketing a product which comes to the end of its lifetime.

Buzz marketing has various risks such as being misleading, coercive and cheating. Therefore, companies are required to search ways to conduct campaigns which are respectful to ethical values.

PORCUPINE METAPHOR

Buzz distribution can be identified with the porcupine metaphor. An impulse or message (buzz) which is received at any hub on the Internet and mobile phone network suddenly draws attention. This rumor, the target of which is not clearly understood, may immediately spread around the close vicinity, and pins (messages) are sent to certain targets. However, this spread is not limited to the close vicinity and it is also distributed to their external peripherals and then to the latter peripherals in the same way. The events and statuses that create a pain or pleasure or similar feelings, are sometimes copied and forwarded to their surroundings and targets in the vicinity are multiplied. And from time to time, they are eliminated by the person and then the distribution continues. Here, the basic information is to acquire information or entertainment; therefore, it does not seek a profit. As it is based on voluntariness, it can indicate a faster performance compared to a profit-aiming flow. It can be considered similar to the abilities of porcupines: their soft wool bristles, short pins, thick and very flexible, long bristles and thick and long pins and the way the porcupines use them (http://tr.wikipedia.org/wiki/Oklu_kirpigiller). The short pins can be considered as those with rather temporary effects, while the soft wool is the soft buzz or connections. However, when the aim is to give harm, then the thick and long bristles are used as an instrument to attack. As they are thick and sharp, they can cause inflammation of the point they cut.

MOBILE LITERACY AND SOCIAL NETWORKS

Literacy is more than just being able to read and write; it includes learning about a wide range of subjects, and to share ideas. It includes cultural knowledge which enables a speaker, writer or reader to recognize and use language appropriate to different social situations (http://www.literacy.sa.edu.au/content.php?page=whatis retrieved 5/8/2009). Mobile literacy, on the other hand, is very important for communication and some interactions in the mobile environment. So, to take part in a mobile society, it is important to learn how to use these devices and to learn to communicate with each other in the society. Using mobile phone requires less talent and knowledge than computers, which have created very dramatic influences about getting prepared for e-life. However, the means included in broadband allows for using many of the functions of Internet, which gradually became wireless, with both mobile phones and handheld computers. This indicates that computers and mobile phones will gradually converge.

Although mobile phones are mostly used by commercial companies in order to reach a customer, and to follow up on the marketed product or marketer, they are also be used for advertising and promotional purposes. From the point of public organizations, in addition to sending announcements and giving information, it can also be used as an important platform for developing economies where functions such as the performance of operations that should be carried out by citizens at those organizations are also included (The 2007 e-Readiness Rankings, The Economist Intelligence Unit, 2007).

The government also has an important role in mobile literacy. An e-government can be defined as a government facilitating the Internet computer and mobile technologies in servicing its citizens. On the other hand, digital divide is the inequality among people in accessing, learning and using such technologies. The digital divide patterns also differ among countries. In some countries, gender may be important; in others, disposable income may be the major source of differences among citizens. However, Internet and mobile technologies are closing the gap between the users and the groups.

The groups that have discovered instant messaging services and enhanced channels such as MySpace (200 million registered users in Holzner, 2009, p.7) on the Internet are now building a very different way of communication via mobile phones with new methods and even new types of correspondence. These are also effective cultural areas to share a lot of information to people in the connection. The social media also leaves its place to social web and ratings, reviews, comments and more could be find within its content (Evans, 2008). Also these kinds of approaches also form around collective interests such as health up to matchmaker's communities. Therefore, these are also important subjects for mobile marketers.

MOBILE DATING

As it has been for centuries, also in our age, building a relationship, going out and dating is a desired phenomena. These are very important but hard to get issues for individuals. In the 1920s, dating or going out was intended to meet social expectations and it was perceived as the manners and attitudes to attract the opposite sex or to organize romantic relationships (Cate & Lloyd, 1992, in Byrne & Findlay, 2004). At that moment, dating or romance relationships were traditionally aimed to conclude in marriage. Now it is seen that, these relationships that begin via phone or face-to-face interaction have changed. Lately, relationships are becoming more virtual due to communication through texting (SMS) or through CMC (computer-mediated communication) (Asia Market Intelligence, 2002; Mori, 2000 in Byrne & Findlay, 2004).

According to Byrne & Findlay (2004), text messaging has an extremely unpretentious meeting structure regarding the acts and behaviors of the first move, beginning with the relationship, dating and coming together (http://en.wikipedia.org/wiki/Mobile_dating retrieved 8/28/2008). Furthermore, it does not increase personal responsibility.

Mobile dating services are also named as *cell dating*, *cellular dating* or *cell phone dating*. These are services provided through text messaging, mobile chatting and web for date chatting, meeting or building a romantic relation.

When they are used for dating or going out, these services develop through the exchanging among users of their short resume which is uploaded as ID on their mobile phones in order to provide information to each other. So, all IDs can find or interrogate each other as online or through a certain telephone number which is directed by a service company. Here, the criteria include, age, sex and sexual preferences and encounters with personal photographs are provided. The use of such sites is free of charge, while a service fee can also be included over the standard text messaging service fee or a service fee per message can be applied.

Mobile dating web sites give emphasis on social network sharing form of location-based services to receive interest. For this purpose, certain companies offer targeting devices that warn the users when the distance of the people to each other is 30 feet in order to facilitate meeting (Kim, 2005). Certain systems use Bluetooth technologies to bring together their users at clubs or bars at certain locations. These meetings are named as proximity dating. These systems are becoming popular as online mostly in Europe and Asia. It is believed that such dating will increase very rapidly through GPS phones and GSM localization.

According to Kim (2005), mobile meeting and dating is one of the big steps among the subjects of online socialization. As of March 2007, 3.6 million mobile phone users logged in such mobile dating sites. Most of the users are male and between the ages 18 and 34 (Divito, 2007). According to M: Metrics, the role of the mobile phone for younger people is different from that for older people (Donovan, in Divito 2007). The mobile phone is an instrument to have fun, connect friends, date, and meet new people for youths. Those who enter in such attempts via Internet are mostly considered as more traditional and although such users prefer Webdate and Match.com, these ventures may shift to mobile devices due to practicality. Here, subscription systems or pricing such as payment per message are performed.

In fact, the replacement of online dating with mobile dating due to practicality has already begun and such encounters and acquaintances have been moved to the mobile platform. Dating website owners describe this as providing people with the opportunity to choose (Kim, 2005).

Here, the leading one of the most important issues is security. All connections are anonymous and the users disclose their real identities only after they choose somebody. Besides, people are recommended to always choose crowded places as the meeting points. Mobile devices enable meeting and beginning to date in a very short time. It is indicated that such meetings and dating have climbed one step further, compared to Internet dating. The reason is mobile dating provides the opportunity to spontaneously meet new people at any location and while moving. This especially increases the requirement for communication and satisfaction. Mobile phone communication becomes influential in satisfying increased entertainment and fun requirements.

REFERENCES

Anderson, C. A. (1997). Effects Of Violent Movies And Trait Hostility On Hostile Feelings And Aggressive Thoughts. *Aggressive Behavior, 23*, 161–178. doi:10.1002/(SICI)1098-2337(1997)23:3<161::AID-AB2>3.0.CO;2-P

Argan, M., & Argan, M. T. (2006). Viral Pazarlama veya Internet Üzerinde Ağızdan Ağıza Reklam: Kuramsal Bir Çerçeve. (Viral Marketing or Word of Mouth Advertising on Internet). *Anadolu Üniversitesi . Sosyal Bilimler Dergisi, 2*, 235–250.

Asia Market Intelligence Limited. (2002). Asia-Pacific Siemens Lifestyle Survey 2002-2003. Singapore.

Balter, D. (2004). *The Word on Word of Mouth.* Retrieved 5/7/2009 from http://www.changethis.com/pdf/7.01.WordOfMouth.pdf

Bensley, L., & van Eenwyk, J. (2001). Video Games and Real-life Aggression: Review of the Literature. *Adolescent Health, 29*, 244–257. doi:10.1016/S1054-139X(01)00239-7

Byrne, R., & Findlay, B. (2004). Preference for SMS versus Telephone Calls in Initiating Romantic Relationships. *Australian Journal of Emerging Technologies and Society, 2*(1), 49.

Cate, R. M., & Lloyd, S. A. (1992). The History of Courtship. In *Courtship* (pp. 13 – 32). Newbury Park, CA: Sage Publications, Inc.

Çete Savaşları Ortalığı Karıştırdı. (2008). Retrieved from http://www.haberturk.com/haber.asp?id=100707&cat=210&dt=2008/10/03

Chard, I. (2008). *Mobile Web 2.0: Leveraging 'Location, IM, Social Web & Search' 2008-2013.* White Paper. Retrieved from http://www.c2mweb.com/bouncer?t=files/Whitepaper_Mobile_Web2.0.pdf

De Souza e Silva. (2006). http://www.modernexistence.net/mobile-phones-connecting-and-disconnecting-people-and-places/ 5/8/2009.

Divito, N. (2007). *Love is Just a Cell Phone Call Away.* Retrieved 12/28/2008 from http://www.theglobeandmail.com/servlet/story/RTGAM.20070711.wgtcelldate11/BNStory/PersonalTech/home

Dye, R. (2000). The Buzz on Buzz. *Harvard Business Review, 78*, 139–146.

Evans, D. (2008). *Social Media Marketing*. Indiana: Wiley Publishing.

Fattah, H. M. (2000). Viral Marketing Is Nothing New. *Technology Marketing Intelligence*, *20*(10), 88–90.

G, S. (2005, March 3). *Social Networking and Music: MySpace Puts It All Together in a Virtual Community.* Retrieved December 13, 2008, from http://ezinearticles.com/?Social-Networking-and-Music:-MySpace-Puts-It-All-Together-in-a-Virtual-Community&id=18424

Gaye, L., Holmquist, L. E., Behrendt, E., & Tanaka, A. (2006). Mobile Music Technology: Report on an Emerging Community. In *NIME'06*, June 4-8, 2006, Paris, France.

Han, I. Kim, M. & Lee, H. (2005). Analysis of Identity and Trust on the Virtual community. *E-Business Review, 5*.

Holson, L. M. (2008). *Text Generation Gap: U R 2 Old (JK).* Retrieved 10/3/2008 from http://www.nytimes.com/2008/03/09/business/09cell.html?_r=1&scp=3&sq=cell%20generation&st=cse&oref=slogin

Holzner, S. (2009). *Facebook Marketing*. Indiana: Que Publishing.

http://en.wikipedia.org/wiki/Mobile_dating (n.d.).

http://en.wikipedia.org/wiki/Mobile_Entertainment#cite_note-6 (n.d.).

http://knowledge.wharton.upenn.edu/index.cfm?fa=viewArticle&id=1105 (n.d.). Retrieved 20/01/2005.

http://tr.wikipedia.org/wiki/Oklu_kirpigiller 1/2/2009 (n.d.).

http://ukfast.net/site/int-news.php?news_id=100&a=1 (n.d.). Retrieved 18/01/2005.

http://www.abqtrib.com/albq/bu_national/article/0,2565,ALBQ_19839_3453265,00.html (n.d.). retrieved 18/01/2005.

http://www.commercialalert.org/index.php/category_id/1/subcategory_id/90/article_id/288 (n.d.). Retrieved 18/01/2005.

http://www.literacy.sa.edu.au/content.php?page=whatis (n.d.). Retrieved 5/8/2009.

http://www.marketingcharts.com/interactive/mobile-social-networking-has-123-million-friends-in-us-and-western-europe-1313/ (n.d.).

http://www.marketingprofs.com/Tutorials/designwordofmouth.asp (n.d.). Retrieved 18/01/2005.

http://www.nucifora.com/art_203.html (n.d.). Retrieved 20/01/2005.

http://www.rechargermag.com/article.asp?id=20031101 0 (n.d.). Retrieved 18/01/2005.

http://www.sloanreview.mit.edu/smr/issue/2005/winter/01/ (n.d.). Retrieved 20/01/2005.

http://www.sovereignmusic.com/viralmarketing.html (n.d.). Retrieved 20/01/2005.

4INFO. (2008). *Marketing Mobile with Mobile: Lessons in Strategy*. Retrieved 5/2/2009 from http://vmware.eyeline.mobi/wp-content/uploads/2008/12/4info_marketing_mobile_20080327.pdf

İnternet Mahir New York Times'da http://www.hurriyetim.com.tr/haber/0,sid~12@nvid~146467,00.asp retrieved 07/01/2005

INTERPHONE. (2009) *Mobile Phone Brain Cancer Study in the UK*. Retrieved 1/2/2009 http://www.mmfai.org/public/docs/eng/MMF_GSMA_ResearchBrief_UK_INTERPHONE.pdf

Karaman, F. (2004). E-Ticaret Ders Notları ppt. sunum (E-business Course notes ppt presentation), Okan University.

Kim, R. (2005). *Hey, Baby, Want a Date? New Mobile Dating Services Allow People to Browse Profiles via Cell Phone and Message Potential Matches -- Even on the Spot Retrieved*. Retrieved 12/28/2008 http://www.sfgate.com/cgi-bin/article.cgi?f=/c/a/2005/07/23/BUGKMDSB4P1.DTL

Kreuz, P. (n.d.). *33 Marketing-Trendchancen, Mehr Markterfolg Im Zeitalter Gesättigter Märkte*. Retrieved 11/30/2004 from www.mkt-trends.com

Leahul, D. (2008). *Brands Need to Listen to Win Over Consumers in Social Networking*. Retrieved from http://www.brandrepublic.com/Digital/News/850927/Brands-need-listen-win-consumers-social-networking/?DCMP=EMC-Digital-Bulletin

Mantovani, G. (2003). Beyond the "Impact" Metaphor: The Mutual Shaping of Psychological Theory and Internet Development. In G. Riva & C. Galimberti (Eds.), *Towards CyberPsychology: Mind, Cognitions and Society in the Internet Age*. Amsterdam: IOS Press. Retrieved 10/2/2008 from http://www.vepsy.com/communication/book2/2SECTIO_03.PDF

Matthews, A. (2001). Computer Games Make Children Anti-Social. *News.Telegraph*. Retrieve 3rd December 2004 from http://news.telegraph.co.uk/news/main.jhtml?xml=/news/2001/08/20/ngame20.xml.

Minney, J., & Lorkin, T. (2007). *Mobile Social Networking Has 12.3 Million Friends in the US and Western Europe*. M:Metrics, Inc. Retrieved 27/04/08 from http://www.mmetrics.com/press/PressRelease.aspx?article=20070815-socialnetworking

Moore, R. E. (2003). From Genericide to Viral Marketing: On Brand. *Language & Communication, 23*, 331–357. doi:10.1016/S0271-5309(03)00017-X

Nucifora, A. (2005). Keep Up With the Buzz On Buzz Marketing. Retrieved 01/20/2005 from http://www.nucifora.com/art_203.html

Peni, H. (2007). *Mobile Music Marketing: Willingness towards Mobile Music Recommendations and Viral Marketing*. Masters Thesis, Lappeenranta University of Technology, Kouvola, Finland.

Preece, J., Abras, C. & Maloney-Krichmar, D. (2004). Designing and Evaluating Online Communities: Research Speaks to Emerging Practice. *Int. J. Web Based Communities, 1* (1).

Schmidt, A. U. (2008). Free-riding and Competition in Network Markets for Digital Goods. In *Proceedings of the 41st Hawaii International Conference on System Sciences* – 2008. Retrieved from http://arxiv.org/PS_cache/cs/pdf/0409/0409028v2.pdf 12/13/2008

Schrader, D. (n.d.). Ending Spam and Other Annoyances. *Privatization Watch, 323.* Retrieved 11/30/2004 from www.privatization.org

Schrader, D. (n.d.). *The End of the Line for Advertising?* Retrieved 30/11/2004 from http://www.teradata.com/t/page/127909/

Solmaz, B. (2004). *Kurumsal Söylenti ve Dedikodu (Institutional Rumor and Gossip)*. Konya, Turkey: Tablet Yay.

Spagnolli, A., Varotto, D., & Mantovani, G. (2003). An Ethnographic, Action-Based Approach to Human Experience in Virtual Environments. *International Journal of Human-Computer Studies, 59*(6), 797–822. doi:10.1016/S1071-5819(03)00120-4

Sproull, L., & Faraj, S. (1997). Atheism, Sex, and Databases: The Net as a Social Technology. In S. Kiesler (Ed.), *Culture of the Internet.* Mahwah, NJ: Erlbaum.

Tactics for Low-Cost Marketing (n.d.). Retrieved 30/11/2004 from http://www.marketingpower.com/

Takeishi, A., & Lee, K. (2006). Mobile Innovation and the Music Business in Japan: The Case of Ringing Tone Melody ("Chaku-Mero"). In S. Barnes, & E. Scornavacca, (eds.), *Unwired Business: Cases in Mobile Business*, (pp. 1-13). Hershey, PA: IRM Press.

The 2007 e-Readiness Rankings, (2007). The Economist Intelligence Unit. Retrieved 5/7/2009 from http://graphics.eiu.com/files/ad_pdfs/2007Ereadiness_Ranking_WP.pdf

Thevenot, C., Watier, K., et al. (2001, May). *Georgetown University, Communications, Culture & Technology Program.* Retrieved from http://www.sovereignmusic.com/articles.html

Turkle, S. (1997). Construction and Reconstruction of Self in Virtual Reality: Playing in the MUDs. In S. Kiesler (Ed.), *Culture of the Internet*, (pp. 143-153). Mahwah, NJ: Erlbaum.

Urry, J. (2002) Mobility and Proximity. *Sociology, 36*(2), 255-274. Retrieved 10/3/2008 from http://www.its.leeds.ac.uk/projects/mobilenetwork/downloads/urry1stpaper.doc

Van Dinther, J. (2005). Emerging Traits of Commercial Online Communities. *E-Business Review, 5.*

Wiertz, C., de Ruyter, K., & Streukens, S. (2002). *On the Role of Normative Influences in Commercial Virtual Communities*. Maastricht, The Netherlands: Maastricht University.

Yamamoto, G.T. (2007). Buzz Gibi Söylenti Pazarlaması (Buzz in Buzz Marketing). *Pazarlama Dünyası, 20*(1).

Yamamoto, G. T. (2008). İnternette İletişim ve Secondlife (Internet Communication and Seconlife). *Pi Dergisi.*

Young, R. (2006). *Of Social Nets & Business Models.* Retrieved from http://gigaom.com/2006/06/19/of-social-networks-and-business-models/

KEY TERMS AND DEFINITIONS

Buzz Hubs: Networks which enable the distribution of information are specialist-centered and social centered, and they can be structured as mobile-based and person-centered.

Buzz Marketing: Buzz marketing is the activity of performing public relations via the customer or turning it into an excitement-creating activity through word of mouth.

Digital Divide: Digital divide is the inequality among people in accessing, learning and using new digital technologies.

Mobile Dating: Going out with the help of mobile systems (mobile dating agencies & devices).

Mobile Family: All family members own and interact with the mobile phone.

Mobile Intrusion: Surrounding the society with the mobile devices.

Mobile Literacy: Being able to read and write, to learn about a wide range of subjects, and to share the ideas from mobile devices.

Mobile Social Networking: The systems that allow for sharing the content through various applications such as chatting, presence, IM and VoIP based on a mobile network created by members.

Chapter 10
Problems

In their Works titled "High Tech/High Touch: Technology and Our Accelerated Search for Meaning" (2001), Naisbitt, Naisbitt and Philips express that technology should be questioned due to its aggressive and brutal speed and that cultural desensitization cannot be prevented, and that it would cause certain social conflicts. Besides, they also draw the attention to the importance of balance in human-technology relationships, and to the issue of balancing material miracles with the spiritual demands of man while creating technology. It is highlighted that due to the continuous tracing caused by technology, both technological immoralities and related personal concerns will continuously increase and spread along society.

With respect to health, high speed technology can cause to extreme stress and tiredness and other problems as its coercive effects may increase because technology proposes eternal series of formations and continuously pushes the limits of mind.

As technology facilitates buying any kind of ready-made solutions, it also damages and loosens certain important human relations such as neighborhood, and relationship; furthermore, it may also increase the dullness of life due to time saving facilities of technology. The person is left alone with the impairment of social-peace providing relations. In order to recover this loneliness, the person either requires developing virtual friendships or counter-reactions which should be evaluated by also taking in to consideration counter-effects.

DOI: 10.4018/978-1-60566-916-8.ch010

Mobile systems may cause the following social symptoms:

- Preferring quick solutions in everything from worshipping to nutrition
- Fear from technology and worshipping technology
- Ambiguity of the real and false
- Taking violence for granted
- Loving technology as if it was a toy
- Leading life into craziness, which should actually have been at arm's length everywhere with everyone, anytime and always
- Intense conflict (both very distant and no distant at all)

Also, humanity may become more susceptible to exploitation through technology. Certain social insensitivities may occur on certain issues due to extreme exploitation. Information will be distributed so long as the information provider allows this; therefore, those who produce technology and expand the information may create a different layer and establish superiority over society. This pressure may increase socially unenviable matters such as exploitation and abasement (http://www.bbc.co.uk/slink/features/iloveme_safety_mobiles.shtml)

In this vicious circle, the most dangerous goofiness of the human is not to know when, where and how to stop. When a situation that can be described as negligence occurs, it always returns to the human being as a loss or a punishment.

The development of technology may from time to time lead to as much bad results as creating Frankensteins. Certain issues such as mobile fraudulences, spam or phishing are continuously expanding, day by day becoming more serious and worse.

Any strong evidence that indicates the unhealthiness of mobile phones has not been found yet. Some researches claim that the use of the mobile has an influence on tumor and cancer while some other researches claim just the contrary. Such a concern was not in question for land lines due to the method of use, and because the frequency and duration of use were within acceptable levels.

At present many headlines related with mobile phone take place in the daily media continuously finding a place in the agenda. It is indicated in certain published news that mobile phones and base stations cause many diseases from cancer to infertility. If all those mentioned negative aspects were taken into account, then the mobile phone should have been totally rejected and not used at all. However, as related information on these issues has not been obtained through scientific ways, it is not possible to say that they are certainly clear. (http://www.bbc.co.uk/sn/tvradio/programmes/shouldiworryabout/mobiles.shtml retrieved 4/5/2008).

Stewart Report, which was published in 2000, explained that a "prudent approach" should be adopted when using mobile phones, and recommended adults to limit the time they use the mobile phone and be more careful about children's mobile phone use. The report recommends that children younger than 16 years old should not use mobile phones at all unless it is necessary.

In 2000, the Ministry of Health of the UK brought in the agenda the issue of turning the recommendations of the report into a leaflet and making them available with all mobile devices to be distributed at sales points by mobile phone vendors. However, whether they were effectively distributed at sales points is suspicious. Also, a conflicting situation is in question in this context. According to Stewart Report, mobile phones do not have any negative effects on health, while it is also expressed in the report that there are many people who do not share the same opinion.

The points related with the dark side of Mobile Marketing can be listed as follows:

- Unwanted calls.
- Unwanted messages.
- Voice calls of previously recorded advertisements.
- SMS advertisements that ring the phone.
- MMS advertisements that ring the phone.
- E-mail advertisements.
- Software and services supported by advertisers.
- Forced messages before using mobile phones.
- Unwanted Location based advertisements (LBA).
- SMS, MMS and videos distributed with buzz.
- Software, e-mail or web-based advertisements that force to call.
- Call recommended advertisements added to phones.

These are also some opportunities for companies to have better relationships with their customers. However if there is not a balance of these implementations, these can also be nightmares to the customers.

PRIVACY

Corporations and marketers gradually better understand the value of location and place for the mobile world. In Nash's practical researches, it is expressed that mobile marketing is scary (Scannell, 2008). Private life or privacy generally means a space, where people stay on their own, where they can freely think and act, and personally decide on where, when and under which conditions to communicate and contact with other people, and the right owned on this space. This right cannot be transferred to others by force or fraudulence.

Personal privacy is among the issues that should be most cared about by online and mobile advertising agencies because in these communications ZIP codes, mobile phone indices, friends and areas of interest of the individuals in the target group can be distributed towards adjacent areas (Scannell, 2008). Intensive aggressive marketing tactics could be seen as a violation of privacy.

Trust and honesty concern millions of communicating people as an issue which gradually gains further importance (Shiffman, 2008, p.11). Similar fraudulences to those frequently committed on the Internet may also be possible via mobile phones or devices. For example, swindling through fake Internet websites can also be realized via mobile environment in a similar way. This method, which is named as phishing, is also possible on the mobile.

Various attacks and tricks can be made to individuals in order to obtain their passwords. For example, personal information can be accessed and card number or passwords and other similar confidential information of individuals can be obtained by the fraudulent person who gives the names of relatives or excuses an emergency or legal issue.

Information acquired through certain software or Trojan horse type of virus software attacks to mobile phones can be used malevolently.

Shopping data can also be copied in a way and used by other people.

Mobile phone fraudulences may increase crime rates.

When using advertisements, blogs, or other marketing instruments, in addition to abstaining from putting the customers into awkward situations, cheating and deluding them, measures to keep the communication environment protected should also be taken (Shiffman, 2008, 11)

The concept of data brokerage should develop in correlation with trust. According to Norcross (2007), data brokerage should not only be based on trust. As methodological access may be more meaningful and important for some people, a controlled approach is essential.

Malevolent Use of Customer Data

Existing mobile phone users can be physically traced at any moment through information obtained from the mobile phone. However, such tracing can be considered as challenging personal privacy (Michael & Salter, 2006, p.58). It brings forward the consideration of privacy and security related legal situations. With such tracing, people's situations or behaviors and their words related with their personal opinions can be illegally identified and malevolently used. As people are not willing to be traced in this sense, a social paranoid may appear. Unauthorized tracing of the individual is not ethically appropriate either. The issues of correcting the concerns and distresses that appear due to this reason come forward. Under mentioned circumstances, a different situation arises with regard to obtaining statistical elements. Furthermore, if such data would fall in the wrong hands, then malevolent uses may occur. Since mobile phone is a personal device, tracing of the individual via this device is a serious violation of personal ethics.

Actually, obtaining the travel models of individuals with a voluntary program would assist in developing better transportation systems. It can help to build a data base for tourism plans. Such information can also assist doctors on issues such as the prevention of epidemics, and the accurate estimation of consequences in many cases (Borenstein, 2008).

In a lifestyle that brings forward individualism, it is foreseen that the individual should not be left alone and accept being held separated from others. In the US, personal privacy was described according to a verdict of the Supreme Court in 1928. Here, the given right is expressed as the most comprehensive of rights and also one of the most value associated rights by the civilized person.

Conceptual privacy is a personal right and it prevents the plundering and disclosure and exposure of private or personal information. This right is just given to the individual. Therefore, the person is set free about when, how and in what way to disclose information related with him or her or his or her private life. In this sense, scanning and recording systems provided through technology can only be considered as the personal authorization based use of local data. Many questions arise regarding the influence of mobile phones on personal privacy, and unfortunately most of these questions remain unresolved yet. As technology develops, attacks to personal rights and the privacy of private life may increase further. However, efforts are also spent in order to obtain solutions to secure them.

Loss of the Private Sphere

Opening the personal information area to some people at the desired place and time may cause the loss of the private sphere. Issues such as the tracing and listening of the individual necessitate taking certain precautions. Certain problems can be overcome through using jammers. Jammers have the ability to deform and mix the communication field. As this field expands, the issue of violating others' communication rights and freedoms may be in question. Actually, the authorities of organizations or individuals to use such devices are limited and they are defined with the law. Their unauthorized use is a crime. The tracing or restricting the communication of an entire society is both impossible and not logical (http://forumturk-talha.azbuz.com/readArticle.jsp?objectID=5000000008550446 retrieved 10/1/2008).

Such listening and tracing activities come on the agenda as phone tapping, and they can cause certain scandals related with people who deal with important assignments. Phone tapping issue may frequently be a news subject. Phone tapping is the enabling of a function or software in the device for unauthorized listening of the individual. These devices can allow transmission of any kind of voice data to any device only in seconds (http://video.haberturk.com/Video.aspx?v_ID=36041&k_A=haberturk 1/4/2009).

Actually, such problems and complaints can be considered as the problem of not only the present time but of almost the entire century. According to Arsan (2008), the word expressed in the US Supreme Court in 1928 proves this fact: "Discoveries and inventions have provided the governments with the opportunity to access the whispers at the backdoors. With these new inventions, espionage has changed form, and became a phone tapping action."

GIS (geographic information system) are the systems which can code and retrieve the addresses of people according to the geographical situation they are in. With the information to be obtained from this system, an opinion regarding the use of credit cards or store discount cards at sales points can be obtained. Graphics and maps, which are produced through these systems, can provide guiding information to those who deal with direct advertising or direct marketing. From the point of view of marketing practices, it can be used as a very important instrument regarding the selection and targeting of place.

Geospatial technologies are also technologies which can be used in order to track record and analyze personal information for commercial applications. These services which can also be partially used in the location-based service sector are obtained through highly developed geospatial technology. Through the evaluation of data that can be retrieved in this way, very important contributions can be obtained in terms of customer profile creation, target market selection and identification. The point that endangers personal privacy is the calculation of potential results of the correlation of geographical and personal information of individuals. Penetration into and violation of an individual's personal life is more possible with these technologies compared to others.

Certain public services can be provided faster and more effectively with the organization of the mobile structure. In certain countries, the distribution of mobile fingerprint identification to the police is planned in order to control the criminal record of suspected people. These devices can be started country-wide in a very short time. The project which allows taking fingerprints on the street aims to support the police. An extensive deployment of these devices which create various opportunities such as accelerating interrogations and savings of both time and money can be expected. Meanwhile, human rights organizations want the police not to archive fingerprints obtained during street controls in a manner similar to listening and monitoring activities, but to destroy them after the control (http://www.bilgicagi.com/index.php?option=com_content&task=view&id=1676&Itemid=137 1/4/2009).

According to Article 12 of the "Universal Declaration of Human Rights" which was accepted by the United Nations in 1948, "No one shall be subjected to arbitrary interference with his privacy, family, home or correspondence, nor to attacks upon his honor and reputation. Everyone has the right to the protection of the law against such interference or attacks." Also, according to Article 8 of the "European Convention on Human Rights", which was adopted by the Council of Europe in 1950 and enforced in 1953, "Everyone has the right to respect his private and family life, his home and his correspondence." In Article 17 of the Constitution of the Republic of Turkey, it is stated that "Everyone has the right to life and the right to protect and develop his material and spiritual entity....." and Article 20 indicates, "Everyone has the right to demand respect for his or her private and family life. Privacy of an individual or family life cannot be violated...." In article 22 of the Constitution, it is stated that everybody has the freedom of communication and secrecy of communication is fundamental. In many countries of the world, personal rights are protected with various clauses in many legal regulations, mainly in Civil Codes and Penal Codes. Some of the principles in the Convention of the Council of Europe express that personal information can only be collected for a certain purpose and it cannot be used for other purposes, and that information should be in accordance with the actual purpose and it should be stored only as long as it is required. Besides, the person about whom the information is collected has the right to learn about this fact and correct the wrong pieces of information when necessary.

According to Cho (2007), the European Union Directive on Data Protection (95/46) also protects the right to use personal and sensitive data with special methods. In this context, sensitive data is considered as those mostly including gender, politics, health, religion or participation in a commercial union.

Here, the important point is the determination of what sensitive information is and what can be disclosed to third people as insensitive information. For a woman who lives alone, the local use of this information may be sensitive because of the fear that she might be teased while some others being known where they are is fearsome. The information in GPS or mobile phones or inclusion in the location based system at various points does not cause any problems. However, the use of data obtained from mobile systems in specific technology to confirm and analyze its accuracy can be considered as the violation of privacy (Cho, 2007).

Giving away the information related with the place of a person is possible through mobile phones, location based services and applications on marketing and sales points.

Violation of location privacy appears through the use of obtained data via local profiling in order to learn about the activities in the living space of the individual. If the data in a certain region related with the activities of the person are collected without permission and not for use as a support for assumptions, then it can be also included in the issue of privacy.

Also a clear distinction should be made between the confidentiality of information and privacy of information. The confidentiality of information is keeping information obtained from someone as special and not sharing it with others. For example, in contracts with banks, age or income information of individuals is not transferred to third parties. Saving personal information is considered protecting and keeping confidential the details in the records of people concerned with law.

Geospatial Applications

The legal protection of the privacy of a person becomes ambiguous due to wrong or different perceptions regarding the concept of privacy. GIS is not a technology that violates personal secrecy, but when it is used with other digital applications such as data mining, there is the possibility of intervention with an individual's secrecy.

When a person uses a credit card or draws money from any ATM, uses a shopping card in anyway, makes a phone call or connects to the Internet or receives an e-mail message, then he or she leaves an "electronic footprint". Such records also feed geospatial applications. However, there are question marks whether the use of such information is ethical or not. Certain industrial policies or laws and rules also come on stream as protective regarding the use of privacy-related information in another field.

Security of information means preventing the acquisition of information by unauthorized people, and allowing people who are authorized to access the information, reaching it at the desired time and with the desired quality. Here, the moral issue is: most users do not realize the danger they are in because most of these situations occur below the awareness limits of people.

The Problem of Protecting Privacy in Public

Certain sensitive and/or private information of people is also normally protected by certain laws and policies. Such information is generally considered as non-public personal information (NPI). However, analysts concerned with privacy dwell on information collected from the Internet and when changed does not have a confidential and private character as social personal information (Tavani, 2005). Really, which information is private for the individual and which information is not? Are they those that cross one's mind, or those implemented, or those given as a secret to a very close friend? In addition to many such questions, both the issues of its use and presentation through technological products such as Internet or mobile instruments, and what can be done to protect that information still remains unanswered.

MOBILE FRAUD

Fraud is the act or group of acts to cheat someone with fraudulent actions against him or her or others', in order to obtain a personal benefit or a benefit for others (http://www.ceza-bb.adalet.gov.tr/makale/162.pdf retrieved 1/8/2009). Fraudulent acts are performed in mobile fraud, while mobile devices are used as the instrument. In these actions, fraudulence can be mostly conducted with methods such as identity hunting and giving appealing news or by creating an emergency feeling.

Mobile frauds generally try to employ the same or similar methods that they developed for Internet also for mobile devices. For example the fraud enters the system with a message which might appear as is if it was sent from a bank, e-sales center or a famous company or operator or any important person so that the message is perceived as a normal instant message or SMS message. This message is generally in a form that could be directly perceived as coming from a particular e-mail address or telephone number of that company. Discriminating an appealing fraud and reality is very difficult. Most of the time, it is even difficult for specialists. Therefore, the best protection method for mobile communication device users is to be careful and pay attention to five simple points. Certain signs that reveal mobile fraudulence are as follows (http://groups.yahoo.com/group/AkadeMilenyum/message/3437 retrieved 1/8/2009).

- Personal information request in an e-mail or mobile message.
- Creation of an emergency feeling with warning messages in order to cause instant answering without thinking much.
- Coercing the victim with several situations.
- Giving devastating offers to a person in general.
- Spelling and grammar mistakes on messages.
- Seeing a slightly changed Web address while doing Internet correspondance. For example, www.microsoft.com may appear as www.micosoft.com, www.mircosoft.com or www.microsoft.com.
- The read text is unbelievably good and full of opportunities

Some other methods that are performed through mobile phones and recorded in the literature as get a call and win fraudulence are as follows. In the simplest of these methods, frauds cause their victims to call themselves for certain reasons, and tell that there is a misunderstanding. They say they would solve the problem if the victim stays on the phone, and cause them to use the phone for minutes and sometimes even for hours, and so put the person into financial trouble (http://www.kom.gov.tr/Tr/HaberDetay.asp?HKey=16 retrieved 1/8/2009). Also, frauds leave unanswered calls to the numbers they randomly choose, or call by introducing themselves as a security officer. These frauds generally say that they are calling from the Internet Crimes Branch or Public Security Branch of Police with certain names. These names are the actual names of the people who work at those units, but they are also unaware of the call. When the victim calls back the given number, he is requested to stay on phone claiming that he teased major politicians, bureaucrats, businessmen or their families or children through text messaging, and for that reason an investigation was started against them. The victim, who is sure that he or she did not commit such a crime, wishes the case to be clarified and closed immediately because he or she is disturbed to hear his or her name mentioned in such a shameful case, and therefore, he or she does everything that is told to him. The victim, who waits on the line for minutes and sometimes even for hours, is without being aware, making the opposite party gain a small interest that is call credits, but at the end he or she has to pay an expensive telephone bill, and therefore he or she is both financially and distressed. In certain cases, when the victim calls back, he or she can be directed to 0900 lines or other similar services by the fraud (http://www.kom.gov.tr/Tr/HaberDetay.asp?HKey=16 retrieved 1/8/2009).

Identity Hunting

Identity hunting is frequently noted as a new type of threatening crime. This phenomenon not only belongs to the information age, but in history, culprits used to create false or fake identities in order to commit similar crimes and introduce themselves as the people of the actual holders of the identity. However at present, in the societies that have faced communication technologies on a global scale, this crime appears with certain new features. Internet, e-mail and mobile phone technologies both influence and change our lives in a positive way, and also in a negative way through facilitating such crime attempts. Together with the new risks, all these changes have also caused new fighting back areas.

Individuals gradually require protecting their identities more and more in relationships they build with others in both real and virtual environments; therefore, identity recognition processes gradually gain more importance in the mobile environment. Also, the identities of individuals are considered as very appealing goods in the world of crime. The importance of identity hunt appeared when personal data in modern society became involved in the crime structure as a problem. Most crimes can be committed over the hunted identity, and put innocent people under unimaginable burdens and accusations.

Various definitions are given for identity hunting. In the widest sense, identity hunting is the acquisition of personal information (e.g. name, address, social security number of credit card details) without permission in order to commit fraudulence and certain other crimes. Therefore, identity hunt is used as an instrument in order to change one's identity to commit the "target crime" and use the information as an instrument in "causing an opportunity for the crime". Here, the "target crime" results in a more complex and organized situation beyond a simple fraudulence and when they turn into crimes committed through technological crime networks, their scopes may increase further (Semmens, 2008).

Upon accepting the existence of two different identity hunting, two different dimensions can be introduced for the subject. Identity fraudulence is considered among the interim crimes, and it is based on using the identities of others for fraudulence. Therefore, identity information can be considered as an instrument which is held in hand until the crime is committed. The situations that activate such crimes also cause using one's identity for borrowing money or drawing loans, and turn into consequences which encumber someone else for the repayments. Some identity hunters can play the entire role of their victims.

Identity hunting frauds try to obtain identity information of their potential victims through various ways. They direct the victim to a fake web page through e-mail services, or mobile phone.

Some frauds, rather than directing to the Web page, tell the victim to call a customer support number. Identity information may be stolen if the victim gives his or her account number, personal ID number, password or other valuable information to the person who replies the call. The account at the bank can be accessed with this information. The person on the other end of the line indicates that the account would be closed or other problems might occur if the questions are not replied; thus causing the caller to panic (http://www.microsoft.com/turkiye/athome/security/email/phishing_with_phones.mspx. 08.01.2009).

In Europe, identity hunting is not considered as a widespread problem yet, and the related data is insufficient. Here, the exceptional data is that the number of such crimes is gradually increasing and turning into a problem. The United Kingdom is one of the countries with the most problems in this regard. According to the earliest related information, every year 137,000 people are affected from identity hunting, and its cost reaches 1.7 billion pound (Semmens, 2008).

From the quality and quantity perspectives, there are two types of victims in fraudulence. The first one is the person who is actually defrauded, that is the person whose identity is hunted while the second is the individuals, organizations or establishments whose names are involved in the fraudulence. Here, the victim who is defrauded is receives damages in various ways. In personal fraudulence, in addition to the damages from money or valuable assets on the personal account, very heavy debiting may also be possible. Fraudulence creates a series of negative effects on the history of every domain and creates a negative influence, which damages confidence. The victim of fraudulence will have to spend time and efforts to recreate his or her identity profile and use the autonomous system.

The researches indicate that loss of personal information decreases the life quality of individuals, and damages the self-confidence of the individual. Secondary victims, although they are not heavily damaged on financial terms, are affected in terms of security and the entire sector is damaged. The most affected sectors from fraud are the operators and finance sectors in the communication environment.

As a country where the utilization of communication technologies became widespread in the mid-1990s, the US officially described identity hunting as a crime. Meanwhile, other countries have been delayed or have fallen behind about taking measures on this issue due to various reasons. Identity hunting did not only take place in the agenda of the governments of the western world, but also became an important issue. In the present situation of technological terrorism, workable processes and methods should be outlined. These problems can be decreased by a large extent with the cooperation of agencies and companies in the world. Applications regarding document exchange between authorized bodies should also be revised (Semmens, 2008).

One of the most important precautions in this regard is the determination of single ID number as the social security number of the individual. Certain drawbacks have also been noted in this coding system which is considered as the key for public and private organizations' change, and certain doubts were indicated claiming that a single identification system would not work in the UK.

Regardless of what is done about identification, fraud will never be totally eliminated. Therefore, people should be educated and informed about such crimes. This may help to adopt a more responsible attitude about personal data. Informing mobile system owners on how to acquire and use the financial means and how to protect detailed information will help solve the problem.

The measures that can be included in the techniques and applications related with mobile devices can be listed as follows (The BlackBerry Enterprise Solution):

The encryption standards of new mobile devices are being developed, and they assist in helping make sure of the security and correctness of the data transmitted as wireless.

- The air wireless IT policy constraints and commands are used in order to find security installations of the devices, and the devices technologically lock when they are lost or stolen, or the location data of the device can be obtained
- Security systems for multi-purpose Internet mail extensions are being developed. Signatures and passwords identified in the wireless environment about personalization and identification should be sent in messaging between the receiver and sender.
- Intelligent card-readers should become readable.
- The current encryption systems should be developed with new and more effective measures.

MOBILE ADDICTION

The mobile use or the games played on mobile phone may turn into a passion for people and cause certain habits. According to the simplest definition, mobile addiction is one's feeling of necessity to always carry the phone with him or her and always keep it available. The issue of what the habits exactly are and the consequences of addiction have not been thoroughly examined yet. Literature related with habits is not very rich, either. Regarding the addiction of teenagers to mobile phone, a reasonable excuse is always shown and the telephone is used due to a countless number of requirements according to the prominent perception. Therefore, it is very important that mobile phone utilization is described as something that makes life more meaningful (Yamamoto, 2008).

Regarding the addiction and dependence of teenagers to mobile, first South Korea developed a program and according to this program, a non-governmental organization based study was performed in order to teach the young generation proper mobile phone use. According to The Korea Times, a pilot study about mobile phone use was performed at 12 elementary, middle and high schools. During the study, the mobile phone user test subjects were examined in terms of how they would behave when they do not have the telephone with them, and the symptoms they experience. Important data were also obtained for the proper use of mobile phones for teenagers.

In order to prevent drawbacks caused by the mobile phone, proper use is considered keeping the mobile phone away from teenagers and children from time to time. Similar to Korea, also in many other countries, there are cupboards for mobile phones, where students voluntarily leave their mobile phones during the lesson.

On the other hand, addiction to many video games or chucking away SMS messages on the mobile is in question and a stronger effort should be spent regarding its influences on teenagers and adults. The opinion that such issues should be inspected and supported is gradually getting stronger (http://www.mobile-weblog.com/50226711/korea_to_combat_mobile_phone_addiction.php retrieved 10/3/2008).

In the addiction related researches, it is understood that mobile phone use of teenagers is gradually increasing. (http://www.impactlab.com/2007/02/28/mobile-phone-addiction-may-cause-psychological-disorders/ retrieved 1/9/2009). The opinion that it would insidiously cause addiction not like alcoholism or other drug taking addictions is gaining prominence. Mobile telephone addiction is not quite understood at first sight and it is regarded as something normal. However, it is stated that 40% of the mobile phone users are young adults who use their mobile phones more than 4 hours a day. These users have low self-esteem. It is noteworthy that these people express that they feel terribly upset when they do not receive a reply to their messages or when they are not called.

Mobile addicts indicate that they pay great attention to the necessity of participating in the social environment such as family's and friend's parties, but they feel totally upset when they are deprived of their mobile phones. These people indicate that switching off their phones causes various disorders such as anxiety, irritability, sleep disorders or sleeplessness, and even shivering and digestive problems.

MOBILE TERROR

Terror is the attacks that do not require making an attack or that are made after an existing attack ends and normal life is reassured. Terror is a process which does not only affect its victims but also the people who watch them when it breaks out. One of the most important goals of terror is to create an environment of insecurity and ambiguity. Insecurity increases the future-related concerns of people who are subject to or who watch the terror attack while the ambiguity equally gives the feeling of a lack of control (http://www.elyadal.org/arge/birinsan.htm 1/9/2009).

From another perspective, mobile terror can be defined as the active or passive use of mobile devices in terror attacks. An attack can be made by using the remote control and similar multi-function features of mobile devices. (http://www.janchipchase.com/ retrieved 1/9/2009). Here, the aim is to hurt people and society, to make them feel pain, and to damage the existing feeling of security. When developing the technical features of existing mobile phones, obtaining good results is always intended; however, individuals or organizations with bad intentions employ this technology in terror attacks and in similar negative practices.

SPAMMING AND JUNK MESSAGES

Spam is a concept which has taken its place in our lives together with Internet and computers. Sending a high number of copies of a message on Internet, and forcing it to people who did not want to take the message, is defined as Spamming. Spam is mostly a commercial advertisement, and they are mostly sent to announce unreliable products, campaigns based on becoming rich very quickly, and semi-legal services. Spam can be sent on a very low cost from the sender's view, and the financial cost is mostly undertaken by the receivers of the message, the conveyor and the service provider (http://www.spam.org.tr/nedir.html retrieved 1/9/2009). With this approach, spams that are sent via mobile phones are named as mobile spams.

According to a joint study conducted by Intrado, University of St.Gallen (Switzerland) and International Telecommunication Association, more than 80% of the mobile phone owners in use received a minimum of one spam text message in 2004. Also, 83% of the participants of the study believe that mobile spam messages will turn into a serious problem for them in the next one or two years (Fındık, 2005).

According to the study of Jupiter Research, mobile spams continue to be a problem for 10% of the users in the US. Manis states that they are surer of the situation compared to 18 months ago that those spam mails were from overseas and the conveyors have moved these messages out of the market as a precaution (Fındık, 2005).

Undesired commercial communication (spamming), and cookies also have a distinct place in mobile phone terror. The Directive on Electronic Communication and Privacy dated December 11, 2003 indicates that a legal infrastructure to identify certain measures regarding location and exchange data, and the public indexes is required in the US (Michael & Salter, 2006, p.97).

MALWARE

Malware is software which is designed in order to penetrate into or damage a digital system without the knowledge (consent) of its owner. The expression malware is a generic term which describes an aggressive, coercive and annoying software or program code (http://doctus.org/bilgisayar-zararlilari-kotucul-yazilim-malware-nedir-t22390.html retrieved 1/8/2009). The main types of malware software can cause the following:

- File Destroying Trojan Horses that can be controlled by the attacker and it can be used for attacking at a certain date and time just like a software bomb.
- Service Preventing Trojan Horses.
- It can turn the computer or mobile phone of the victim into services equipped with different authorizations.
- There may be Trojan Horses which are received through File Transfer Protocols.
- Preventers of software detection (They prevent the functioning of popular antivirus and firewall software which protect the computer or mobile phones, and enable the attacker to penetrate into the victim's system via a personal computer or system).
- Worms can be sent.
- The control of mobile devices can be taken over through different ways.
- Advertising supported software can be realized with spywares.
- Advertising-effect messages can be distributed with shared spywares.
- They can be used in order to earn money without selling product or service to the user.
- By depleting the processor's power, annoying situations can be created at inappropriate times.
- The settings can be changed.
- A security threat and a difficulty of cleaning the conditions can be created on the system.

Additional precautions and costs may be required in order to prevent the occurrence of these computer and Internet environment situations also in the mobile environment. Wolverton (2008) indicates that the next war field of the malware software after the PCs will be mobile phones, which are most frequently used by individuals and especially Smartphones in this regard. Because the Smartphone can perform all the functions of a PC, it can also face similar problems. Although it is currently considered as a long shot, it is obviously an intuitive probability in the future. However, at present, losing mobile phones can be considered as a more important problem. Hazelton security experts state that they have already identified 500 viruses or similar malwares targeting mobile phones (Wolverton, 2008). Although the ratio of mobile malware is very low compared to PC malware, they are still very important in order to prevent personal damages.

So far, the most well known virus is Commwarrior-A, a malware which was spread on mobile phones in 2005. They are distributed through text messages on Symbian operating systems.

The hacker threat is expected to pass on GPS features. Also, a new type of spyware which is defined as "snoopware" is being developed. Snoopware allows tracing through the microphone or camera of mobile phones, and it can be used by malevolent people.

Expecting such malware to increase further in line with the increase in smart-phone use would not be wrong. It can be expected that the users of the products such as iPhone will also be affected by this situation. Here, because there is less control on the downloaded and uploaded applications such attempts and new applications can be expected on android platforms.

The experts recommend the following precautions in order to be protected from dangerous software (http://www.mercurynews.com/ci_10558223?source=rss retrieved 1/9/2009).

- To turn off the Bluetooth when it is not functioning
- To be careful when adding various applications to mobile phones, and to ensure that the source of the received messages are secure
- To seek digital signature which is used to approve correctness
- To be careful when opening the attachments of e-mails or text messages
- To use antivirus spyware type of protective programs which are developed for mobile phones.

COCOONING

As life is full of minor troubles which are sometimes manageable and which are not a continuously carried mobile device can also create such effects. Although each new generation enters into different interactions, the discomforts related with technological factors are noticed in similar ways. These troubles can arise not only from mobile phones, but also from any other discomforting situation in the vicinity. In order to move away from disturbing situations, some individuals return to their cocoons and cover their environment with an invisible shield. At present, technology develops innovations to assist in preventing this and similar troubles. For example, people can block the telephone calls or numbers that they do not want to answer their batteries can run out if they are not willing to talk long with the other side, or filter unnecessary mails and spams when they attend a meeting. The situation caused by these troubles is considered as an opportunity by telephone marketers.

The environment, which is covered with excessive advertising data, can gradually become more and more distressing for customers as explained in the mobile advertising section. Similar to an automobile commercial in the middle of a favorite TV series, an inconvenient telephone call or received text messages may also cause a similar and even more troublesome situation. Customers are willing to get rid of and isolate themselves from advertisement bombarding.

According to the cocooning trend, the home and surroundings of a person is gradually netted with different systems and the living environment turns into a shielded cocoon. The passwords for undesired data on mobile phones or systems which prevent understanding whether someone is at home or not and even access systems in the form of home offices provide opportunities to make business. It almost results in locking one's self at home. The questions below are coming on the scene "How should marketing be in an environment which helps use the time more economically, yet restricts the ability to move, and how should marketers benefit from that?" As potential customers use technology in order to build a cocoon around them, they begin to employ the logic of being interested in only the products or services in their areas of interest and in their advertisements. This gradually developing trend requires making use of cocooning in marketing. Therefore, a passage from the generic to specific should be designed.

According to the answers obtained from the question about this issue, a system with suitable conductivity for the company can be created. This system can convey only the values which preferred by the customer, and provide data access without damaging the cocoon.

It is also possible to allow customers to take control of the cocoons of product and service providers (by presenting options in elements such as mobile publications and the web). Such customers are offered with product and service recommendations that they would actually like to hear, and marketing with effective information becomes possible such as mobile vacation or myhotelguide.com for those who would like to go on a vacation. The development of technological products which strengthen the cocoon also supports those who present products and services in this field.

This trend can be sustainable through correct offerings to prevent customer distress and discomfort, and to strengthen the feeling of protection and happiness.

Also in mobile phone structure, certain efforts can be exhibited within the scope of premium services for those who would like to cover similar needs and build their own cocoons. Actually, although certain services offered by existing mobile operators provide consumers with certain options such as blocking unwanted calls or not sending advertisements, new methods should be sought in order to provide the suitable service for the customer group which needs a denser need for cocooning. Such ventures where the CRM concept can also be used more frequently will also facilitate the customer's need to isolate himself or herself from others.

BELLEROPHON METAPHOR

In life, as the balance of binary human-technology is lost and technology becomes heavier, its impacts on human relationships may also be felt strongly. Technology may cause a person to face certain negative consequences or barriers due to its potentials, and turn into a source of certain problems. The mythological tales mention such issues. After birth, Pegasus lives on Helicon Mountain. One day, a hero named Bellerophon appears and acquires the ability to control Pegasus everywhere by taking advantage of its speed. He thinks that one day he could fly to Olympus, the place where Gods live, and gain very large powers after reaching to the level where the biggest god is situated. The God Zeus does not need to do much to prevent Bellerophon and just sends a horsefly to bite Pegasus. Bellorophon loses control of the winged horse and falls on the ground.

The things that happened to Pegasus may also happen in our lives where both mobile phone and the Internet made pretentious entries. The great technology behind a tiny device, which is called mobile phone allowing it to be anywhere with a speed and size that almost challenges all of the antecedent communication technologies. That is, in addition to providing an ability to communicate and send messages everywhere, it also climbs up towards the peak in terms of transferring all aspects such as color, shape and sound. However, a small virus or a buzz would create a strong impact similar to the fly that bites Pegasus, and mess up the entire attempt all of a sudden like the power of the winged horse becoming uncontrollable. Therefore, users' problems may increase further when they lose control of their mobile phones which abusers aim to do.

DIFFERENCES & AFFECTS OF MOBILIZING

The legendary influences of the impressive voice on people has been narrated from generation to generation and made it to the present. Everybody expresses and accepts that as one of the most astounding voices a person can ever hear. The new legend is the communication environment which has been strengthened with mobile phones and similar instruments. The weakened person, the human and the society under the influence of momentary changes in the mobilizing environment really tries to coalesce with the fast-decision-making trend. At present, almost everybody agrees that the mobile phone is one of the handiest instruments people use.

Economic crises do not influence developments in the mobile field very much, and while industrial production decreases, the sectors that contain informatics and mobile-based technologies can remain standing. It is believed that this sector will regain its growth tempo in a very short time with minor adjustments, and its strongest aspect is to integrate it with communication instruments required for creating new businesses or sustaining the existing ones (Çalışkan, 2008).

The future of wireless technology will go much beyond laptops, mobile phones or MP3 players. Moving people or devices with this technology will be able to wirelessly communicate with each other. The sensor networks between cars and road equipment will further enable to acquire and use more personal and automatic traffic-related data (Ad hoc wireless and sensor networks. http://www.bbc.co.uk/worldservice/programmes/digital_planet.shtml 4/5/2008)

TECHNICAL PROBLEMS

In societies where technology enters, certain ideas and conditions that are required by technology should also enter. Such means organize ideas related with the change and create the classical conflict of ideas and attitudes between generations. Even the technical means that have been obtained through transfer create influences on the social intellect, the state, government and live styles, etc. (Ülken, 2005, s.35).

High-technologies constitute the shell of the technical spiral in our era. Communication technologies –with an important part on this shell- have certain self-specific technical problems. The transfer of a web site or a mobile sales project with a previously constructed technical structure from one place to another during the realization phase or in other phases after the completion of the project requires an extra cost and effort.

Creative ideas are required during the production of mobile devices, yet it does not prevent the occurrence of certain problems. There are also certain problems related with the supply of mobile devices and configuration of mobile systems. Besides, the brains suitable to use the technology are required when the devices are put in use. Learning how to use the devices, and the ability to perform the actions required on the system is also an important problem. The user manuals of mobile phones or the consulting services to be provided in this regard should be structured so as to address even to the most novice person. This is also required for marketers.

Therefore, any kind of mobile and web-related data should be followed up and analyzed. For the user, web site or mobile site is a problem which should be made suitable in terms of content, design and optimization. Turning the sites into a better and more effective form is also an important issue. However, most companies fail to measure the mobile traffic received from mobile devices at present. According to the research of Omniture', 71% of the companies do not count the mobile based traffics of their web sites (http://ekinozcelik.com/?p=197 retrieved 1/6/2009).

Also, certain problems may also appear because most mobile-phone accessible web sites are not optimized in order to fully comply with the changes in mobile technology, and that the content and design is not adapted to mobile devices. The use of mobile phone will increase further day by day, and those who fail to meet that demand will be condemned to disappear. Mobile marketing methods which draw the attention of mobile users are required, but it is noted that sufficient attempts are not taken in order to obtain scientific information in this regard. In this system, first of all, mobile based traffics should be measured and those measures should be interpreted. Correct interpretations will allow finding out the correct paths, and directing to the correct destinations.

REACHLESS TECHNOLOGY

The use of new products is proposed with the very rapid advancement of technology and the continuous introduction of new and higher features to mobile phone users. It is seen that because most users were born at a time without mobile phones and as they are used to the classical communication style; they do not show interest in most of these things proposed. Most of the time, they cannot use these devices due to lack of knowledge or failure to understand. Although every new development triggers a following one, it is possible to say that most users are not ready for such innovations. Even though groups that allow the immediate adaptation of societies pioneer for the innovation-based development offered by the mobile sector, in the overall society, the adaptation of such changes is not common enough to support fast development. It is obvious that new structures and customs in the society will also influence other developments. Having achieved a great progress in building superficial communication, will mobile technologies also guarantee building real communication?

MOBILE USE SIDE PRODUCTS

For example, cleaning materials for mobile phones have become a new sector, similar to those for computer keyboard cleaning. Also the accessories and apparatuses related with mobile phones have begun to develop a new production area. iPhones several applets and appliances also give as an example.

Instructors, who are to train the people to be equipped with the information based on new technologies, will create another group to develop new methods related with the application. Universities and certain research institutes will deal with these subjects further in order to obtain academic information.

INFRASTRUCTURE

The development of mobile communication infrastructure is utmost essential. Although wireless communication providing infrastructure can be offered in the metropolises of certain developed countries, developing countries adopt a different attitude. For example in Turkey, service providers adopt the objectives such as including every settlement unit with a population of over 2000 to serve a wide mass with a wide coverage area (Turkcell, 2008). In countries with a fewer population, this can be obtained much easily while the very high mobile communication taxes in developing or underdeveloped countries prevent the volume from rising.

MOBILE RECYCLING

Since the quick spread of mobile instruments in the global market and its rapid consumption with continuously changing devices due to new technologies, new mobile instruments are used within very short lifespans, and the old devices are disposed or sold in a second hand market if they still have any buyers. Considering the number of individuals who replace their mobile phones every two years on average, it is obvious that a serious issue of waste is in question. Having gone to waste, mobile phones and PDA etc. devices cover a place equal to their average size and height. For example, if 6 million mobile phones are disposed in a country on average (the average volume of a mobile phone is 50 cm3) approximately 300,000 m3 volume of waste is in question.

Furthermore, efforts should be spent in order to protect the environment from the damages of batteries and pollutant materials in the mobile devices.

In a world which has turned into an electronic market due to technological developments, computers are dumped as electronic waste in 2 to 5 years, mobile phones in 18 months and durable consumer goods in 15 years. The human kind produces 20 to 50 million tons of e-waste per year, even continuing to increase this rate by 5 to 8% every year.

According to the data of Gartner, Inc., global mobile phone sales in the first quarter of 2008 were 294.3 million units, indicating a 13.6% year per year increase. Compared to the first quarter of 2007, mobile phone sales in Western Europe decreased by 16.4%, indicating the first decrease in the region. (http://www.gartner.com/it/page.jsp?id=680207 retrieved 8/28/2008).

SOCIAL PROBLEMS

The leading problem is certain behavioral addictions. New addictions mean brand new cases for psychiatry clinics because new groups and concepts have arisen now regarding technological dependencies. Text messaging is shown as its best evidence. It is stated that certain related patients stay in Internet chat rooms or text message for more than seven hours a day (Michael & Salter, 2006, p. 9).

LAST WORDS

As people, we love making a story out of various things and turning them into mythological elements. It was probably the reason for writing the last words with certain mythological elements. Yin and Yang, one of the oldest Chinese mythologies, is considered a theory that explains the fundamental laws of nature and universe, basing their observation on a geography that covers a wide part of Asia for thousands of years. In its essence, the theory identifies that everything in nature and universe is conducted in opposition relations. Breeding, growing up and transformation are performed with the dynamism of the continuous struggle of the counterparts. Nothing can be stable, permanent and absolute. Yin and Yang is, rather than being the names of any objects, a concept that describes counter relations. In this context, a continuous development and dynamism also takes place in mobile marketing, with the use of mobile instruments, and as much as its pluses it also has minuses in our lives. While it is one of the channels to be most commonly used in the future which opens the gate for brand new instruments and occupations equipped with features of realizing social interactions, it will also cause elimination of certain occupations that became classical.

So, users in the future will decide whether mobile phone and new technologies are a gift or a punishment. For better or worse, one thing is sure that it introduces many new applications to both the marketer and the target consumer of marketing.

REFERENCES

Åkesson, J. (2004). The Birth of an Idea. In S. Gratton & D.A. Gratton (eds.), *Marketing Wireless Products*. Oxford, UK: Elsevier Ltd.

Arsan, L. (2008). Kişisel Gizliliğimiz Beklemede (Our Personal Privacy on Hold). *Bilgi Çağı, 4*(5), 51, 32.

Belek, I. (1999). *Post-Kapitalist Paradigmalar*. İstanbul: Sorun Yayınları.

Bilgisayar 5, (n.d.). 'Cep' 1,5 Yılda e-Atık Oluyor (Computers Turn into e-Waste in 5 years, Mobile in 1.5), *Hürriyet*, 10. Ağustos Pazar, s.12)

Bland, E. (2008). *Minority Report-Style Bots Coming Soon*? Retrieved 5/24/2009 from http://foorum.lcfk.org/viewtopic.php?p=1409&sid=35ca97fe42f081970f49f85a3dacfded

Borenstein, S. (2008). *Cell Phone Users Secretly Tracked*. Retrieved 8/3/2008 from http://dsc.discovery.com/news/2008/06/04/cell-phone-track.html

Çalışkan, N. (2008). Bilişim 6 Ayda Toparlanır (Informatics recovers in 6 months). Cumhuriyet, 12/30/2008, p. 10.

Cho, G. (2007). Privacy. In *Encyclopedia of Geographic Information Science*. Thousand Oaks, CA: SAGE Publications. Retrieved 1/4/2009 from http://www.sage-ereference.com/geoinfoscience/Article_n163.html

Ersoydan, A. (2007). MoPromo - Bluetooth Tabanli Reklam Agi Projesi Hakkinda (About Mopromo Bluetooth Based Advertising network) 12/10/ 2007 dated e-mail attachment Mopromo ppt.

Fındık, H. K. (2005). *Mobil Spam: Yeni Bir Sorun mu? (Mobil Spam: Is It a New Problem?)*. Retrieved from http://www.tele.com.tr/blog_comment.asp?bi=1402

Gratton, S., & Gratton, D. A. (Eds.). (2004). *Marketing Wireless Products*. Oxford: Elsevier Ltd. http://ekinozcelik.com/?p=197 (n.d.). http://dsc.discovery.com/news/2008/05/02/army-tiny-robots-02.htmlretrieved 8/3/2008 (n.d.). http://en.wikipedia.org/wiki/James_Bond#Films (n.d.). Retrieved 4/1/2008.

http://en.wikipedia.org/wiki/Privacy (n.d.).

http://en.wikipedia.org/wiki/Private_sphere (n.d.).

http://en.wikipedia.org/wiki/Spam_(electronic) (n.d.).

http://forumturk-talha.azbuz.com/readArticle.jsp?objectID=5000000008550446 (n.d.).

http://groups.yahoo.com/group/AkadeMilenyum/message/3437 (n.d.).

http://video.haberturk.com/Video.aspx?v_ID=36041&k_A=haberturk (n.d.).

http://www.bbc.co.uk/slink/features/iloveme_safety_mobiles.shtml (n.d.).

http://www.bbc.co.uk/sn/tvradio/programmes/shouldiworryabout/mobiles.shtml (n.d.).

http://www.bbc.co.uk/worldservice/programmes/digital_planet.shtml (n.d.).

http://www.bestkeptsimple.org/index/2005-1-12.php (n.d.). Retrieved 8/10/2006.

http://www.bilgicagi.com/index.php?option=com_content&task=view&id=1676&Itemid=13 (n.d.).

http://www.ceza-bb.adalet.gov.tr/makale/162.pdf (n.d.).

http://www.elyadal.org/arge/birinsan.htm%20 (n.d.). Retrieved from 1/9/2009.

http://www.gartner.com/it/page.jsp?id=680207 (n.d.).

http://www.impactlab.com/2007/02/28/mobile-phone-addiction-may-cause-psychological-disorders/ (n.d.).

http://www.janchipchase.com/ (n.d.).

http://www.kom.gov.tr/Tr/HaberDetay.asp?HKey=16 (n.d.).

http://www.microsoft.com/turkiye/athome/security/email/phishing_with_phones.mspx (n.d.).

http://www.turkcell.com.tr/bireysel/servisler/mobilinternet/cepteninternet (n.d.). Retrieved December 17, 2007.

Jensen, T. (2007). Retrieved 3/30/2008 from [REMOVED HYPERLINK FIELD]http://trekmovie.com/2007/11/07/star-trek-begins-shooting-today/

Korea to Combat Mobile Phone Addiction. (2007). Retrieved from http://www.mobile-weblog.com/50226711/korea_to_combat_mobile_phone_addiction.php

Michael, A., & Salter, B. (2006). *Mobile Marketing: Achieving Competitive Advantage Through Wireless Technology*. Oxford: Butterworth Heinemann.

Mobilcüzdan ile Kredi Kartı Cebe Giriyor (Credit Cards getting into pockets with Mobile wallet) (n.d.). Retrieved from http://www.tele.com.tr/blog_comment.asp?bi=2487&m=6&y=2008&d=1&s=search

Naisbitt, J., Naisbitt, N., & Philips, D. (2001). *High Tech/High Touch: Technology and Our Search for Meaning*. London: Nicholas Brealey Publishing.

Norcross, T. (2007). *Why Data Brokering is Losing Ground.* Retrieved 9/13/2008 http://www.consumer-preference.com

Postman, N. (1995). *The End of Education*. New York: First Vintage Books Edition, p. ix.

RB27036 / 25 October 2007 / Dutch Mobile Promotions 2007-Q3 in www.telekompaper.com

Scannell, T. (2008). *GPS-Based Mobile Ads: Where Does Privacy Fit?* Retrieved 10/24/2008 from http://news.earthweb.com/mobility/article.php/3780591

Semmens, N. (2008). Identity Theft. In *Encyclopedia of Social Problems*. Thousand Oaks, CA: SAGE Publications. Retrieved Nov. 18, 2008 from http://www.sage-ereference.com/socialproblems/Article_n276.html

Sheth, J., & Sisodia, R. (1999). Revisiting Marketing's Lawlike Generalizations. *Journal of the Academy of Marketing Science*, *27*(1), 71–87. doi:10.1177/0092070399271006

Shiffman, D. (2008). *The Age of Engage*. Ladera Ranch, CA: Hunt Street Press.

Smyth, P. (Ed.). (2004). *Mobile and Wireless Communications: Key Technologies and Future Applications.* London: BT Communications Technology Series 9.

Spam, M. *Yeni Bir Sorun Mu? (Mobile Spam: Another Problem?).* (n.d.). http://www.tele.com.tr/blog_comment.asp?bi=1402

Tarhan, B. (2008). Retrieved 6/13/2008 from http://www.marketingturkiye.com/BilgiBankasi/Detay/?no=479

Tavani, H. (2005). *International Review of Information Ethics, 3*. Retrieved from http://www.i-r-i-e.net/inhalt/003/003_full.pdf#page=41

Tekin, M. ve Çiçek, E. (2006). *Bilgi Çağında Bilgi Toplumu ve Bilgi Ekonomisi.* Retrieved from http://bilgitoplumu.blogspot.com/2006/10/bilgi-anda-bilgi-toplumu-ve-bilgi.html#_ftn3

The BlackBerry Enterprise Solution was Created with Corporate Data Security in Mind. (n.d.). Retrieved from 3030_BlackBerry_-_Developing_an_Application_Mobilization_Plan_for_your_Business.pdf

Toffler, A. (1981). *Üçüncü Dalga (Third Wave)*. Istanbul: Altın Kitaplar.

Turkcell (2008). Kapsama Alanı Yüzde 98'i Geçti (Coverage Area Exceeds 98%). *Cumhuriyet,* (23.08.08).

Ülken, H. Z. (2005). *Türkiye ve Çağdaş Düşünce Tarihi*. İstanbul: Ülken Yay.

Wolverton, T. (2008). *New Worry for Mobile Phones: Malware.* Retrieved from http://www.mercurynews.com/ci_10558223?source=rss

Yamamoto, G. T. (2008). Mobil Telefon Alışkanlığı ve Psikolojik Etkiler (Mobile Phone Addiction and Physicological Impacts). Retrieved from http://goncatelli.blogspot.com/2008/11/mobil-telefon-alkanl-ve-pskolojik.html

Yüksel, M. (n.d.). Mahremiyet Hakkı ve Sosyo-Tarihsel Gelişimi (Right of Privacy and its Socio-Historical Development). SBF Dergisi. Retrieved 1/4/2009 from http://dergiler.ankara.edu.tr/dergiler/42/458/5202.pdf

KEY TERMS AND DEFINITIONS

Fraud: On a simple level, it can be described as any activity by which service is obtained without intention of paying.

Malware: The word Malware is composed from the words "malicious" and "software". This is a collective term for all types of damaging software such as Trojans, Spyware or Viruses (http://www.emsisoft.es/es/kb/articles/tec080424/ 1/22/2009).

Mobile Recycling: The recycle of mobile devices to protect the environment.

Mobile Fraud: Mobile frauds generally try to convince and engage in cheating activities without intention of paying.

Spam: Spam is the abuse of electronic messaging systems to indiscriminately send unsolicited bulk messages. While the most widely recognized form of spam is e-mail spam, the term is applied to similar abuses in other media: instant messaging spam, Usenet newsgroup spam, Web search engine spam, spam in blogs, wiki spam, Online classified ads spam, mobile phone messaging spam, Internet forum spam, junk fax transmissions, and file sharing network spam (http://en.wikipedia.org/wiki/Spam_(electronic) retrieved 1/22/2009).

Privacy: Privacy is the ability of an individual or group to seclude themselves or information about themselves is revealed selectively. The boundaries and content of what is considered private differ among cultures and individuals, but share basic common themes (http://en.wikipedia.org/wiki/Privacy retrieved 1/22/2009).

Private Sphere: Private sphere is a certain sector of societal life in which an individual enjoys a degree of authority, unhampered by interventions from the government or other institutions. Examples of private sphere are family and home. Martin Heidegger argues that it is only in the private sphere that one can be one's authentic self (http://en.wikipedia.org/wiki/Private_sphere retrieved 1/22/2009).

Selected Readings from the Author

Chapter 11
E-HRM as a Reality in Virtual World

Gonca Telli Yamamoto
Okan University, Turkey

Ahmet Özbek
Iremis Consultancy, Turkey

INTRODUCTION

The human resources (HR) profession has had a rich history: that is, from the concept of labor unions in 1930 to the creation of selection tools in 1940 then to the reward systems in 1970 shifting to teamwork and then to partnerships in 1990. After the 1990s, like all other businesses, HR also influenced the Internet revolution. Classical human resources management (HRM) concerns the recruitment, selection, development, compensation, retention, evaluation, and promotion of personnel within an organization (Bernardin & Russel, 1993), which could be mostly transferred to the virtual world. Therefore, the HR profession faces an important challenge with new e-HRM and virtual actions encountered in the business world. This chapter examines the e-HRM phenomenon in the virtual world in detail and tries to bring new angles to e-HRM, and then manifests new definitions for the new reality while changes and interactions take place in the virtual world.

BACKGROUND

The estimations of access, usage, acceptance, and communication in the Internet have tripled or more in the last years (Hall, 2001; Hays, 1999; Murphy, 1999) and recruitment stages and career opportunities have changed (Kumar, 2003) after Internet has come into the stage. E-HRM fast break has already started within the 2000's. According to Rüel, Bondarouk, and Looise (2004), e-HRM is a relatively new term for the IT supported HRM, especially through the use of web technology. Some authors have singled out mastery of HR technology as one of the five competency domains for HR along with strategic contribution, personal credibility, HR delivery, and business knowledge (Schramm, 2006). E-HR approaches use both internal corporate intranets and the public Internet (Business Europe, 2000). According to Rüel, Bondarouk, and Vandervelde (2007), since the mid 1990s, organizations have increasingly introduced electronic HRM. It has also different names, for example e-HRM (Ulrich, 2000), virtual HRM (Lepak & Snell, 1998) Web-based HRM (Rüel et al., 2004), digital HRM, computer-based human resource information systems (CHRIS), and HR portals (Rüel, 2002).

However, e-HRM has been known as its position as the dynamic version of HRM. E-HRM functions are similar to HR management functions. These are planning for organizations and jobs for people, acquiring human resources, building individual and organizational performance, rewarding employees, maintaining human resources (Fischer, Schoenfeldt, & Shaw, 1996). The distinction between HRM and e-HRM functions is their process location. West and Berman (2001) also dwelled on differences between traditional HR and virtual HR.

E-HRM is also the cause of the start of the controversial part of the Internet revolution, which achieved dramatic improvement in critical areas of performance such as validity, credibility, compliance, proficiency, speed, etc. Within today's rapid developments, HRM departments must be strategy-focused, flexible, efficient, and client oriented, and all must be so at the same time (Lepak et al., 1998).

E-HRM is a platform that we use and follow. HR functions with desktop architecture, intranet, and Internet. E-HRM can also be defined as the information flow system pending from one user to a network of HR. Biesalski (2003) describes e-HRM as a Web- based tool to automate and support HR processes. E-HRM is also defined as a way of implementing HR strategies, policies, and practices in organizations through a conscious and directed support of and/or with the full use of Web-technology-based channels (Rüel et al., 2004). According to Rüel (2002), e-HRM is not a specific stage in the development of HRM, but a choice for an approach to HRM.

E-HRM is the planning, implementation and application of information technology for both networking and supporting at least two individual or collective actors in their shared performance of HR activities (Strohmeier, 2007). Karakanian (2002) explains e-HRM as "the overall HR strategy that shifts HR out of the HR Department and isolated HR activities, and redistributes it to the organization and its trusted business partners old and new." This new strategic approach also needs new tools and systems.

E-HRM TOOLS

There are four types of tools for e-HRM. These tools are functional HR tools, selection tools, development tools, and user tools. E-HRM tools for HR functions can compile career planning, job analysis, recruitment information systems, measurement and statistical records, selection, assessment, performance and compensation, training & development tools which an be used online or mobile. E-HRM selection tools are effective online and are legally useful concerning application blanks, bio-data, experience, records databases and tests, which can be suitable for completion using technological devices, reference and background checks from early records, e-HRM evaluation designs, and employee development tools.

The e-HRM developer tools are the software tools that give rise to make E-HRM software. Oracle E-BUSINESS SUITE (www.oracle.com), SAP - ERP HCM: human capital management (HCM) | HR management (www.sap.com), and IAS CANIAS ERP (www.ias.com.tr) are some examples. These are mostly modular applications that automate the entire recruit-to-retire process, so firms can align their

Table. 1. e-HRM flow schema (Source: Yamamoto, & Ozbek, 2007)

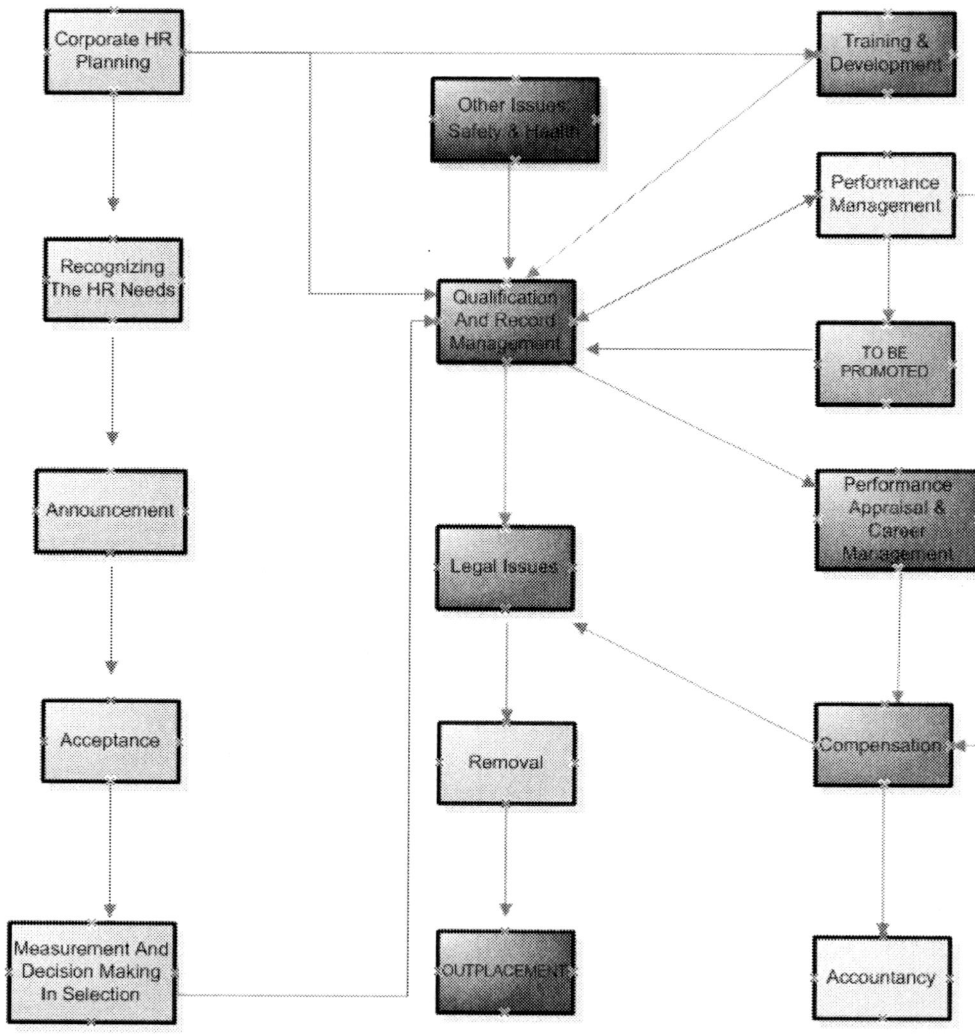

workforce with strategic objectives. User tools are mostly creative tools for career pursuits, which could employ new marketing techniques for self for new types of admissions, and follow-ups.

E-HRM AS A SYSTEM

The e–HRM system is a system that human resources experience is trying to deliver and interact within virtual world. Characteristically, this means that human resource management does what it thinks is right, as do the recruitment and the Web channels and all other HR work. Furthermore, brilliant technologies lead to fantastic products that create new systems that consist of producers, suppliers, and users of HRM which are named as the e-HRM macro system.

Table 2. Stages of e-HRM transformation and virtualization

Stages	E-Transformation Progress	Virtualization
1	Human Computer Interaction.	Computer as a medium for collecting and using some information.
2	Internet comes into circuit which means Web-presence HR Technological and systematical developments.	First step in the virtual world: e-HRM applications debut.
3	Functional movements through Internet which mean Web-enabled HR.	Virtual organizations and teams, and some functions of HR done by some software, some technological devices, a lot of new jobs and job definitions.
4	e-HRM fully implemented and intensively used by workers and companies turn to e-HMR.	Multiple virtualization-virtual HMR (human managed relationships) - post bureaucratic organizations.

Mostly for producers and suppliers, which could also intervene as a micro system and the reviews of Knouse, Heneman, and Greenberger (2004) and Heneman and Greenberger (2002), these address the crucial question of: how human resource (HR) functions are carried out in virtual organizations. This question is especially significant considering that most traditional HR functions—hiring, training, evaluating performance, and rewarding for performance (or disciplining)—assume there will be face-to-face interactions as the basis for carrying out that function.

This system is mainly of three parts: application, retention, and abruption. The system also comprises the back stage. Table 1 shows the flow diagram of e-HRM. Corporate HR planning is the first conduct for preparedness. After the HR need is identified, it should be announced. Then appeals are taken. Finally, the selection and replacement of workers are completed.

In the retention period, performance and credentials management maintain its importance, and legal reports and announcements are also created from performance management.

While carrying out performance training and development, balanced scorecards, scores, payroll management, and accountancy are kept up-to-date and used.

E-HRM TRANSFORMATION

E-HRM transformation begins with human and computer interaction, which is the basic step, and the primitive usage of hardware and software mostly constrained by laws or stressed by government applications.

In the second stage of transformation, intranet or Internet is used as a negotiator in internal relationships to collect information from all departments and partners. There is the need to negotiate better specifications and to add value to something that is negotiated from within and with out firms. Software is developed or adopted for HR at this stage. The reinforced technical infrastructure and technical support have been accomplished. Tabular matters and job descriptions are made while a business could also be opened to the public with a Web site and work applications are taken. This stage could also be defined as the Web presence of HR that means stepping into the virtual world.

In the third stage, the time of realizing Internet stream and opportunities are surfacing as a HR company in the Internet. This needs to be ready for user interaction and acknowledgment. Career centers and the other functions of HR that come out of the Internet means steps to improve HR services. Knowledge management comes into prominence. To create better harmony with collaboration and integration also maintains their importance. There will be examples of some successful e-HRM applications on the waves of change. How to forge great and matching workers to their jobs with e-HRM implementations will also be questioned in this stage. According to Adler and Heckscher (2006), a growing group of theorists have been exploring the possibility of a new form of trust that would enable interdependent activity in the more fluid, open contexts characteristic of knowledge production, reconciling choice with community. The past few years have seen a proliferation of work based on non-traditional forms of trust: "studied," "deliberate," "swift," and "reflective." And an emerging body of research focuses on new forms of organization among professionals and in "post-bureaucratic" firms and markets. This stage could be defined as Web-enabled HR with all parts of the HR solutions present and accessed online which could let firms, organizations and people enter into the virtual world.

The last stage is the total transformation of knowledge being used by artificial intelligence and systems that reach the total efficiency of human beings which also take into account human senses and human-managed relationships given the acronym of e-HMR (E-human managed relationships) what means multiple virtualization.

VIRTUAL WORLD, VIRTUAL ORGANIZATION, VIRTUAL TEAMS, VIRTUAL HR

A virtual world is an environment in which people use computer-screens and/or akin technological devices in front of them to access to Web pages to share common knowledge. There are possibly thousands of virtual worlds on the Web (Guittet, Loekkemyhr, & Schickel, 1998). Virtual world is an environment where people can share knowledge, do business, and study together by using new technological devices such as PCs, laptops, mobile phones, PDAs, PDA phones and other devices by accessing to common networks such as the Internet (Yamamoto, Telli, Featherstone, & Borstorff, 2006). Glaser (1997) discussed some of the limitations of virtual reality with reference to socio-technical systems.

Virtual teams and virtual work could be done as a source for differentiation and market responsiveness in this world, which uses outsourcing (Kakabadse & Kakabadse, 2005). E-business and virtual enterprises are being set up and promoted with or for outsourcing.

Virtual teams can be identified as their characteristics of working teams and leisure teams according to their functions or functional teams and dysfunctional teams according to their availability, etc. Virtual teams could be composed in the virtual world and these teams will be supplanted for real teams in businesses. These teams could easily bring together and add correct team members with e-HR for a significant project or work. But, this kind of collaborative work also requires a level of personal familiarity and trust.

Virtual teams are also teams of people who primarily interact electronically and who may meet face-to-face occasionally (http://www.seanet.com/~daveg/vrteams.htm). Virtual teams are defined as groups that have "gone digital" in order to function as a team (Borstorff, Featherstone, Camp, & Newton, 2006). Virtual team can be explained as a holistic approach integrating three key performance drives: people, process, and technology (Chase, 1997). Virtual teams are also members of a new and very creative virtual world which are composed of a certain virtual organization which can even be formal or informal. Helms and Raiszadeh (2002) explore the challenges of managers in establishing virtual offices and teams.

A virtual organization has been defined as a temporary network of independent companies, suppliers, customers, and erstwhile rivals—linked together by information technology to share skills, cost, and access to one another's markets (Byrne, 1993). Virtual organization is an organization distributed geographically and whose work is coordinated through electronic communications (http://www.skyrme.com/insights/2virtorg.htm).

According to Jones (1998), virtual HR relies on technology, ranging from multimedia personal computers to voice-response communication systems to incorporate Intranet that brings employees into direct contact with their organization's human resource system. In this system many of the typical HR tasks are accomplished without the direct involvement of the HR department. Virtual HR is handled by Bartol, Lepak, and Gardner, (2001). According to the Subramanian concept of Virtual HR, it relies on technology, ranging from multimedia personal computers to corporate intranet that brings employees into direct contact with their organization's human resource system (http://www.expressindia.com/fullstory.php?newsid=81617#compstory). Virtual HR also uses the latest technology to bring employees into direct contact with HR systems (HR Focus, 1994). When companies and people use virtual HR, there needs to be a system that meets HR supply and demand together. Virtual e-HRM hubs could be used for this connection.

A hub is the irreducible core of an organization (Dickerson, 1998). Virtual e-HRM hubs are the repository for the central core competencies of HRM in digital conditions and additionally, include the whole basic functional infrastructure necessary to coordinate and control workflow as it progresses between value chain and/or alliance partners as organizational hubs. Virtual hubs are also explained as the meeting points of virtual career centers (VCCs).

VCCs (career sites/career portals) are the Web portals that present global work opportunities. These career centers supply information regarding improvements of work life and competencies for people who are involved or connected to them.

These kinds of centers could also be concerned with applications, the results of the applications, and the relationships with companies. The criteria and details should be taken from both suppliers and workers within the application period. With VCC, comparisons among candidates should be done with competitions, exams, and projects. CV admissions are easier, detailed, and assorted by desired criteria from both the supply and demand side. VCCs are creating not only daily work opportunities, they also create punctual work opportunities for their applicants both on the supply and demand side. Virtual career centers could also segment the works and workers in niches. Career portals can also help streamline these efforts. Virtual career portals should also be concerned with ethical practices and the continuity of right work processes. National/international central data base operations are needed for the legal infrastructure of these systems.

Career Mosaic (http://www.careermosaic.com), E-Span (http://www.espan.com), the Monster Board (http://www.monster.com), and Virtual Job Fairs are some examples of these centers.

VCCs offer improved career services, on-line interest, work values assessments, e-mail VCC resumes for employers, search for occupations matching work skills and abilities, extensive labor market information in customized formats, expanded skills analysis; and saves information for on-line career planning, explores training options, and stores multiple resumes.

These virtual hubs are also defined as the channels for delivering information relevant to HR and the location for the meeting point of e-HRM market. These channels can be described as interdependent organizations involved in the process of making digital HR services available for utilization. Virtual career markets are the markets where e-HR solutions are done. These are markets where employees and employers encounter in virtual market space.

FUTURE TRENDS

There is a need for new tools and techniques for e-HRM for the realization of using the virtual world. There will be approaches for some useful e-HR tools for users, applicants, acceptors and deliverers. It is also a need for studies of the differences of e-HRM and virtual world. The studies for structural and environmental factors in e-HR formation are also another subject. Extemporary virtual reality opportunities and threats should also be discussed for e-HR. When applying e-HR, people and their behavior in the virtual communities should be focused. In adoption and adoption studies, achievements in stages and other circuit elements should be revised sequentially.

CONCLUSION

This study tries to construct a road map for e-HRM and e-HRM systems through virtual world which would be a substitute of reality for the near future. First the e-HRM concept is scrutinized in detail with several definitions. Then, e-HRM tools are explained as functional HR tools, selection tools, development tools and user tools. Furthermore e-HRM is analyzed to convey it as a system of macro and micro base. After the explanation of the stages from HRM to e-HRM, the relationships of the virtual world and e-HR are shown and the virtual e-HRM hubs and virtual career center concepts and future trends are discussed as a part of the e-HRM market.

REFERENCES

Adler, P. S., & Heckscher, C. (2006). Towards collaborative community. In C. Heckscher & P. S Adler (Eds.), *The firm as a collaborative community: Reconstructing trust in the knowledge economy* (pp. 11-106). Oxford, UK: Oxford University Press.

Anonymous. (2000). Virtual HR. *Business Europe, The Economist Intelligence Unit*, *8*(3), 1-2.

Anonymous. (n.a). *Virtual HR: A boon for booming companies*. Retrieved October 15, 2007, http://www.expressindia.com/fullstory.php?newsid=81617#compstory

Anonymous. (n.a). *Work force Santa Cruz county*. Retrieved October 15, 2007, from http://www.workforcescc.com /3/5/2007

Anonymous. (n.a). *Empower your people with smart e-HR tools*. Retrieved October 15, 2007, from http://www.epower.be/nl/weblog.html 3/2/2007

Bartol, K. M., Lepak, D. P., & Gardner, S. (2001, August). *Virtual HR: Information technology upgrades and HR effectiveness*. Paper presented at the 61st Annual Conference of the Academy of Management, Washington, D.C.

Bernardin, J. H., & Russel, J. E. A. (1993). *Human resource management: An experiential approach*. Singapore: McGraw Hill.

Biesalski, E. (2003). *Knowledge management and human resource management*. Retrieved October 15, 2007, from http://km.aifb.unikarlsruhe.de/ws/LLWA/fgwm/Resources/FGWM03_08_Ernst_Biesalski.pdf

Borstorff, P., Featherstone, M. D., Camp, L., & Newton, S. R. (2006). Virtual teams: The challenges of communication. *E- Business Review*, *6*, 20-23.

Bsat, M. Z., & Beckers, A. M. (2006). A comparison between electronic human resource systems. *E-Business Review*, *6*, 20-23.

Byrne, J. (1993, February 3). The virtual corporation. *Business Week*, 98-102.

Chase, R. L. (1997). Knowledge management benchmarks. *Journal of Knowledge Management, 1*(1), 83-92.

Dickerson, C. M. (1998). Virtual organizations: From dominance to opportunism. *New Zealand Journal of Industrial Relations, 1*, 35-46.

Fischer, C. D., Schoenfeldt, L. F., & Shaw, J. B. (1996). *Human resource management* (3rd ed.). Boston: Houghton Mifflin Co.

Fry, K. (2000). E-Learning markets and providers: Some issues and prospects. *Education & Training, 43*(4/5), 233-239.

Gardner, S. D., Lepak, D., & Bartol, K. M. (2003). Virtual HR: The impact of information technology on the human resource professional. *Journal of Vocational Behavior, 63*, 159-179.

Glaser, S. (1997). The real world and virtual worlds. *Internet Research, 7*(4), 246-251.

Guittet, C., Best, C., Loekkemyhr, P., & Schickel, P. (1998, December). *A shared virtual world: The virtual European statistical laboratory*. Online Information 98 International Conference, London. Retrieved October 15, 2007, from http://www.geocities.com/christian.guittet/papers/online_information_98/

Hall, J. R. (2001, April 9). Recruiting via the Internet. *Air conditioning, Heating and Refrigeration News, 212*(15), 26.

Hays, S. (1999). Hiring on the Web. *Workforce, 78*(8), 76-84.

Helms, M. M., & Raiszadeh, F. M. E. (2002). Virtual offices: Understanding and managing what you cannot see. *Work Study, 51*(5), 240-247.

Heneman, R. L, & Greenberger, D. B. (2002). *Human resource management in virtual organizations*. Greenwich, CT: Information Age Publishing.

Hooi, L.W. (2006). Implementing e-HRM: The readiness of small and medium sized manufacturing companies in Malaysia. *Asia Pacific Business Review, 12*(4), 465-485.

Jones, J. W. (1998, June). *Virtual HR: Human resources management in the digital age*. 1998 IPMA Assessment Council Conference, Chicago. Retrieved March 2, 2007, from http://www.ipmaac.org/conf/98/jones.pdf

Karakanian, M. (2002). *Are human resource departments ready for e-HR, architectures for e-business system building. The foundation for tomorrow's success.* Sanjiv Purba: Auerbach Publications.

Hooi, L.W. (2006). Implementing e-HRM: The readiness of small and medium sized manufacturing companies in Malaysia. *Asia Pacific Business Review, 12*(4), 465-485.

Knouse, S. B., Heneman, R. L.., &, Greenberger, D. B. (Eds.) (2004). *Human resource management in virtual organizations (book review)*. Retrieved March 3, 2007, from http://www.accessmylibrary.com/comsite5/bin/comsite5.pl?page=document_print&item_id=0286-21823792&purchase_type=ITM&action=print

Lepak, D. P., & Snell, S. A. (1998). Virtual HR: Strategic human resource management in the 21st Century. *Human Resource Management Review, 8*(3), 215-234.

Murphy, H. L. (1999). Top job sites. *Marketing News, 33*(16), 13-17.

Rüel, H. (2002). *E-HRM: Disaster or destiny?* Case-study research in five large companies on Web-based HRM. Retrieved October 15, 2007, from http://www.nwo.nl/nwohome.nsf/pages/NWOP_6D4HE8

Rüel, H., Bondarouk, T., & Looise, J. K. (2004). *E-HRM: Innovation of irritation.* Utrecht: Lemma.

Rüel, H., Bondarouk, T., & Looise, J. K. (2004). E-HRM: Innovation or irritation. An explorative empirical study in five large companies on Web-based HRM. *Management Revenue, 15*(3), 364-380.

Rüel, H., Bondarouk, T., & Vandervelde, M. (2007). The contribution of e-HRM to HRM effectiveness. *Employee Relations, 29*(3), 280-291.

Schramm, J. (2006). *HR technology competencies: New roles for HR professionals.* Retrieved October 15, 2007, from http://www.shrm.org/research/quarterly/2006/0406RQuart.asp 3/2/2007

Strohmeier, S. (2007). Research in e-HRM: Review and implications. *Human Resource Management Review, 17*, 19-37.

Ulrich, D. (2001, June). *What's next for HR?* Paper presented at ICEDR Meeting in Ann Arbour. Retrieved October 15, 2007, from http://www.daveulrich.com/pdf/papers/whatsnext.pdf 02.03.2007

Ulrich, D. (2000). From e-business to e-HR. *Human Resource Planning, 23*(2), 12-21.

Van den Bos, M., & Versteeg, R. (2007). *E-HRM: 'Enabling' HRM! E-Tools.* Voor Competentiemanagement. Retrieved October 15, 2007, from http://www.managementsite.nl/content/articles/303/303.asp 3/2/2007

West, J. P., & Berman E. V. (2001). From traditional to virtual HR. *Review of Public Personnel Administration, 21*(1), 38-64.

Yamamoto, G. T. (2004). Your new career: e-Learning process manager. *Training & Development, 58*(4), 66-68.

Yamamoto, G. T., Telli, A., Featherstone M., & Borstorff, P. (2006, July). *Virtual environments, virtual works, virtual lives?* Paper presented at PICMET '06, Portland International Conference on Management Engineering and Technology, Istanbul.

KEY TERMS

E-HRM: A platform that we use and follow HR functions with desktop architecture, intranet, and Internet / defined as the information flow system pending from one user to a network of HR/ defined as a way of implementing HR strategies, policies, and practices in organizations through a conscious and directed support of and/or with the full use of Web-technology-based channels

E-HRM Tools: The tools that used for E-HR functions, selection, software development for E-HR, and give convenience to users.

E-HRM Transformation: A modulation development from HRM to e-HRM.

Vırtual Career Center: Web portals that present global work opportunities.

Virtual Market: Markets where e-HR solutions are done.

Virtual Organization: A temporary network of independent companies, suppliers, customers, and erstwhile rivals—linked together by information technology to share skills, cost, and access to one another's markets

Virtual Teams: Defined as groups that have "gone digital" in order to function as a team.

Virtual World: An environment where people can share knowledge, do business, and study together by using new technological devices such as PCs, laptops, mobile phones, PDAs, PDA phones and other devices by accessing to common networks such as Internet

This work was previously published in Encyclopedia of Human Resources Information Systems: Challenges in e-HRM, edited by T. Torres-Coronas; M. Arias-Oliva, pp. 286-292 , copyright 2009 by Information Science Reference (an imprint of IGI Global).

Chapter 12
Business Ethics and Technology in Turkey:
An Emerging Country at the Crossroad of Civilizations

Gonca Telli Yamamoto
Okan University, Turkey

Faruk Karaman
Okan University, Turkey

ABSTRACT

IT ethics cannot be analyzed without assessing business ethics in general and the cultural environment. This study is based on the Turkish case. Turkey lies at the crossroads of civilizations, making it hard to define a generally accepted set of ethical principles. Western, Islamic, and Turkish cultures are in competition with each of them, and a synthesis has not yet been achieved. Therefore, a common identity and common ethical standards cannot be acquired. In fact, such a synthesis could be categorized as a new civilization. This disagreement causes proliferation of unethical behaviors such as the illegal copying of software. The majority of highly educated technical people in Turkey approves of the illegal copying of software, if it is necessitated by the interests of the country. This shows that we have a long way to go to reach global ethical standards, and country-specific differences cannot be eliminated in the short term.

INTRODUCTION

While the IT sector is growing exponentially and converging with every aspect of our lives and work, it also has effects on business ethics, and various moral or ethical problems can arise. The Internet also presents us with utterly new ethical challenges for which we have no precedents in our efforts to struggle (Bynum, 2005; Ess, 2002).

There are several definitions made through IT ethics. Some defined computer ethics; others defined information ethics. According to Bynum (2001), Deborah Johnson (1985) defined computer ethics as one that studies the way in which computers "pose new versions of standard moral problems and moral dilemmas, exacerbating the old problems, and forcing us to apply ordinary moral norms in uncharted realms" (Johnson, 1985, p. 1). Moor (1985) also defined computer ethics as a field concerned with policy vacuums and conceptual muddles regarding the social and ethical use of information technology.

On the other hand, information ethics is characterized as a biologically unbiased extension of environmental ethics based on the concepts of information object/infosphere/entropy rather than life/ecosystem/pain (Floridi & Sanders, 2002).

In Figure 1, we present a conceptual map of IT ethics. Business ethics is a subset of ethics, and IT is a subset of technology. The intersection of IT and business ethics is IT ethics. Business or IT ethics develops within the context of culture; geography; history; legal, economic, and political environments; and so forth. Unethical behavior in one country or civilization may be seen as ethical in another. For example some highly educated technical people in Turkey approve of the illegal copying of software, if it is necessitated by the interests of the country. A common example is cracking the Hotmail e-mail account in order to use a 250 MB quota. In fact, such high quotas are available only to American citizens.

For IT ethics, this presents an important obstacle, since IT products are used worldwide and cultural disparities hinder the establishment of globally accepted, standard IT ethics. The solution may be that each major culture and civilization should be analyzed from the perspective of individual businesses and IT ethics, and then, common denominators should be found.

The ideal solution would be the global acceptance of a particular country's environment and reducing the cultural differences to a minimum. In fact, at the start of the 19th century, Eastern civilization came close to that point; however, it could not succeed. In the 19th century, a great majority of the world was ruled by Western countries. Exceptions were China and the Ottoman Empire (Huntington, 1997; Lewis, 2002). Today, globalization again has become a major trend. However, Eastern cultures' contributions are much more significant. This is due to the relative economic and cultural strength of the East compared to the situation in the 19th century.

That does not mean that a clash among civilizations will take place, as proposed by Huntington (1997). Rather, it simply means that Eastern cultures also should contribute to the formation of a global culture. A globally accepted set of cultural values should not be determined solely by the West. This would diminish the attractiveness of the global culture for Eastern citizens. Global culture and ethics should embrace the best practices of the East. Huntington (1997) argues that differences among civilizations would not diminish but, rather, would widen. He thinks that this eventually would cause global-scale

Figure 1. The conceptual map of IT ethics

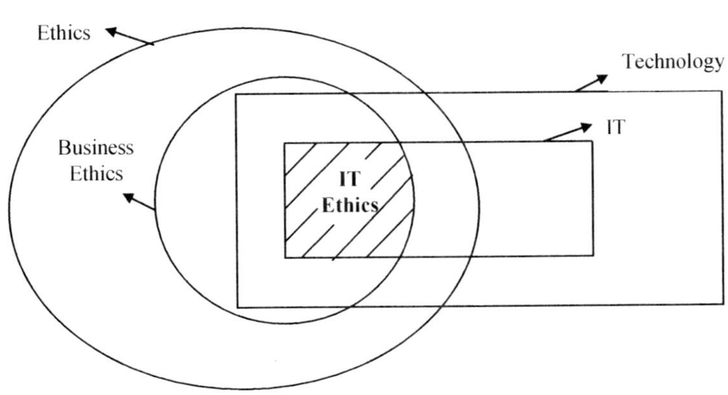

wars. It should be noted here that, although a thought-provoking theory, Huntington's views cannot be seen as Newton rules of global politics, and they are severely criticized. We think that "the clash of civilizations" was an important theory to explain global events, but we also recognize other forces and factors, such as globalization, nation-states, ethnic groups, and so forth. Regional economic pacts such as the EU and NAFTA also should be included. However, we do not see the clash of civilizations theory as totally useless.

Globalization can be seen as an antithesis of the clash of civilizations. However, we do not embrace globalization, too. The ideal mix is in between. Increasingly, the value of cultural diversity is becoming stressed. Cultural differences may pose problems for establishing global standards, but a strictly standardized world may hinder innovativeness, and the diversity of ideas and perspectives can be lost.

In Figure 2, we place the clash of civilizations at one extreme and pure globalization at the other. By *pure globalization*, we mean an ultimate level of globalization in which no nation-state or civilization exists other than the global state and civilization.

In this chapter, after analyzing the improvements and effects of IT all over the world, we will try to introduce the Turkish case for IT ethics and, thus, contribute to the global understanding of the cultural differences affecting the approaches toward business and IT ethics.

IMPROVEMENTS AND THE EFFECTS OF IT

There are several discussions about technology and business improvements and their effects on IT. According to Martin and Freeman (2004), the traditional way of technology posits two alternatives at either end of a spectrum. These alternatives are technological and social determinism. Technological determinism incorporates "the idea that technology develops as the sole result of an internal dynamic and then, unmediated by any other influence, molds society to fit its patterns" (Winner, 1986, p.21)

Such control requires technology to exist outside the construction of society and relies on an artefact's essence, as described by Heidegger (1977). A prevalent metaphor for technological determinism is a railroad line in which "technology is recognized as a separate entity that follows a linear path. Social determinism refers to 'what matters is not technology itself, but the social or economic system in which it is embedded'" (Winner, 1986, p. 20). Technology can be viewed as value-laden with inherent abilities to influence or direct its uses and users all around the world (Martin & Freeman, 2004).

Figure 2. The clash of civilizations vs. pure globalization

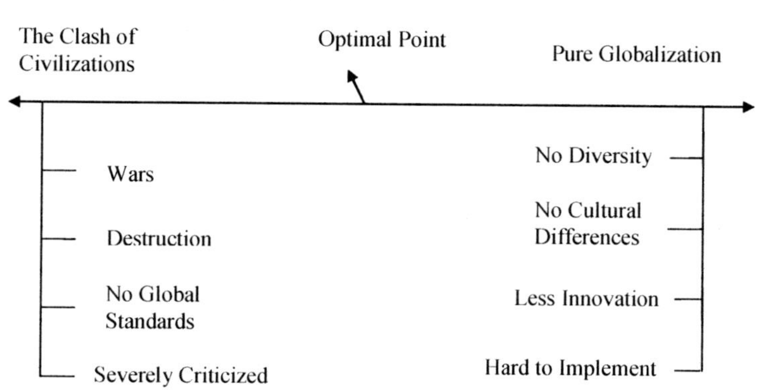

Our view of social determinism and technological determinism is a little bit more sophisticated than that of Heidegger's. We described our view in Figure 3. We think that Heidegger, as a philosopher, presented an oversimplified model for technology and society. We present a more sophisticated model in the figure. Even so, this is also a very simplified view.

Considering an interview with Brenan and Johnson (2004) in an article titled "Social Ethical and Policy Implications of Information Technology," information has been a key driver of globalization. Individuals as human beings and groups as a society gradually are relying on information and information technologies that make technology more prominent in their social and business lives. Intellectual property rights and security issues in information ethics also are concerns (Peace & Freeman, 2005).

Information and ethics should be studied in detail. Convergence of the technologies is an important concept that should be taken into account for predicting the future of IT ethics. The widening generation gap is particularly important.

Unethical and often illegal behaviors in the use of IT seem to be commonplace all around the world. At the same time, however, another study—software piracy and unauthorized access to computing facilities—are illegal acts in most countries, even in developing and developed countries (Phukan, 2002). Issues such as software privacy, virus development, and illegal system access, which once were viewed as an annoyance, now are considered major problems for organizations (Gattiker & Kelley, 1999). Crime and moral systems adoption should be asserted comprehensively, and emerging new ethical issues should be given more public attention.

Ethical questions arise when issues involve moral standards (Fritzsche, 1990). Therefore, unemployment, changing work life, and marketing styles should be emphasized in an emerging country context. The validity of ethics is based upon the acceptability of the reasons that support and justify them. In other words, emerging ethical issues should be explained to the ordinary citizen.

IT Converges with Everything

Technological revolutions are becoming more frequent, and technological changes are accelerating. Computers, telecommunications, consumer electronics, and entertainment all are converging. Mechanical parts are replaced by electronic parts, and electronic parts are replaced with software. With the manufacturing of cheaper and cheaper computer chips, computing power is now available for a wide

Figure 3. Technological determinism, social determinism

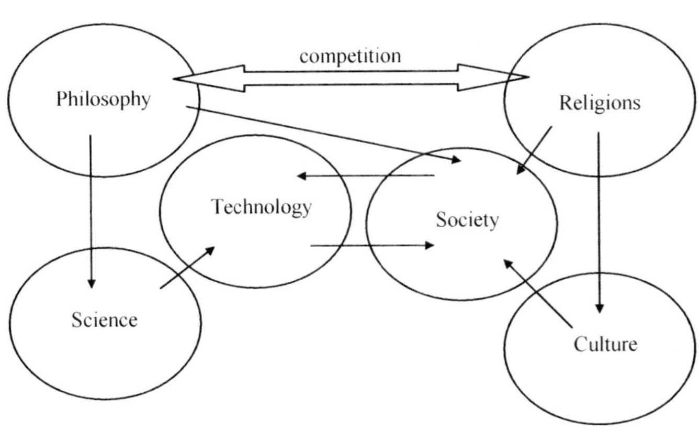

range of consumer goods, ranging from refrigerators to automobiles to MP3 players. As Moore's Law is expected to hold for two more decades, the future likely is to be characterized by more intelligent consumer products. According to Moore's Law, the number of transistors on a computer chip doubled every 18 to 24 months, and the corollary is that the speed of microprocessors, at a constant cost, also doubles every 18 to 24 months (Hulbert, Capon, & Piercy, 2003).

Digitally converging technologies already have been revolutionary technologies. Examples are the personal computer, the Internet, the television, video games, and mobile technologies. The term *new economy* frequently was used to emphasize the importance of these new products. It is a frequently criticized idea that that these products were revolutionary. Especially the dot.com crash in U.S. markets supported the views of the critics.

Even so, we reckon that these products were revolutionary. In our view, they have very deep impacts that can be visualized only after a long period of time. This is not to say that we adopt Heidegger's technological determinism as our point of view, but rather, we think that second- or third-order effects also should be considered before declaring these products as not revolutionary. Society may control these secondary or tertiary effects, provided that these effects can be discovered.

Therefore, it will have effects and side effects while it has a hold on everything without borders. Their convergence will have a synergistic effect, augmenting their revolutionary nature. This also makes their impact more complex, and second- and third-order issues should be considered.

Of course, it may seem rather strange to talk about the problems of such exciting new technologies. However, technology changes people's lives, and they will find themselves in a much different world in only a few years.

IT and the Generation Gap

As technological revolutions accelerate, the generation gaps widen. While parents can use the gadgets used by their children, usually parents are less tech-savvy than their children.

The term *digital divide* commonly is used to describe the generation gap in the use of new technology. Older people have difficulty adapting to new technologies. In fact, they need to be retrained in the use of new technologies. Their resistance to the idea of change needs to be decreased if they are to adapt successfully to new technologies. Therefore, special adult education programs are needed.

Information Technologies make new products more complex and harder to handle. In fact, most people cannot use the full capabilities of the products they use, but they pay for the unnecessary features, nonetheless. Companies benefit from this behavior in that R&D investments for the new and unneeded features are financed by such customers. Some of these features become mainstream as society learns how to use them. Younger generations are more accustomed to these kinds of new technologies and implementations.

IT and Crime

New technologies also bring new ways to conduct new types of crimes, since new technologies bring previously unknown, new capabilities. Existing measures against crime assume that criminal people have access only to old technology and can conduct crimes by traditional ways. Legal and regulatory systems are slow to adapt to new technologies, and such adaptation can be made only after a significant number of people are hurt by technology-driven crime.

The challenge of advanced IT products is that such products augment the capabilities of individual technologies, making them hard to regulate. The easiest solution apparently is banning the new technology; however, it is hard to persuade people not to use such attractive technologies. Given the slowness of the regulatory process, predicting the effects of technology on society will help the adaptation of society and its legal system.

IT and the Distribution of Income

In the past, technological changes were not so fast. Societies had more time to adapt to changes. Even then (e.g., in the period of the Industrial Revolution), the social fabric was severely harmed, crime exploded, and the world saw the rise of socialism and communism as a response to these problems. New technologies are eroding the value of some traditional professions and are giving birth to new professions. Thus, some specializations gain value, while others lose value. Those having rising professions will enjoy rising earnings, and those with falling professions will witness declining earnings.

Advanced IT products are much more capable than products of the previous technologies, and they offer a more suitable substitute for human capital. Also, IT products offer more functionality per dollar (i.e., technological capabilities can be purchased at a lesser cost). These factors have caused structural unemployment and a widening income gap. According to the State Institute of Statistics (SIS) data, unemployment rate remained as high as about 10% in spite of the high economic growth in the last three years. The most widely accepted explanation for this contradiction is the view that productivity investments prevented the translation of high economic growth into lower unemployment.

Myopia of the Marketing Function

Changing trends and technology have altered the traditional way of doing business. These trends have caused modifications in the marketing world. We are all concerned with the effects of successful marketing applications associated with the wireless world and new technologies (Yamamoto & Telli, 2005). Technological advances come with a cost. Human society faces increasing difficulty in amassing new waves of technology. In the face of fierce competition, high technology companies continually offer new and more capable products. These products address new and sometimes unknown needs of customers. These products should be transformed based on needs and/or should increase their usage for markets. When customers are not aware of these needs, high technology companies with excellent marketing efforts will stimulate their buying activities. Communication will gain more importance due to the application of these efforts, since interaction with company stakeholders will increase. Therefore, digital and/or mobile communication environments increase the speed and effectiveness of the marketing strategy.

Actually, marketing became the main function for companies with management in developed countries. In fact, the marketing function is more important than R&D functions for many companies. However, in practice, the marketing function by itself can be considered shortsighted, because usually the only important thing for companies is the sale of the product. The shortsightedness of the marketing function is emphasized by the introduction of complex IT products. These products have complex side effects, and in fact, the time of the market stage should include the analysis of such effects. For many companies, however, societal effects of the products they develop and market are not the primary concern.

Strategic thinking is the starting point of change. Changes also depend on the size and age of the company and other environmental factors. Focusing on the future is analyzing the risks and threats in the existing market and trying to forecast probable opportunities in the future. These risks are not only financial or market risks but also political, and technological risks. Even a change in the lifestyle/demographic characteristics of customers will influence customer purchases and the customer decision-making process. The R&D function should associate much more with marketing in order to create and fill the needs and wants of the consumer. The importance of R&D function should be understood by marketers, and a new synergy should result from that kind of implementation.

IT and New Ethical Issues

In brief, as these various elements show, projected complex IT products make problems stemming from technology more severe. Thus, the previous ways of technology production must be revised, and societal effects and possible problems should be predicted beforehand. This absolutely will lengthen the time-to-market period and add to the cost of the new product development, but this is exactly what society needs; namely, a slower and more responsive process of new technological development.

Complex use of several technologies makes the resulting technology highly complex and powerful. The problems caused by a complex technology are also complex. Such problems are different from the problems caused by separate problems caused by individual problems of the initial technologies that are converging. For example, the digital camera and mobile phone are two distinct products. Consumers can buy them separately. However, advanced mobile phones include a built-in digital camera. Since mobile phones have become indispensable, they are allowed in many places, such as classrooms. However, students may record some parts of the lecture and may distribute the recording via the Internet. Instructors are not protected effectively against such an incident by law or by generally accepted ethical standards. This example is taken from a real experience we faced in our lectures. Apparently, the ethical use of mobile phones with cameras in the classroom setting is unclear, and students have leeway to use their expensive gadgets to harm their instructors.

It is nearly impossible to predict all such second- or third-order impacts. Some of these only will be visible many years after the first introduction of the new technology. One recent example is the mobile phone with built-in digital cameras. Such cameras have unpredicted uses and may be used for spying, for example. This was not apparently the intended usage when these products first were introduced.

A person who is armed with several such new hybrid products enhances his or her capabilities. He or she may use them for terrorist or life-threatening activities. Even worse, most people are not educated in using such tools properly, and society-wide education is ignored due to its cost.

To conclude, the advancement of more complex IT products is not solely a technological issue. It has implications in a wide array of areas ranging from legal and ethical to social. Besides the new horizons brought by the convergence of the technologies, problems also are likely to arise. If we can identify such problems at an early stage, we can get ready for them.

THE TURKISH CASE

With its Westernization efforts, Turkey presents a very special case for analyzing IT ethics. In spite of the great efforts to become part of the Western civilization, Turkey is still struggling to decide to which civilization it wants to belong—Western civilization or Islamic or Eastern Civilizations. Some even argue the importance of establishing a unique Turkish civilization together with the Central Asian Turkish countries.

Business ethics, in general, are not established in emerging countries like Turkey. Although Turkish culture has its own traditional way of dealing with business ethics, the complex issues of the modern world are covered better by Western practices. The reason behind this is that traditional Turkish business ethics of the Lonca-Ahi system of the 15th and 16th centuries could not cope with the social, technological, and demographic changes throughout the world. The Lonca-Ahi system was a voluntary civil organization monitoring its members' ethical behaviors. Members with unethical conduct lost their membership, and nonmembers could by no means practice their occupations.

In other words, at the climax of the Ottoman Empire, Turkish people had developed good practices in many areas of business. Even today's Western civilization is affected partly by these good practices. This is inevitable, since all great empires in history have made their contributions. One example of the Ottoman influence is the Lonca-Ahi system. In this system, a craftsman should become a member of the relevant Lonca. Today, in the U.S., engineers and accountants are required to become members of regulating bodies in order to provide professional services. These were self-regulatory bodies in the Ottoman Empire, similar to today's U.S. practice.

However, as the Ottoman Empire declined, the Ottoman civilization gradually lost some of its qualities, including a high level of business ethics as well as economic power and military strength. On the other hand, Europe and North America both prospered economically and established higher business ethics practices.

In Turkey, in the traditional sectors, there is still some inheritance of good old business practices that are supported by the Central Asian Turkish practices and the Islamic view of business ethics. In the sectors using advanced technology, however, Turkish people can not accumulate comparable good practices yet.

When a civilization or culture does not develop technology, it cannot shape the technology in accordance with its values and cannot adapt itself to new and complex ethical problems forced by technology. In other words, business and IT ethics are high-level achievements for a society. To establish an advanced ethical framework and practice, a country first should solve its economic and security problems. Thus, Maslow's Hierarchy of Needs also can be applied to countries, cultures, and civilizations. When a country or civilization faces economic decline, high-level social qualities such as high ethical behavior also deteriorate, since low-level needs stimulate urgent action without considering far-reaching consequences.

For example, in a country in which software development is minimal, it is hard to persuade people to pay for software, since that society does not even know the meanings of software or similar technological concepts. There is also difficulty in the development process, and society does not give value to those efforts. In fact, software is not seen as property similar to a building or land. It is intangible, and people experience hardship visualizing why it should have a value and why copying it should be illegal and unethical.

This is exactly the situation in Turkey. Even the highly educated, Western-oriented segment of the Turkish population cannot see the impact or ethical problem of copying intellectual property. Society as a whole could not develop an understanding of the software industry, since the only interaction with software is at the user level. In contrast, there is little problem in the case of medical doctors and lawyers, because these are traditional occupations. Society has experience in these fields and respects the value of medical doctors' or lawyers' knowledge. People pay for these traditional services without hesitation but do not want to pay for software. Probably in the future, software also will gain the status of these types of services. People also will recognize software to be as valuable of a service as a doctor or lawyer.

Naturally, in the course of time, Turkish people also will get used to the concept of the value of an intangible asset such as software and will be more willing to pay for it. Intellectual property laws will speed up the process. However, another key issue is that the Turkish regulatory system is caught off-guard by rapid technological changes. A very severe problem is the lack of human capital having expertise both in laws and technology. Only such multidisciplinary-oriented people can solve the complex ethical and legal problems bought by IT technology. Fortunately, regulatory bodies do recognize the problem and take action accordingly.

The development of a high level of IT ethics in Turkey will take time and should be viewed as a process rather than a one-time, big step. This technology alters all societies, developed or underdeveloped. However, in emerging countries like Turkey, the results are much more catastrophic. Old institutions became obsolete in a short time and there is little idea about how to establish new ones or reform the existing ones. Obviously, such a transformation cannot be performed overnight.

To understand Turkey, one should understand the roots of the Turkish culture. Modern Turkey's culture can be examined looking closely at the following dimensions:

1. Central Asian Turkish Culture
2. Arab-Islam Culture
3. Western-European Culture
4. Migration to Big Cities and Modernization
5. The European Union (EU) Accession Process
6. A Synthesis or Dissolution

Central Asian Turkish Culture

Ethical behavior in Central Asia was the norm. However, since it was a nomadic civilization, the range of ethical behavior did not include respect for others' land or buildings. Even today in Turkey's central regions, some nomadic families have difficulty understanding why a certain land belongs to a particular person. Today, such families are rare; however, a substantial number of families live in suburban houses built on lands they do not own and, thus, create slum areas. They apparently see the land as public good and think that those who use it also own it. The ownership of real estate is apparently much more systematic in Western countries, given their people's settlement in the same cities for several centuries.

The weak understanding of the rights and privileges of a settled civilization becomes a more severe problem in understanding intangible assets such as company shares, bonds, and so forth. When it comes to intellectual and property rights, it becomes very hard to understand why one should pay for a software CD, especially in the amount of US$1,000 instead of copying and using it. Since it is a conceptual value with little physical attribute other than being a software CD, it is seen as having little value, if not zero. Needless to say, patents and copyrights are not recognized and, thus, cannot be protected.

Of course, that behavior harms the country, because intellectual production is not rewarded. Thus, the country cannot use its full potential and produce scholars, writers, computer programmers, inventors, artists, and so forth. In fact, this is a common problem for most developing countries. In Turkey, however, the problem partly can be attributed to the nomadic roots of the country's people. Although nomadic cultures are strong on building new states and even empires, they are not powerful in science and technology, as settled civilizations are.

Arab-Islam Culture

Turkey's second root is Islam, although lessened in importance after the modernization efforts of the past two centuries. Turkish Islam has differences from its original Arabic versions. Turks are generally strong practitioners of the religion. However, according to the ethical framework in Turkey, the effects of Islam are very deep.

The afterlife concept is stressed very much in Islam. Muslims think that they will be punished for their nonreligious and unethical behavior and be rewarded for their correct behavior. The problem is that the right or wrong behaviors do not match those of the Judeo-Christian Western civilization. Thus, some unethical behaviors in a Western country may be proper in an Islamic country. For example, a divorced woman is rejected by society in some parts of Turkey. In the Western sense, such a behavior does not violate any law, but it is unethical. The Muslim society uses this behavior to discourage a woman from divorcing. In this way, families and children are protected. However, Western civilization gives more value to the woman's freedom than to protecting families and children.

The interesting point is that even well-practicing Muslims may see nothing wrong with copying a book or software and would be reluctant to pay for it unless he or she has no other choice. Those who would consider intellectual property theft to be forbidden are rare. The argument can be that "Western countries sell us expensive goods and enjoy unfair advantage. Those countries are rich, and we are poor. We do not have disposable income to pay for those original books and software. Also, in the past, Westerners have stolen our wealth, so we now steal their wealth in response, and we will not be responsible for that behavior after death." We should note that Turkey is not an Islamic country according to the Arab-Islam culture, but its people have Islamic roots, and some of them are well-practicing Muslims.

Thus, with this kind of reasoning, the strong prohibition in Islam against stealing becomes ineffective in practice when it comes to the theft of intellectual property. This behavior also is attributed to the first factor; namely, nomadic roots. Intellectual property is not real tangible property and has no value—so it cannot be stolen, and its copying cannot be forbidden.

The argument that Westerners have stolen our wealth in the past has a pitfall in that the producers of current intellectual property are not the same Western thieves and most probably are not their grand-children. Even so, such arguments are seen as strong enough to see the copying activity as an ethical activity, hence pointing to the huge gap between the ethical understanding of a Western country and a non-Western country like Turkey. So, recently, copying has been seen as an opportunity to put the copiers in a retaliation mode.

Western-European Culture

Turkey's interaction with Western civilization started in the Ottoman Empire (ca. 1300-1922), especially in the 19th century. However, the greatest advancements toward Westernization were made by Turkey's founder, Mustafa Kemal Ataturk (1881-1938). In Turkey, Ataturk has a legendary position, and his views have affected Turkish people deeply. His reforms effected change directly in society, largely without any turbulence and disturbance of the community.

Ataturk aimed to modernize and westernize Turkey and succeeded to a great extent. After his death, however, the reforms lost their pace, and Turkey became a country of an eclectic culture. Today's modern Turkey cannot be understood fully by understanding only its cultures based on its major roots. In other words, Turkey has acquired many traits of Western culture, including Western ethics.

Individuality can be described as the quality of being an individual. A human being regarded as a unique personality is one of the strongest Western values (Wel & Royakkers, 2004). This probably empowers some of the highly pro-Western and modern segments of the population; it is common practice to copy books, music, and software. Thus, even the Western culture and ethical environment did not stop Turkish people's copying attitudes from taking root.

Migration to Big Cities and Modernization

In addition to the effects on the identities of the people living in Turkey, a very significant force affecting all the country—namely, migration to big cities—should be examined.

Although Turkey has a presence in industries worldwide, such as textiles, iron, steel, construction, white goods, and so forth, its economy is still dependent on agriculture, with 40% of its people working in agricultural jobs. Even so, that rate is steadily declining, and the recent economic crisis sped up that process.

In other words, Turkey only recently has experienced a transformation comparable to the Industrial Revolution of 19th-century England. The newcomers to big cities experience a culture shock, and their values and beliefs are under attack. The first generation keeps its culture and values taken from the rural areas of Turkey, and they identify themselves with the Turkish and Islamic identities, with the West being unknown or little known. By contrast, it is certain that city inhabitants have adopted Western culture and values as part of their behavior as they lived through the transformation that took place much earlier.

The second- and third-generation newcomers were born in the big cities or have lived for a large portion of their lives in those cities. Thus, they are more open to the effects of the cosmopolitan cultural atmosphere of the big cities. They also are affected by their families, and they live with the conflict of two different worlds. Identity crises are common among these people, and most of them are unsure about the set of values to which they belong. Naturally, ethical problems are not their primary concerns, as they are struggling with their basic needs.

The European Union (EU) Accession Process

In the case of Turkey, the European Union (EU) accession process is likely to fill the gap between Turkey and Europe. So Turkey will become part of a Western culture. There is an interpenetration argument concerning the concordance among Turkish, Islam, and European concepts. Göle highlights the need for a change in the minds of the members of these societies, because there is a horizontal relation taking place between the EU and Turkey (Tepeli, 2005). However, Europe as well as Turkey will be different than the Europe of today and will be less Westernized. This even may increase the differences between the two major powers of the world: the United States and the EU.

Turkish citizens living in foreign countries are known for their obedience to rules and regulations, although cultural and ethical adaptation takes time. The hesitation of the EU regarding Turkey is understandable, since Turkey also will affect the EU. If the accession is not finalized, the cultural problems in Turkey will deepen. If another set of countries is selected to form an alternative alliance, Turkey's dominant set of ethical values will be reshaped accordingly.

In fact, Turkey can be seen as an important case for setting global ethical standards. Success of such standards in Turkey will be a strong sign for global success, too, since Turkey is at the crossroads of civilizations and is still trying to establish the practice of the right set of ethical standards. The issue is which set of standards will be accepted as right by virtually all citizens. Such a set apparently will depend on the common denominators among people with diverse backgrounds.

Synthesis or Dissolution

Anatolia, which contains most of rural Turkey, is a very fertile land as well as a passage in all directions and, thus, affected by a lot of cultures. These kinds of natural environmental conditions and cross-cultural relations have a lot of influence on the ethical values. On the other hand, looking at Islamic culture, Central Asian Turkish culture, and Western culture in isolation does not define Turkish people. It also should be noted that as a successor of the Ottoman Empire, Turkey also accepts many subcultures. But the cultural, religious, and environmental factors are important and are analyzed in this study.

Lewis (2002) examined the position of the Ottoman Empire as ahead of Western Europe. Up to a certain point, the Ottoman Empire totally ignored the West and saw Western culture as inferior. After increasing military losses, however, Western culture is recognized gradually as superior, and the East is rejected. However, Lewis states that even today, Turkey is not a member of the West. Neither does it belong to the East. In his classic book, Huntington (1997) also defines Turkey as a thorn country together with Russia and Mexico. Such countries are at the borderline of civilization and are defined by distinctive identity crises. Those countries find themselves as fully encapsulated in neither of their distinct neighbors, while such countries nonetheless incorporate much of the cultural elements of their neighbors in their own identities. These cultural mixtures make them less clear and certain as to their national identity, at least compared to their neighbors.

As already seen, each component of the modern culture of Turkey has aspects that support ethical behavior; however, these are disregarded, and ethical issues are paramount. This can be attributed mainly to the remaining differences between Western culture and Turkish culture. Although Ataturk stressed Western civilization, he also emphasized the importance of the local culture and aimed to reach a synthesis of Western culture and local culture.

The problem may be the incompleteness of that synthesis. Turkish society today cannot be deemed homogeneous. There are several subcultures remaining, and naturally, each has different ethical values and different views about right and wrong. A good example is the obedience to traffic laws. In some developed western provinces such as Eskişehir, people do not cross the road when the traffic light is red, even if there is no vehicle passing. However, in Istanbul, pedestrians cross the road whenever they see that traffic slows a little bit.

Even if the ultimate synthesis is achieved one day in the future in the way Ataturk dreamed, it still will be a different culture than Western culture. So, the problem remains. The issue is again the problem of constituting one global set of ethical standards. However, the local differences will hinder such efforts.

In the pessimistic scenario that no such synthesis can be achieved or no one single identity and one single set of ethical standards can be adopted and applied by the great majority of Turkish people, the country may even eventually dissolve and not keep its unity.

THE INTERNET AND ETHICS IN THE WORKPLACE

The Internet is cited as a powerful technological revolution that affects all aspects of business. Early research on the Internet was interested mostly in finance and marketing functions, and the scope was the company as a whole. After extensive research in such tangible and measurable aspects of the firm, attention now turns to intangible assets such as human capital.

In Turkey, the Internet revolution is felt in a delayed fashion as compared to other countries, such as the U.S, Scandinavian countries, and certain Far Eastern countries in which Internet penetration reached high levels earlier. This delay can be attributed to the lack of telecommunication infrastructure, low per-capita income, and low PC ownership and literacy rates in Turkey.

Whatever the reason, Turkey also felt the changing forces of the Internet in a delayed fashion. This is not true for all segments of society, however. High-income segments are highly educated, they follow the global trends more closely, and the Internet penetration in these segments followed a more similar trend to those in the developed countries. Thus, the digital divide became a much more severe problem in Turkey.

Some segments of society, especially those living in the more developed Aegean and Marmara regions, left the agricultural sector early enough to build factories. Thus, people living in these regions became aware of the industrial-age concepts of company shares, capital markets, marketing, production, finance, and so forth, due to this development.

It is then not surprising to see that these segments of society had no difficulty learning to use and adapt to the Internet. Their Internet experience is very similar to that in developed countries and presents little research interest, given the extensive research coverage of the behavior of people in developed countries. Few studies have examined technology adoption and usage behavior in developing countries (Anandarajan, Igbaria, & Anekwe, 2002), few have examined the technology acceptance model across cultures (Straub, Keil, & Brenner, 1997), and few have examined the differences in Internet adoption and usage in developing countries (Brown & Licker, 2003). However, there is not enough documentation about the behavioral experiences in developing countries.

Apparently, the major adaptation problems are felt by low-income segments of society. These people are characterized by the crowded, low-educated families whose members either are still struggling with the traditional way of agricultural life or recently have left their villages and migrated to big cities. They are in transition from the Agricultural Age to the Industrial Age. In a way, they face the problems witnessed by the mid-19th-century England peasants or craftsmen.

In other words, while a segment of Turkish society faces the challenges of the transition from the Industrial Age, the great majority still tries to adapt to the Industrial Age. Even so, information age scientists who presented technologies such as the Internet, mobile communication, and ICT in general are not aware of this situation. There is no restriction for people to buy and use these post-industrial products. But can they use the full potential of them or use them properly? The answer is certainly not affirmative.

In particular, the Internet gives too much freedom and too much enthusiasm for individual space where these are otherwise largely unknown in a society such as Turkey's. In Turkey, the Islamic tradition still has powerful effects, and interpersonal relationships are subject to tight rules. That is changing fast, however. For example, partners increasingly are found on the Internet, and marriages based on love between individuals rather than as arranged by families become more frequent. The impact is profound among younger generations. Wheeler (2003, 2005) has some studies about Internet and age and gender relationships in Arab countries. It should be noted, however, that Turkey was the only country in the region with a secular system. Therefore, the experience in Arab countries is different than the Turkish case. Women have much more freedom in Turkey, especially in the developed western parts of the country.

In the early stages of the Internet, the primary access in Turkey was the workplace, and Internet connection was used for purposes other than the performance of tasks assigned by employers. In a way, efficiency was brought by the Internet and captured by workers, and they started to finish their tasks in a shorter period of time and use the rest for leisure purposes.

The issue was not of concern until the deep 2001 economic crisis in Turkey, and inefficient working practices prevailed. Until 2001, productivity became a top concern among companies, and they examined the daily lives of their workers more closely. The result was a wave of layoffs and restrictions over the use of the Internet in the workplace.

Nevertheless, home connections grew rapidly, thanks to DSL promotions supported by the local telecom monopoly, Turkish Telecom (TT). Naturally, unlimited DSL connections from home were not limited by supervisors or employers, and the use of P2P applications, chat rooms, MP3 sites, and dating sites exploded.

After the initial prohibition, even the employers were at the stage of accepting the power of the P2P application that they once prohibited and cautiously allowed their use for professional work. Even e-mail was not an acceptable formal way of communication until very recently. However, things are changing fast, especially among the young generations who are friendly to the Internet and joining the workforce.

The current stage in the transformation of the workforce after the Internet is the explosion of distance work. Traditional-style, office-based jobs are highly appreciated in Turkey, and temporary or flexible jobs are not respected at all. This can be attributed again to the late industrialization of the country. Apparently, industrial-age professions were very prestigious among the majority of peasants, and offices were considered as holy places compared to the fields they work.

Apparently, even the highly educated segments of society are affected by the old-fashioned ideas and beliefs of the majority about what an ideal job is. It is easier to follow the crowd, and it is difficult to challenge it. But transformation is inevitable. This may sound like technological determinism. Although we do not strictly follow Heidegger's view, in countries such as Turkey, society has less control over technology. This is related directly to educational levels and low income per capita.

This is a fast, difficult, and bitter experience for Turkish society as a whole. It is one matter to stay alive and another to be swept away by the waves of change. Who will survive and who will not is yet to be seen.

In brief, the Internet does transform Turkish society and the workplace. It is the driving force that is changing the culture and values of its users. The use of P2P programs and decreasing productivity are problems brought with the widespread use of the Internet by employees.

CONCLUSION

From the view of Western IT companies, emerging countries are problematic in terms of intellectual property laws and IT ethics. In these countries, such laws are nonexistent or are not applied strictly. Unethical behavior is also more common.

However, in order to discover the real reasons for intellectual property infringements and unethical business practices, the historic, cultural, geographic, economic, and political environments are analyzed. This can be done best by the scholars of that particular emerging country. In this study, we tried to analyze the root causes of the low standards of IT ethics in countries like Turkey. Turkey especially is a very complex case and cannot be understood without considering its history, culture, and neighbors.

In most of the developing countries, per capita GDP is only a small fraction of that of a developed country. However, software and hardware companies set flat prices globally, irrespective of the economic situation of a particular country. Especially for software and content, these disparities among the incomes of people become a motive for using very cheap illegal copies. Such people do not accept the view that they have not acted ethically, and they defend themselves by arguing that their lower disposable income justifies their copying. Thus, for IT companies, new and flexible pricing schemes in accordance with the individual characteristics of the companies might help to alleviate the problem. Illegal copying is also a problem in the developed world. However, these countries are better equipped to fight against it than are the developing countries.

Establishment of global standards in business and IT ethics is an ideal goal. However, there is much to do to reach that point. For many years to come, emerging countries will be headaches for developed countries in terms of IT ethics. However, the solution should start first from understanding and recognizing the situation of these countries. Once root causes are discovered, the rest will be much easier.

REFERENCES

ABI/INFORM Global. (2004). Social, ethical, and policy implications of information technology. *Information Management, 17*(1/2), 30.

Anandarajan, M., Igbaria, M., & Anekwe, U. (2002). IT acceptance in a less-developed country: A motivational factor perspective. *International Journal of Information Management, 22*, 47-65.

Brown, I., & Licker, P. (2003). Exploring differences in Internet adoption and usage between historically advantaged and disadvantaged groups in South Africa. *Journal of Global Information Technology Management, 6*(4), 6-26.

Bynum T. W. (2001). Computer ethics: Its birth and its future. *Ethics and Information Technology, 3*(2), 109-112. Retrieved September 21, 2005, from en.wikipedia.org/wiki/Business_ethics

Bynum, T. W. (2005). *Computer ethics: Basic concepts and historical overview.* Retrieved November 18, 2005, from http://plato.stanford.edu/entries/ethics-computer

Ess, C. (2002). Computer-mediated colonization, the Renaissance, and educational imperatives for an intercultural global village. *Ethics and Information Technology, 4*(1), 11-22.

Floridi, L., & Sanders, J. W. (2002). Mapping the foundationalist debate in computer ethics. *Ethics and Information Technology, 4*(1), p. 1-9.

Fritzche, D. J. (1990). Emerging ethical issues in international business. *S.A.M Advanced Management Journal, 55*(4), 42-46.

Gattiker, U. E., & Kelley, H. (1999). Morality and computers: Attitudes and differences in moral judgements. *Information Systems Research, 10*, 233-254.

Heidegger, M. (1977). The question concerning technology. *The question concerning technology and other essays*. New York: Harper & Row.

Hulbert, J. M., Capon, N., & Piercy, N. F. (2003). *Total integrated marketing*. New York: Free Press.

Huntington, S. P. (1997). *The clash of civilizations and the remaking of the world order*. New York: Touchstone.

Johnson, D. G. (1985). *Computers and ethics*. Prentice Hall.

Lewis, B. (2002). *What went wrong? Western impact and Middle East response*. New York: Oxford University Press.

Martin, K. E, & Freeman, R. E. (2004). The separation of technology and ethics in business ethics. *Journal of Business Ethics, 53*, 353-364.

Moor, J. H. (1985). What is computer ethics? *Metaphilosophy, 16*(4), 266-275. Blackwell.

Peace, A. G., & Freeman, L. (2005). *Information ethics: Privacy and intellectual property*. Hershey, PA: Information Science Publishing.

Phukan, S. (2002). *IT ethics in the Internet age: New dimensions*. Retrieved from http://proceedings.informingscience.org/IS2002Proceedings/papers/phuka037iteth.pdf

Straub, D., Keil, M., & Brenner, W. (1997). Testing the technology acceptance model across cultures: A three country study. *Information and Management, 33*, 1-11.

Tepeli, S. (2005). Interview with Nilüfer Göle, author of interpenetration l'Islam et l'Europe. *Yeni Aktüel, 11*, 48-50.

Wel, L., van, & Royakkers, L. (2004). Ethical issues in Web data mining. *Ethics and Information Technology, 6*(2), 129-140.

Wheeler, D. (2003). The Internet and youth subculture in Kuwait. In C. Ess & F. Sudweeks (Eds.), Technologies of despair and hope: Liberatory potentials and practices of CMC in the Middle East. *Journal of Computer-Mediated Communication, 8*(2). Retrieved August 20, 2006, from http://jcmc.indiana.edu/vol8/issue2/wheeler.html

Wheeler, D. (2005). Gender matters in the Internet age: Voices from the Middle East. In M. Thorseth & C. Ess (Eds.), *Technology in a multicultural and global society: Programme for applied ethics, publication series no. 6*. (pp. 27-42). Trondheim: Norwegian University of Science and Technology.

Winner, L. (1986). *The whale and the reactor: A search for limits in an age of high technology*. Chicago, IL: University of Chicago Press.

Yamamoto, G. T., & Karaman, F. (2005). A road-map for the development of the content protecting technologies (CPT) for the content based e-business models. *E-Business Review, 5,* 226-232.

Yamamoto, G. T., & Telli, A. (2005). High technology for marketing: New applications & integrated circuits. In *Proceedings of the 4th International Conference on Science Marketing,* Pretoria, South Africa.

This work was previously published in Information Technology Ethics: Cultural Perspectives, edited by S. Hongladarom; C. Ess, pp. 184-199, copyright 2007 by Information Science Reference (an imprint of IGI Global).

Chapter 13
Barriers to E-Procurement Adoption:
The Turkish Case

Gonca Telli Yamamoto
Okan University, Turkey

Faruk Karaman
Okan University, Turkey

ABSTRACT

E-procurement practice is not well-established in emerging countries. There are barriers in terms of transportation, financial, telecommunication, and legal infrastructures. Also, a lack of a qualified workforce, cultural barriers, and security problems hinder the development of e-procurement activities. These are not such significant problems in the developed countries. In this study, we examined the Turkish practice. We gave country background information including macroeconomic data and Internet penetration data, and we gave a picture of the current situation of Turkey. We then discussed how these barriers can be overcome in Turkey. Finally, we gave predictions about how m-procurement may alter the situation and its potential for the Turkish market.

INTRODUCTION

E-procurement is the use of electronic technologies to flow and enable the procurement activities of an organization. It is used particularly within the range of the operational, professional purchase. E-procurement is also defined as the business-to-business purchase and sale of supplies and services over the Internet. Typically, e-procurement Web sites allow qualified and registered users to look for buyers or sellers of goods and services.

E-procurement is a concept strongly related to concepts such as logistics, supply chain management (SCM), and even e-commerce. Although, some definitions are suggested to distinguish between these concepts, the problems faced are similar. They all require financial, transportation, legal, and communication infrastructures (Ohmae, 2000). If a country is weak in one or some of these infrastructures, then e-procurement activities are destined to fail. In addition to these four factors, education of the e-procurement personnel, security issues, societal readiness to use information and communication technologies (ICT), and the impact of the wireless technologies should also be considered.

In countries like Turkey, huge infrastructure investments are the norm. However, since these investments are heavily affected by short-term political concerns and voter influence, projects function most of the time at a sub-optimal rate. A planned and systematic approach is rarely found. Even so, as the economy grows and international trade rises, Turkish companies are facing the pressure of global competition. To stay alive in this environment of fierce competition, efficient tools such as e-procurement are being introduced. However, this is an ongoing process, and fast improvements should not be expected. Problems are still paramount and can slow down the widespread adoption of the e-procurement practices.

In this study, we try to assess the e-procurement activities from a developing country's perspective, namely Turkey. Turkey's unique problems and strengths are also examined. In addition, given the widespread use of wireless technology in Turkey, we have also elaborated on the concepts of m-procurement. The role of ICT is well-documented in the literature, but the possible outcomes of the widespread use of wireless technology in procurement activities are rather an uncharted territory. These issues are discussed in the "Future Trends" section of this chapter.

BACKGROUND

Changes and developments in technology and macro trends such as globalization also affect the micro world of enterprises and functions in the companies. In this respect, purchasing has morphed into procurement (Kotler, 2004). In the past, the purchasing function was seen as a way to execute a transaction between a buyer and a seller. Nowadays purchasing is a function that needs to be executed on a broader level. This means connecting different partners and helping them come up to the mark (Shah, 2002). Pride and Ferrel (1989) classified the three types of organizational purchases as new task, modified rebuy, and straight rebuy purchases.

Internet procurement has the potential to redesign and streamline procurement processes by shaping electronic markets that are tailored to reflect the contracts, purchasing rules, and business workflow of an individual buying organization (Aberdeen Group Profile, n.d.).

E-procurement has a multi-layered body and is connected to ancillary industries and distribution channels. For example, vehicle routing problems are part of the general purchasing system (Emel, Taskin, & Deniz, 2004). These problems are also part of e-procurement. Stanton and Stanton (2002) have created a model of Internet purchasing to show the link between personality, predisposition towards innovativeness, and adoption. However, there is not an adequate solution in the monetary value of face-to-face bargaining.

Tomkins (2000) proposed the supply chain synthesis (SCS) which he claimed to be the next step to SCM. If SCS can be implemented, procurement activities throughout the supply chain can be streamlined. Raisch (2001) saw the future of SCM and e-procurement as B2B marketplaces and stressed the importance of content management and community building. Bovet and Martha (2000) use the term "value net" as almost the equivalent of B2B e-marketplaces. Kuglin and Rosenbaum (2001) and Guinipero and

Sawchuk (2000) are among others who have discussed how the Internet has revolutionized the supply chain and procurement process.

Croom (2000) recognized the operational and strategic benefits and their influences on Web-based procurement systems. There is also a global procurement bias not only for goods but for the service activities. Global procurement has received an increasing amount of managerial attention in recent years for service activities (Kotabe & Murray, 2004). Trends in global sourcing, emphasis on time to market, customer uncertainty, and the need to improve bottom line costs affects the procurement processes (Kalakota & Robinson, 2001), and this conveys e-procurement's very important role.

Vigoroso (1999) considered the pros and cons for buyers and sellers by migrating business onto the Internet. E-procurement pros include saving sourcing time, receiving accurate information, having new sources of supply and having the ability of comparing different sellers, lowering overall operating costs and prices paid, optimizing the supply base, and having more control over spending and inventory for sellers. However, lack of security is a major problem concerning e-procurement.

Generally, studies on e-procurement report large efficiencies regarding process and procurement costs (Gebauer & Segev, 1998). Puschmann and Alt (2005) discussed the use and success of e-procurement in supply chains.

According to Neef (2001), e-procurement becomes the catalyst that will allow companies to finally integrate their supply chains from end to end, from sales to supplier, with shared pricing, availability, and performance data that will allow buyers and suppliers to work to an optimum level with mutually-beneficial prices and schedules.

The UND (International Transportation Association of Turkey) officials state that logistics in general and procurement in particular is highly crucial for Turkey, given Turkey's heavy dependence on the textile sector and fast-growing international trade volume (E-mail correspondence of Dr. Yamamoto with UND officials, July 15, 2005).

It is stated that the sector representatives were aware of the Internet's potential; however, they were hesitant to make purchases over the Internet for the following reasons:

1. Companies and state enterprises in Turkey are not ready to share data over the Internet or by classical means; they see data sharing as a threat to their competitive position in the industry.
2. For large-scale purchases over the Internet, companies want legal protection; although e-signature regulations have recently been passed, there is still time for e-signature applications to mature.
3. E-procurement requires a large-scale implementation of a network of companies with seamless integration of hardware, software, firm data, and legal, regulatory, and supervisory organizations; this means there needs to be a very serious dimensional integration in developing countries.
4. Companies do not want their procurement activities to stop because of electricity shortages, software or hardware problems, or viruses; purchasing everything online will make companies dependent on the Internet, and they will be much more vulnerable to its problems.
5. Especially for international procurement activities, there are numerous official documents in use, and they should be transferred to the electronic medium; such a study is under way in cooperation with the United Nations, and pilot work is currently being done for the Gemlik Industrial Zone of the Bursa province of Turkey.

There is a huge potential for e-procurement in Turkey; however, a turnkey solution is needed which would encompass the hardware and software needs, and those regulatory issues which are specific to

Turkey. It can be concluded that Turkish companies are aware of e-procurement, but they are far from implementing it fully. However, they are likely to adopt e-procurement at a slow pace. The rate of its acceptance depends on the work that will be done by nongovernmental organizations (NGOs), state organizations, and private-sector unions.

TURKEY: BASIC COUNTRY INFORMATION

Turkey is a mid-sized developing country located between the Middle East and Asia. Turkey also meets the Slavic countries with the Arab world, and Caucasian countries with the Mediterranean countries. This highly strategic geographical position gives the country the role of a mediating bridge and a facilitator of regional trade. However, this fact also means that Turkey needs to spend heavily for its defense given the conflicts abundant in the region.

According to the State Institute of Statistics (SIS), the surface area of Turkey is 769,604 km^2. In the year 2000, the Turkish population was counted as 67.8 million. The population growth rate per year averages 1.8% and the life expectancy from birth is 70 years. The city population is 64.9% of the total population. The total literacy rate is 87.3%: The male literacy rate is 93.9%, and the female literacy rate is 80.6%. Higher education levels are still insufficient, however, given the total rate of 7.8%, with the male higher education rate of 10.2% and the female rate of 5.4%.

The SIS also reported that, in 2004, Turkey's gross national product (GNP) reached $299.5 billion. The gross domestic product (GDP) growth rate was calculated to be 9.9%, a record level for the last 10 years. Per capita GNP was reported as $4,172. Foreign debt at the end of 2004 was $161.8 billion, which was 54% of GNP.

In the year 2004, total exports were $63.1 billion, while imports reached $97.5 billion. These numbers translates into a trade deficit of $34.4 billion. Another problem is unemployment, which averaged 10.5% in the years 2003 and 2004 despite the high growth rate of GNP.

In brief, Turkey has a very dynamic young population eager to use new technologies and to take new approaches. The EU accession process gives a reliable road map for the infrastructure upgrade in technological-enabled procurement. Turkey's geographical position and its globally-integrated economy make it a country that needs great improvements in its procurement processes.

ISSUES AND PROBLEMS RELATED TO THE BUSINESS ENVIRONMENT

These problems and issues have secondary or indirect effect over the process of e-procurement activities. However, in the long run, they are as important as the factors related to its infrastructure.

- EU accession and its implications over the procurement process
- The interrelationship of e-procurement and e-governance activities
- Small and medium-sized enterprises (SMEs)
- International procurement
- Organizational resistance
- Lack of qualified workforce
- Widespread use of mobile technology

EU Accession and Its Implications on the Procurement Process

Turkey has had a long history with the European integration process. She made her first application in July, 1959. At that time, EU was called the European Economic Community (EEC). As of the end of 2003, Turkey was EU's 7th biggest trading partner (up from 9th in 1990). It was also the 13th biggest exporter to the EU (up from 17th in 1990) (EU - Turkey relations, n.d.).

Turkey is a European Union (EU) candidate and despite skeptics, she has a great chance to gain accession by the year 2015. Thus the EU factor must be taken into account in any analysis made about Turkey. After the December 17, 2004, decisions of the EU, Turkey was categorized as a "converging country" rather than as an "emerging country." In the year 2005, the explosion in foreign direct investments (FDI) is a direct result of that changing outlook. Even so, membership has not been finalized yet, and Turkey still shows most of the traits of an emerging country; therefore, Turkish practice is valuable to understand business procurement activities in an emerging country.

Throughout the EU process, it is most likely that Turkey will be forced by the EU to upgrade her infrastructure to EU standards. At that point, professional e-procurement and m-procurement applications in the developed countries will be very relevant for Turkey. However, the greatest challenge is managing the fastened process of migration from an emerging country to an EU-country.

The Interrelationship of E-Procurement and E-Governance Activities

According to the UK Office of Government Commerce Manual (2005), e-procurement reduces paperwork and speeds up payment. This is especially important in government services, since it is known that these services are highly inefficient. It is also applicable to Turkey and other developing countries.

Although e-procurement is very essential in all sectors beginning with the public sector, its implementation is not so easy. The reasons are:

- **More efficient processes will mean less need for labor:** In Turkey, unemployment is a significant problem; it averaged around 10-11% after the 2001 crisis. The public sector was used as a vehicle to circumvent the unemployment problem by hiring unneeded employees. In spite of the fact that the International Monetary Fund (IMF) wants Turkey to shrink the government sector, little progress can be made. Therefore, efficiency tools such as e-procurement are hard to be accepted by the public sector workers in Turkey.
- **Public sector workers are undereducated:** The private sector in Turkey is very dynamic and competitive, and its workers are young, dynamic, and highly educated. The opposite is true for the public sector. In these enterprises, job security is very high, and it is hard for the government to lay off these workers. This makes it unnecessary for the public workers to retrain themselves in new technologies and management techniques. The public sector also faces the probable resistance to these new methods and technological developments in Turkey. Also, the recruitment process in the public sector is not efficient. The result is an undereducated, noncompetitive employee with outdated knowledge and skills. In other words, the human capital of the public sector is much lower and less valuable than that of the private sector. This is an obstacle for the effective implementation of e-procurement projects in the Turkish public sector in spite of the enthusiasm presented by some of the politicians.

Small and Medium-Sized Enterprises (SMEs)

Yanıklar (2003) states that 99.5% of the enterprises in Turkey are SMEs. They are the primary source of employment and the basic driver of economic growth. Yanıklar also adds that 23% of the Turkish SMEs use computers; however, only 2% have e-commerce activity. Recalling that e-commerce is a broader concept than e-procurement, we can deduct that e-procurement is practically nonexistent in Turkish SMEs.

These numbers are not so surprising since SMEs have weak financial resources to implement e-commerce or e-procurement projects. They cannot employ the talented workforce needed in these projects. The solution may come from a collaboration of such companies and establishment common e-commerce and e-procurement platforms.

The five forces of competition, as proposed by Michael Porter (1998), require SMEs to make collaborations to become competitive players in the global marketplace. Among the five forces, the bargaining power of suppliers and bargaining power of customers especially apply to SMEs. When SMEs come together, new and advanced management techniques including e-procurement can be afforded just as large companies. The Internet makes it easier and cheaper for the SMEs to cooperate.

International Procurement

International trade exploded after the fall of communism and the victory of the liberal economic system. In fact, such a high degree of globalization was also witnessed in the 19th century, but the 20th century witnessed a world divided into two camps. Even so, international trade is a very old phenomenon which gave rise to well-known business practices in this area. However, the developments in the ICT field and the widespread use of the Internet changed the whole paradigm, and old practices became out-of-date.

Previously, international trade could be performed by highly specialized professionals and agent firms. The profit margins and earnings were very high, and such professions were very prestigious. Today's Internet world dimmed their shiny outlook, and these professionals are struggling to survive by means of their dying profession. Their specialized expertise is becoming worthless in over the Internet. Even a teenager can order a product from abroad and pay for it.

In Turkey, the situation is even more dramatic in that such professions denied adopting the change for a long time, and now they face a much more accelerated and devastating situation in a short period of time. The understandable psychological reaction is total refusal of the merits of e-procurement in international procurement activities. They still dream about a world of published special legal documents, obscure trading regulations, clumsy transactions, and guru-like status stemming from their specialization in them.

Technology is known to be killing some professions and giving birth to new ones. The industrial revolution has made craftsmen unemployed, and that has given rise to the communism in the end. However, the information revolution occurred in a much shorter time period than the industrial revolution. People having classical occupations could not find time even to understand that they needed to retrain themselves in the new ways of doing business. In Turkey, the change was even more dramatic, in that Turkey marched into the information age without first being an industrial country. The effects on the personal lives of people are catastrophic. The solution is to retrain people. But first they must accept that they need to be retrained.

Organizational Resistance

Similar to international trade specialists, employees in the classical procurement function could not adapt to changes brought by the Internet. In other words, they resist against e-procurement. As for m-procurement, they have little idea about it and they cannot even visualize it.

Companies mistakenly delegate the transformation of e-procurement to the classical procurement personnel. Understandably, they resist against the establishment of the online purchasing infrastructure. They claim that technology is not yet ready to perform the purchasing activity that they used to do with the telephone and the personal contact.

A purchaser, just like salespeople, depends on his or her personal contacts in an organization. In other words, the social network he or she possesses is his or her most valuable asset. The Internet allows the establishment of such social networks in a different way. Online communities, e-mail groups, forums, and even friendship sites are places to develop such a network. Some of the classical purchasers and salespeople deny the value of such virtual networks. They claim that there is nothing that can match human contact. According to them, Turkish culture values personal communication. In this view, the Internet will never replace the human contact.

In fact, the Internet is still an evolving technology and is merging with other technologies such as voice and video technologies. Using voice over IP (VoIP), people can also talk to each other. Skype is the industry leader in this field; other Instant-Messenger (IM) applications such as Yahoo Messenger and MSN Messenger are also modifying their products in order to catch up. Internet contact will resemble classical human contact with growing broadband Internet access technologies such as video conferencing.

In Turkey, broadband came late but diffused fast. DSL membership increased in the last two years thanks to the subsidies offered by Turkish Telecom. This is due to the fact that the Turkish population is young and dynamic and ready to accept technological innovations. Of course, old workers are still skeptical about new technology and believe that technology is not adequate. This point is further analyzed in the next subsection.

Lack of Qualified Workforce

Since the professional implementation of procurement is rather a new concept in Turkey, employees do not have formal training in this field and are unaware of formal approaches. The rapid explosion in e-procurement activities has worsened the problem. Even though new schools are opened to address this issue, their full effect cannot be seen in the near future. Newly-established non-governmental organizations (NGOs) take a greater and greater role and try to speed up the learning curve by means of collaborating with all of the stakeholders.

Another important trend for Turkey is the proliferation of private universities, especially after the year 1990. Private universities have closer ties with the industry and respond to their needs in a faster manner. Logistics, procurement, and supply chain training programs are offered, and this will help alleviate the problem of the lack of a qualified workforce for procurement activities. As these more qualified employees replace the ones with no formal education, more sophisticated procurement systems can be established.

It is often stated that Turkey lacks the human capital in terms of specialized blue-collar workers. However, an even more severe problem is at the managerial level. Turkish companies are quite conservative and young managers, such as those seen in American companies, are rare. Thus, Turkey cannot make full use of its young, educated, and technologically-savvy population. Older people may not fully appreciate the impacts of new technologies and may even fear them. Thus they may postpone the adoption of new technologies. Therefore, even top management should be retrained in new technologies and new methods such as e-procurement.

Widespread Use of Mobile Technology

Wireless technology has already made great impacts, but in the near future that impact will largely be amplified (Gratton & Gratton, 2004). In fact, we are in the midst of the wireless revolution which extends to the boundaries of the Internet revolution. Since the Internet and mobile technologies are integrated and interrelated technologies, before analyzing the state of the wireless technology in Turkey, it will be helpful to look at the Internet penetration in selected countries.

Internet usage in the Middle East was reported to be 8.3%. Global penetration was given as 14.9%. Apparently the Middle East lags the world in terms of Internet usage. The reason may be closed societies and governments found in the region. Turkey's penetration was given as 9.9%. The highest rate of penetration was in Israel with 44.8%, and the lowest was Iraq with 0.1%. The highest penetration rates were given in Sweden (73.6%), Hong Kong (70.7%), Denmark (68.7%), and United States (68.5%). Given these rates, Turkey has a serious competitive disadvantage in terms of Internet usage (http://www.internetworldstats.com).

A World Bank study indicates that among 115 countries, Turkey is 50th in terms of Internet penetration, with 85 users in 1,000 in 2003. Similarly, the penetration of mobile phones and wired phones was 662 in 1,000 in Turkey in the year 2003. With this penetration rate, Turkey was 46th among 115 countries (Türkiye Cep Zengini, n.d.).

According to the State Institute of Statistics of Turkey, in the year 2004, 92.19% of the households of Turkey had TV sets, 53.64% had cell phones, 9.98% had personal computers, 5.86% had Internet connection, 2.85% had game consoles, 0.85% had laptops, and 0.13% had PDAs. Although laptops, PDAs, game consoles, and cell phones are all new products, Turkish people have heavily invested in one of them, namely the cell phones. We do not need these data to reach such a conclusion. By making a short trip in one of the big cities of Turkey, one can easily observe Turkish people's attraction to the cell phones. Of course, this phenomenon is not unique to Turkey; however, the change in Turkey is highly visible because of the late industrialization of the country.

In Turkey, mobile phones gained very favorable customer acceptance, and diffusion of mobile technology occurred much faster than that of the Internet. Therefore, we may conclude that with the advances in wireless technology, the mobile phone can become a primary Internet access tool for Turkish people. Thus, in Turkey, m-procurement has a better chance of success than e-procurement.

Countries like Turkey can attain the most recent technology advancement without following the same long path of the learning curve that developed countries have followed. For Turkey, this means that mobile technologies and m-commerce will become the most successful applications.

Thus for Turkey, directly developing m-procurement and m-business applications is much more reasonable than trying to transfer the e-procurement and e-business practices of the developed world. Other developing countries may also choose to jump directly to mobile technologies. However, in the Turkish case, this is a necessity rather than a means of choice, given the large established base of mobile users in Turkey.

ISSUES AND PROBLEMS RELATED TO INFRASTRUCTURE

The previous section has analyzed the business environment in Turkey. We will now focus on the infrastructure-related issues and problems. These can be listed as: legal and regulatory environment; security issues; transportation infrastruture; financial infrastructure; and telecommunication infrastrucure.

Legal and Regulatory Environment

The legal and regulatory environment in developing countries is far behind technology. One major problem is that these countries lack human capital with expertise in diverse fields such as information and communication technologies (ICT), domestic law, international law, international trade, logistics, and procurement. Without such experts, the regulatory environment cannot cater to the new and sophisticated needs of the industry.

Although Turkey was slow in adopting a regulatory framework, the EU process, the exploding foreign trade, and foreign direct investment (FDI) through privatization activities speeded up the adoption of the legal infrastructure derived from the best practices around the world. According to the official e-government Web site of Turkey (Vatandaşın Sadece %3'ü e-devlet hizmetlerinden yararlanıyor, n.d.), e-signature is now formally accepted as a legal equivalent of the classical signature. This will positively affect the spread of e-procurement activities without doubt. However, security issues related to e-signature should be resolved before widespread use.

Security Issues

Security is a major issue for all kinds of e-business and m-business applications. In emerging countries like Turkey, this problem is exacerbated by the inefficient and out-of-date legal system and unknowledgeable technology users who can easily be victims of technology-related crime or fraud. Cases in the popular press make people hesitant to use technology.

Security complaints in Turkey are on the rise recently. This can be attributed the fact that cheap DSL Internet access by Turkish Telecom (TT) increased the number of Internet users is Turkey in a short time (ADSL Abone sayısı 900bini Geçti, n.d.). In 18 months, the number of subscribers rose from 56,000 to 901,000. New subscribers are less experienced over the Internet, but stay connected for a long duration due to the unlimited hours of access offered by DSL service. As a result, such users become easy victims of hacking, spying, and phishing activities.

There is no established insurance system to alleviate victims' problems and to foster trust on advanced technology. In Turkey, with the spread of technology, online and mobile crime exploded unexpectedly. Thus, the security issue will be among the most important issues in the years to come.

Transportation Infrastructure

Although e-procurement is a revolutionary step over classical procurement, goods purchased are still delivered by classical means. In other words, transportation and logistics services have not changed much. Of course, information content is added, and transported goods can now be tracked down to some degree. However, we authors still need a good transportation infrastructure, which is expensive and slow to build.

Emerging economies often face financial and economic difficulties in building their transportation infrastructures (TI). Such projects span many years, and weak governments cannot persuade society to make short-term sacrifices for long-term benefits. TI projects are often financed with high-interest foreign loans, and high sovereign debt levels of such countries put a limit on the opportunity to borrow additional debt.

TI can be divided to five categories:

1. Surface transportation infrastructure (STI)
2. Railroads transportation infrastructure (RTI)
3. Marine transportation infrastructure (MTI)
4. Airline transportation infrastructure (ATI)
5. Transportation infrastructure integration and management (TIIM)

The first four categories are obvious. The fifth one, that is, the TIIM category, is a newly-developing meta-component that uses information and communication technologies (ICT). Transportation in Turkey is mainly conducted by trucks on motorways and highways. Railroads and marine transportation have not been fully exploited yet. Especially in the late 1980s, superhighways have been built throughout the country. Air transportation within Turkey has only been recently opened due to the competition from the private sector. Previously, the only carrier was the state-owned Turkish Airlines. The lack of an infrastructure in Turkey may seem like a handicap, but it is also an opportunity for both Turkey and its trading partners.

Financial Infrastructure

Banks have had a central role in trading or business activity in the past. This was due to the fact that processes of trading and commerce were inefficient, and businessmen needed trust supplied by banks. However, given the exploding security problems, banks are not as secure and trusted as before. Also, technology makes it possible for large companies to perform some of the functions of banks within their organizations. In the future, banks may even cease to exist in today's terms. There is one big central global bank which brings together all sides, and it is called the Internet.

In the year 2001, Turkey fell in a deep financial crisis due to the weaknesses of the banking sector. Since then, with the help of the IMF, the financial strength of Turkish banks improved with the help of foreign investments in the sector. Although it gave rise to some criticisms from globalization opponents in Turkey, the increasing share of foreign banks in the sector also means that banking services in Turkey will meet global standards.

Even before the entry of foreign banks, the Turkish banking sector was using the most advanced technology in its operations. However, these investments were not so efficient, and choosing the wrong technology was a common mistake. The development of the sector will no doubt enhance e-procurement activities in Turkey. However, in the future we will be able to use the development of the banking sector and the development of the Internet interchangeably. When the legal infrastructure and the use of e-signature applications advance, the banking services will also develop more in depth and breadth.

Telecommunication Infrastructure

The Turkish Telecom (TT) has been recently privatized after decades of debate. In those countries with privatization and deregulation, their telecommunications sectors are reported to have made great advances in ICT-related industries. Thus, for the development of e-procurement activities, this should be considered as positive news. Although the debate was still continuing when we were writing this chapter, we believe that the privatization of TT was finalized during this time after several years. But unfortunately, according to predicted analysis, its value will decrease substantially, and new technologies will erode much of the competitive power of TT.

Huge public companies are known to be slow to adapt to new technologies, and they are found to be irresponsive to the needs of the consumer. A private TT, therefore, will need to increase its investments

Figure 1. Procurement/public procurement process in Turkey (Source: Elektronik İhale Forumu, n.d.)

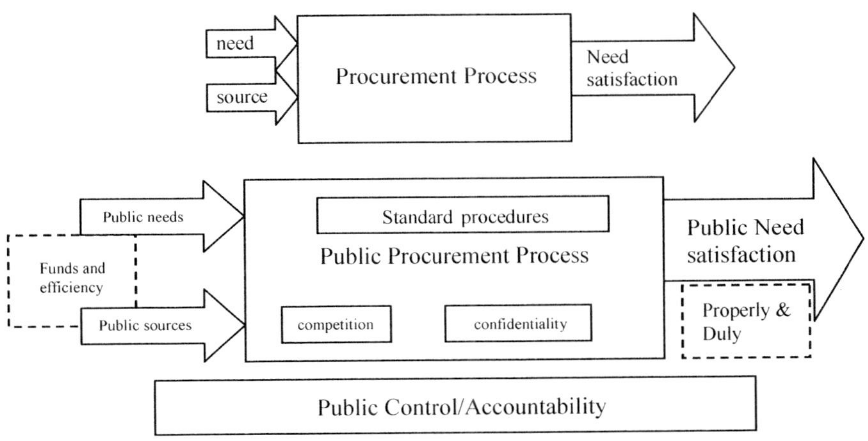

Figure 2. E-bid reference model for public purchasing (Source: Elektronik İhale Forumu, n.d.)

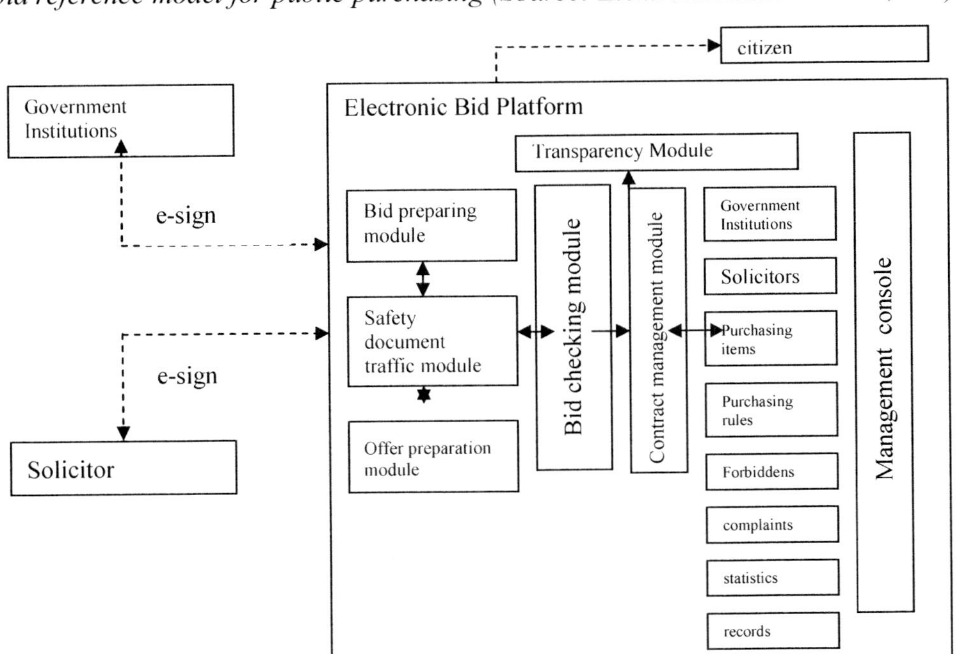

Figure 3. Migration from procurement to m-procurement

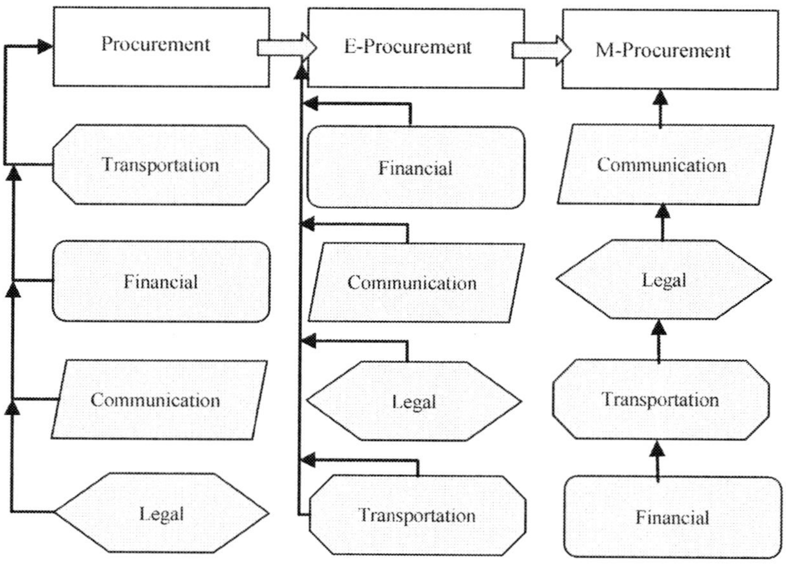

in new technologies. One possible problem is that privatization does not always mean deregulation and more competition. If competing firms are allowed to enter into the arena, then the telecommunication infrastructure of Turkey may be upgraded in a few years. That will mean a friendlier atmosphere for e-commerce in general and e-procurement in particular. The widespread use of broadband applications will also be valuable.

CASE STUDY: PUBLIC PROCUREMENT AUTHORITY OF TURKEY

Until recent privatization efforts, the public sector has dominated the Turkish economy. Thus the best examples of e-procurement in Turkey are from the public sector. Also, more information is available for these enterprises as they are public. Thus we have taken Public Procurement Authority (PPA) of Turkey for our case study.

The procurement process in Turkey is shown in Figure 1 from the perspectives of both the public sector and the private sector. The public procurement process is more complicated than the private sector's procurement process.

In the procurement process, first the needs are identified and then the available resources are determined. The ultimate goal is the satisfaction of each of the stakeholders throughout the process. For a successful public procurement process, the state authorities should treat all parties equally. The confidentality of trade secrets is also criticial.

The PPA is taking an active part in e-government activities in Turkey. The intention of PPA is to achieve the establishment of a clear, transparent system for public procurement. In the past, corruption in the public sector was one of the biggest problems. E-procurement will bring openness, accountability, and efficiency in this field.

As it can be seen in Figure 2, there is an electronic bid platform where government institutions, solicitors, citizens, and so forth, meet. In this electronic platform, buyers and sellers are able to make transactions in a faster, easier, and open manner. There is also a documentation system for technical specifications. An important goal is to build a database that can be shared with all other governmental institutions. Standardization in procurement is also an important objective.

As a basic e-marketplace characteristic, price/quality competition is achieved by giving an opportunity to every solicitor for bidding. In other words, openness and easiness for bidding fosters competition. This system also contains computer-assisted bid assessment, that is, the e-bid reference model maintains its status as a reliable assessment system.

As it can be seen, the PPA platform does not feature a built-in m-procurement component. This can be attributed to the fact that even the idea of the importance of the Internet has only recently been disseminated by the public sector. Mobile commerce or m-procurement sound like futuristic ideas at this stage. However, it can be predicted that in the future, the e-procurement platform of PPA will evolve into an m-procurement-based exchange mechanism.

FUTURE TRENDS

So far we have talked about business- and infrastructure-related issues and barriers to e-procurement adoption. In this section, we present some future trends and discuss how technology such as mobile can be an enabler for Turkey.

Migration to M-Procurement

The future of e-procurement is apparently m-procurement. However, mobile technologies are still at their development stage, and their full potential is yet to be exploited. Especially, hybrid products such as merging cell phones and PDAs are candidates to be the enabling devices for m-procurement.

In Figure 3, we tried to give a simple model of the migration from the classical procurement to first e-procurement and then to m-procurement. Among the infrastructures, the communication infrastructure gains more importance as this migration occurs. With the new technology, the legal system soon becomes outdated, and it also gains importance to have a robust and up-to-date legal infrastructure. Transportation and financial infrastructures are still important, but they are much more mature than communication and legal infrastructures.

Classical procurement, e-procurement, or m-procurement (mobile procurement) each require transportation, communication, financial, and legal infrastructures. Emerging markets are known to be underdeveloped in all of these dimensions. On the other hand, Turkey is closing the gap with developed countries.

The Changing Face of the Marketing Function

Migration from classical procurement to m-procurement will change the way the marketing function operates in a firm, in that e-procurement will make the sales and procurement departments shrink as less employees will be needed. The emphasis will shift from sales to the analysis of the data obtained from digital sales-procurement activities of companies.

Thus the profile of the people in the marketing and procurement departments will be altered, and market researchers and analysts will gain value. Also, such workers will need to have expertise in using and modifying software packages. These traits are much different from those needed in the age of classical procurement.

In fact, even the marketing function is new compared to production and finance, for example. It evolved from sales and became a discipline using quantitative methods known as market research. In the sales era, the desired attribute was being presentable. However, market research needs an analytical mind.

M-procurement will be very much like the classical procurement in terms of human contact. However, since this contact is completely digitalized, it will be tracked, analyzed, stored, and data-mined. We may even see the rise of virtual selling or purchasing agents. Thus, the seemingly visible human contact will be performed by software. If even sales activities are performed by software, then what kind of jobs will be left to human beings other than computer programming?

Certainly, the final stage data analysis will still need human intellect, and employment will continue there. This will make the marketing function a part of the strategic management function; marketing and strategic management will come to be in apposition, in which both will be used interchangeably. Analysis will be based on the data collected from em-procurement activities. With the exception of some multinational firms, most of the firms consider the production and marketing functions as separate. However, e-procurement tools necessitate an integrated approach.

CONCLUSION

Emerging countries are different from developed countries, and even emerging countries have differences among them. In this chapter, we tried to give a picture of Turkey in terms of m-procurement, e-procurement and em-procurement activities.

In general, the e-procurement infrastructure in Turkey is developing but still insufficient. It should also be noted that even a weakness in only one of these areas of infrastructure will hinder the total e-procurement process. To date, the most significant problem is the security problem, and the most advanced area can be seen as the financial area. However, security problems weaken the strengths of the financial infrastructure.

Given the low Internet penetration but a wider mobile customer base, we see potential in Turkey first for m-procurement and then for integrated em-procurement. Of course, em-procurement requires an even more sophisticated infrastructure than just a wide customer base.

Obviously, m-procurement is still an evolving concept. Its major advantage is the freedom from the need of wired connection. In fact, in the future wireless technologies and the Internet technologies are likely to merge and become one single integrated technology. Thus we prefer the term em-procurement instead of m-procurement.

Em-procurement can be seen as the most advanced form of procurement given the available technologies. With em-procurement, the exact position of a purchased product can be tracked down, and even possible problems with the product can be detected and reported to sales.

Em-procurement is also good for tracking the performance of logistic personnel. Thus, they will be more accountable, and supervision will become easier. Also, in countries where mobile technologies are more widely used, employee training costs will be lower, since they are already accustomed to them.

The most important problem with em-procurement is that the technology is still too underdeveloped to handle the procurement needs of the companies. Cell size and life, memory size, bandwidth, and high communication costs prevent their use by all employees. At the present, only technologically-savvy middle or top managers have the privilege to use them.

For mobile technologies to make their full impact, they should become much cheaper and much more capable. After cheap broadband services are widely available, people will be able to handle their procurement activities with almost perfect human contact. At that point, the concept of m-procurement and em-procurement terms will become more meaningful.

REFERENCES

Aberdeen Group Profile. (n.d.). *elcom: Delivering the benefits of Internet procurement automation without the burdens.* White paper. Retrieved July 29, 2005, from www.aberdeen.com

ADSL Abone sayısı 900bini Geçti (ADSL Subscriber's are more than 900 thousand). (n.d.). Retrieved August 8, 2005, from http://www.ntvmsnbc.com/news/336004.asp

Bovet, D., & Martha, J. (2000). *Value nets: Breaking the supply chain to unlock the hidden profits.* New York: John Wiley Sons Inc.

Croom, S. (2000). The impact of Web-based procurement on the management of operating resource supply. *The Journal of Supply Chain Management, 36*(1), 4-13.

Emel, G. G., Taskın, C., & Deniz, G. (2004). Tedarik Zinciri Yönetimi: Otomotiv Sektöründe Bir Araç Rotalama Uygulaması. *Öneri, 16*(21), 59-70.

Gebauer, J., & Segev, A. (1998). Assessing Internet-based procurement to support the virtual enterprise. *Electronic Journal of Organizational Virtualness, 2*(3), 30-43. In T. Pushmann & R. Alt (Eds.), Successful use of e-procurement in supply chains. Retrieved July 29, 2005, from verdi.unisg.ch/.../$file/Successful%20Use%20of%20eProcurement%20in%20Supply%20Chains.pdf

Giunipero, L. C., & Sawchuk, C. (2000). *E-purchasing plus: Changing the way corporations buy.* JGC Enterprises. Retrieved July 29, 2005, from http://www.sharemax.com

Gratton, S. -J., & Gratton, D. A. (2004). *Marketing wireless products.* Oxford, UK: Elsevier.

Hawking, P., Stein, A., Wyld, D. C., & Foster, S. (2004). E-procurement: Is the ugly duckling actually a swan down under? *Asia Pacific Journal of Marketing and Logistics, 16*(1).

Elektronik İhale Forumu. (Electronic Bid Forum). (n.d.). Retrieved January 14, 2006, from http://www.kik.gov.tr/eihalecs/eihale.ppt

EU - Turkey relations. (n.d.). Retrieved August 10, 2005, from http://europa.eu.int/comm/enlargement/turkey/eu_relations.htm

Johnson, P. F., & Klassen, R. D. (2005). E-procurement. *MIT Sloan Management Review, 46*(2), 7-10.

Kalakota, R., & Robinson, M. (2001). *E-business 2.0: Road map for success.* New Jersey: Addison-Wesley.

Kalakota, R., & Robinson, M. (2002). *M-business: The race to mobility.* New York: McGraw-Hill.

Kotabe, M., & Murray, J. Y. (2004). Global procurement of service activities by firms. *International Marketing Review, 21*(6), 615-633.

Kuglin, F. A., & Rosenbaum, B. A. (2001). *The supply chain network @ Internet speed: Preparing your company for the e-commerce revolution.* New York: Amacom.

Kumar Nirmalaya foreword by Philip Kotler. (2004). *Marketing as strategy* (p. vii). Boston: HBS Press.

Neef, D. (2001). *Procurement: Present and future.* Retrieved July 30, 2005, from http://www.informit.com/articles/article.asp?p=24124&seqNum=7

Ohmae, K. (2000). *The invisible continent: Four strategic imperatives of the new economy.* New York: HarperCollins Publishers.

Pride, W. M., & Ferrel, O. C. (1989). *Marketing concepts and strategies.* Boston: Houghton Mifflin.

Porter, M. E. (1998). *On competition.* Boston: Harvard Business School Press.

Puschmann, T., & Alt, R. (2005). *Successful use of e-procurement in supply chains.* Retrieved July 30, 2005, from http://www.alexandria.unisg.ch/EXPORT/DL/28242.pdf

Raisch, W. D. (2001). *The e-marketplace strategies for success in B2B e-commerce.* New York: McGraw Hill.

SAP.com. (2002). *E-procurement. Glossary.* Retrieved from http://searchsap.techtarget.com/gDefinition/0,294236,sid21_gci214418,00.html

Shah, J. B. (2002, April 1). Three steps to better procurement—through a mix of strategies, innovative companies are making strides to cut costs and improve inventory management. *EBN, Manhasset,* (1306), 29.

Stanton, A. D., & Stanton,W. W. (2002, February). The link between personality, innovativeness predisposition and adoption: A model of Internet purchasing. In *Proceedings of the American Marketing Association Winter Educators' Conference* (published as abstract, p. 42).

Tompkins, J. A. (2000). *No boundaries: Moving beyond supply chain management.* Raleigh, NC: Tompkins Press.

Turkish Statistical Indicators. (n.d.). Retrieved August 4, 2005, from www.die.gov.tr

UK Office of Government Commerce. (2005). *Overview to selling to government.* Retrieved June 13, 2005, from http://www.businesslink.gov.uk/bdotg/action/detail?r.l3=1074033478&r.l2=1073858827&r.l1=1073861169&r.s=sc&type=RESOURCES&itemId=1073792572

Vatandaşın Sadece %3'ü e-devlet hizmetlerinden yararlanıyor (Only the 3% of the citizen are benefited from e-government services). (n.d.). Retrieved August 8, 2005, from www.turkiye.gov.tr

Vigoroso, M. (1999). Buyers prepare for brave new world of e-commerce. *Purchasing, 126*(6), 4-12.

Wheatley, M. (2003, June 15). How to know if e-procurement is right for you. *CIO Magazine*. Retrieved August 8, 2005, from http://www.cio.com/archive/061503/eproc.html

Türkiye Cep Zengini (Turkey is a mobile paradise). (n.d.). Retrieved June 12, 2005, from http://www.hurriyetim.com.tr/haber/0,,sid~1@w~5@tarih~2005-06-11-m@

Yanıklar, M. (2003, December 19-21). Bilişim (E-ticaret). In *The Published Papers of the Second SME Summit Meeting of Turkey* (p. 177).

APPENDIX

The following is the list of the definitions and abbreviations used throughout this chapter. They are listed here for quick reference and to facilitate a clearer understanding of the concepts developed.

ATI: Airline transportation infrastructure

e-procurement: The use of electronic technologies to flow freely and enable the procurement activities of an organization

em-procurement: The procurement activities conducted by the integrated use of both classical and electronic means (the Internet, extranets, etc), and mobile technologies

FDI: Foreign direct investment

ICT: Information and communication technologies

IMF: International Monetary Fund

m-procurement: The procurement activities conducted by the use of mobile technologies such as cell-phone, PDA, and so forth

MTI: Marine transportation infrastructure

PPA: Public procurement authority

RTI: Railroads transportation infrastructure

SME: Small and medium-sized enterprises

STI: Surface transportation infrastructure

TI: Transportation infrastructure

TIIM: Transportation infrastructure integration and management

TT: The Turkish Telecom Company

VoIP: Voice over Internet protocol

YTL: New Turkish Lira

This work was previously published in E-Procurement in Emerging Economies: Theory and Cases, edited by A. Pani; A. Agrahari;, pp. 101-125, copyright 2007 by IGI Publishing (an imprint of IGI Global).

About the Author

Gonca Telli Yamamoto is an associate professor in the School of Applied Sciences at Okan University, Turkey. She is the head of Information Sciences and Technologies Department. She is also coordinating Okan University Distance Education Center at present. She was formerly the founder and director of Social Sciences Institute, Business Admisitration Department and several other departments of Okan University. She currently teaches, consults and conducts research on mobile and integrated marketing, and new learning technologies in business. She has been studying integrated marketing, technological developments and customer value more than a decade in the academic field. She is also interested in broader implications associated with emerging information society. She has some books and chapters related in sales, integrated and mobile marketing and IT Ethics. Her several articles were published in several national and international journals. She has also received two outsanding paper awards from the International Academy of E- Business.

Index

D

E

H

I

N

O

P

Q

R

S

Y

Z